TISSUE OPTICS

Light Scattering Methods and Instruments for Medical Diagnosis

SECOND EDITION

TISSUE OPTICS

Light Scattering Methods and Instruments for Medical Diagnosis

SECOND EDITION

Valery Tuchin

PRESS

Bellingham, Washington USA

Library of Congress Cataloging-in-Publication Data

Tuchin, V. V. (Valerii Viktorovich)
 Tissue optics : light scattering methods and instruments for medical diagnosis / Valery V. Tuchin.
-- 2nd ed.
    p. ; cm.
 Includes bibliographical references and index.
 ISBN-13: 978-0-8194-6433-0
 ISBN-10: 0-8194-6433-3
 1. Tissues--Optical properties. 2. Light--Scattering. 3. Diagnostic imaging. 4. Imaging systems
in medicine. I. Society of Photo-optical Instrumentation Engineers. II. Title.
 [DNLM: 1. Diagnostic Imaging. 2. Light. 3. Optics. 4. Spectrum Analysis. 5. Tissues--
radiography. WN 180 T888t 2007]

 QH642.T83 2007
 616.07'54--dc22

                                                                            2006034872

Published by

SPIE
P.O. Box 10
Bellingham, Washington 98227-0010 USA
Phone: +1 360 676 3290
Fax: +1 360 647 1445
Email: spie@spie.org
Web: http://spie.org

Contents

Part II

Light Scattering Methods and Instruments for Medical Diagnosis

6

Continuous Wave and Time-Resolved Spectrometry

The specificity of optical spectral diffusion techniques is discussed in this chapter. As usual for this tutorial, two types of instruments and measuring techniques are presented: spectroscopic, used for tissue local parameters monitoring, and tomographic, used for tissue pathology imaging. Some of them are based on CW light source tissue probing. A few examples of CW measuring and imaging instruments and results of clinical studies are presented. Time-resolved techniques and instruments, which are the most promising for an accurate *in vivo* measurement, are also analyzed. In accordance with the basic principles discussed in Chapter 1, three types of time-resolved techniques and instruments are considered: the time-domain technique, which uses ultrashort laser pulses; the frequency-domain technique, which exploits an intensity-modulated light and narrowband heterodyne detection; and the phased array technique, which utilizes an interference of photon diffusion waves.

6.1 Continuous wave spectrophotometry

6.1.1 Techniques and instruments for *in vivo* spectroscopy and imaging of tissues

For the *in vivo* study of thick tissue (for example, the female breast), the collimated light transmittance can be described by an exponential law such as Eq. (1.1), taking into account that due to multiple scattering, the effective migration path of a photon before it is absorbed should be larger than the thickness of the tissue.[288] For a slab of thickness d, the diffusion equation can be used to calculate a mean path length L of the photons as[272]

$$L = \frac{\mu_{eff}}{2\mu_a \mu_s'} \frac{(\mu_s'd - 1)\exp(2\mu_{eff}/\mu_s') - (\mu_s'd + 1)}{\exp(2\mu_{eff}/\mu_s') - 1}, \tag{6.1}$$

where μ_{eff} is defined by Eq. (1.18). Using Eq. (1.1) for the matched boundaries ($n = 1$), the collimated transmittance can be written in the form[288]

$$T_c(\lambda) = x_1 \exp[-\mu_a(\lambda)L(\lambda)x_2], \tag{6.2}$$

where $L(\lambda)$ reflects the wavelength dependence of $\mu_a(\lambda)$ and $\mu_s'(\lambda)$; x_1 takes into account multiply scattered but not absorbed photons, which do not arrive at the

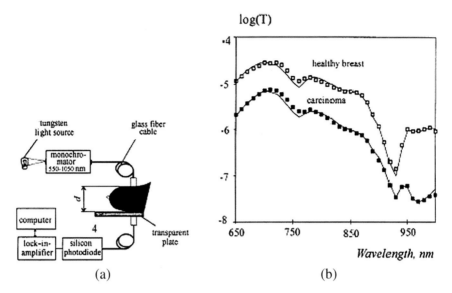

(a)   (b)

Figure 6.1 (a) Schematic setup of a spectrophotometer system used for *in vivo* measurements of breast tissue spectra.[288] (b) Spectra and respective fits of a breast cancer patient (56 yrs., breast thickness of 60 mm) within the area of carcinoma and for a healthy breast and the similar localization as carcinoma.

detector, and the measurement geometry; and x_2 compensates for measurement error of the thickness d and inaccuracies in the reduced scattering coefficient μ_s'.

The semiempirical Eq. (6.2) was successfully used for fitting the *in vivo* measurement spectra of the female breast and estimation of the concentrations of the following absorbers: water (H_2O), fat (f), deoxyhemoglobin (Hb), and oxyhemoglobin (HbO_2) as[288]

$$\mu_a = c_{H_2O}\sigma_{H_2O} + c_f\sigma_f + c_{Hb}\sigma_{Hb} + c_{HbO_2}\sigma_{HbO_2},\qquad(6.3)$$

where σ_i is the cross section of the absorption of the ith component.

By varying the concentrations of the four tissue components, the measurement spectra could be fitted well by Eq. (6.2); the correlation coefficients were better than 0.99 in all cases.[288] Figure 6.1. shows the spectrometer for *in vivo* measurement of the collimated transmittance spectra of a female breast and some examples of measured and fitted spectra for normal and pathological (cancer tumor) tissues. Typically, most carcinoma spectra exhibit a lower transmittance than the reference spectrum (for the same breast thickness). The fits show that this is generally due to an increased blood perfusion (higher Hb/HbO_2 values of the carcinoma curve). In the wavelength region between about 900 and 1000 nm, spectra are quite different; this is clearly due to the altered water and fat content of carcinomas compared with that of the healthy breast. The majority of mastopathies and carcinomas show a higher water concentration and a higher blood volume at the lesion site. A comparison of healthy and cancerous sites yields a slightly lower concentration of oxyhemoglobin for the tumor. Unfortunately, the specificity is not good enough because it is not possible to discriminate between benign mastopathies and malignant carcinomas by means of water content, blood volume, and oxygenation.[288]

Transmittance NIR spectrometry for measuring oxygenation has had the most success to date in the newborn infant head, largely because of the small size of the head, the thin overlying surface tissues and skull, and the lower scattering coefficient of the infant brain.[55] The development of the cooled CCD and time-resolved and spatially resolved techniques and instruments has proceeded rapidly, and they are increasing the area of NIR spectroscopy investigations and applications. At present, there are more than 500 commercial clinical NIR spectroscopy instruments for monitoring and imaging the degree of oxygenation in tissues, the concentration of oxidized cytochrome, and tissue hemodynamics.

For many tissues, *in vivo* measurements are possible only in the geometry of the backscattering. The corresponding relations can be written on the basis of a diffusion approximation. For a semi-infinite medium and source and detector probes separated by a distance r_{sd}, normal to the sample surface (see Fig. 6.2), and optically matched (so that specular reflectance at the surface can be neglected), the reflecting flux is given by Eq. (2.17).[685] A more general expression valid for refractive index mismatch conditions on the boundary is given by Eq. (1.27).[205,206]

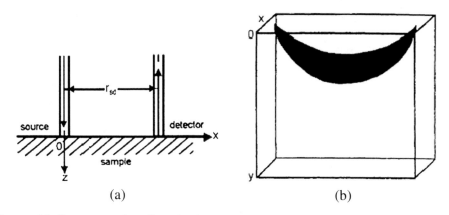

Figure 6.2 Geometry of a fiber backscattering experiment for investigation of (a) a semi-infinite medium, and (b) a "banana" shape region of photon path distribution.[1120,1121]

For backscattering optical spectroscopy and tomography, in addition to the measured coefficient of reflection defined by Eqs. (1.27) and (2.17), we have to know from what depth the optical signal is coming. That depth is defined by the photon-path-distribution function for the photons migrating from a source to a detector.[1120,1121] This spatial distribution function for a homogeneous scattering medium has a "banana" shape [see Fig. 6.2(b)]. In the weak absorption limit, the modal line of the banana region (the curve of the most probable direction of a photon migration) is given by[1120,1121]

$$z \approx \left[\frac{1}{8} \left(\left\{ \left[x^2 + (r_{sd} - x)^2 \right]^2 + 32s^2 (r_{sd} - x)^2 \right\}^{1/2} - x^2 - (r_{sd} - x)^2 \right) \right]^{1/2}, \quad (6.4)$$

where $0 \le x \le r_{sd}$. At $x = r_{sd}/2$, the modal line of the banana region reaches a maximum depth,

$$z^{max} \approx \frac{r_{sd}}{2\sqrt{2}}. \quad (6.5)$$

Instead of Eq. (6.2), used for *in vivo* study in transillumination experiments, using Eqs. (2.17) and (6.4), we can write a modified Beer-Lambert law to describe the optical attenuation in the following form:[1120,1121]

$$\frac{I}{I_0} = \exp(-\varepsilon_{ab} c_{ab} r_{sd} DPF - G_s), \quad (6.6)$$

where I_0 is the intensity of the incident light, I is the intensity of the detected light, ε_{ab} is the absorption coefficient measured in $\mu mol^{-1} cm^{-1}$, c_{ab} is the concentration of absorber in μmol, r_{sd} is the distance between the light source and detector, DPF is the differential path length factor accounting for the increase in the photons' migration paths due to scattering, and G_s is the attenuation factor accounting for scattering and geometry of the tissue.

When r_{sd}, DPF, and G are kept constant (for example, during the estimation of the total hemoglobin or degree of oxygenation), then the changes in the absorbing medium concentration can be calculated using measurements of the changes in the optical density (OD), $\Delta(OD) = \Delta[\log(I_0/I)]$ as

$$\Delta c_{ab} = \frac{\Delta(OD)}{\varepsilon_{ab} r_{sd} DPF}. \quad (6.7)$$

In optical imaging, the changes in optical density are measured as[1120,1121]

$$\Delta(OD) = \log\left(\frac{I_0}{I_{test}}\right) - \log\left(\frac{I_0}{I_{rest}}\right) = \log(I_{rest}) - \log(I_{test}), \quad (6.8)$$

where I_{rest} and I_{test} represent the light intensity detected when the object is at rest (brain tissue, skeletal muscle, etc.) and being tested (induced brain activity, cold or visual test, training, etc.), respectively. For example, based on the OD changes at wavelengths of 760 and 850 nm, one can get either the absorption images for these two wavelengths or functional images (oxygenation or blood volume) within the detection region of study as

$$\Delta(OD)_{oxy} = \Delta(OD)_{850} - \Delta(OD)_{760}, \quad (6.9)$$

$$\Delta(OD)_{total} = \Delta(OD)_{850} + k_{bvo} \Delta(OD)_{760}, \quad (6.10)$$

where $(OD)_{850}$ and $(OD)_{760}$ are the optical densities measured at the wavelengths 850 and 760 nm, respectively, and k_{bvo} is the modification factor for reducing the cross talk between changes of blood volume and oxygenation. This factor is determined by calibration on a blood model.

NIR absorption spectra of oxy- and deoxyhemoglobin and water are presented in Fig. 6.3.[4] The water band at about 980 nm can be used as an internal standard for the evaluation of the absolute concentrations of the blood components in tissue *in vivo*.[1122]

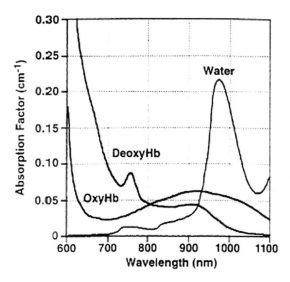

Figure 6.3 Near-infrared attenuation $[\log_{10}]$ for 1-cm depth deoxyhemoglobin (DeoxyHb), oxyhemoglobin (OxyHb), and water; hemoglobin concentration, 210 μM in water.[4]

6.1.2 Example of a CW imaging system

The whole-spectrum NIR spectroscopy system described in Ref. 1123 uses illumination of the subject's head with light from a halogen lamp emitting a continuous spectrum [see Fig. 6.4(a)]. The back-reflected light is detected and spectrally analyzed by a commercial grating spectrograph equipped with a liquid nitrogen cooled CCD detector. The system provides a spectral resolution of 5 nm in the range 700–1000 nm; spectra were collected every 100 ms. Figure 6.4(b) illustrates the image received using this optical instrument and the testing algorithm described [see Eqs. (6.9) and (6.10)]. It shows a focal increase in total Hb in response to stimulation with a stationary multicolored dodecahedron. The area of the peak response is clearly focused and it is about 0.5×0.5 cm in size.

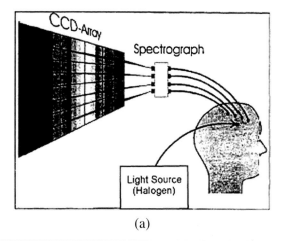

(a)

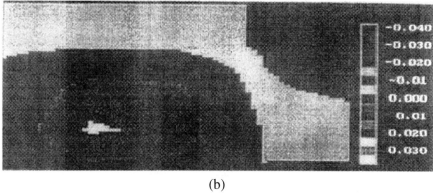

(b)

Figure 6.4 A CCD-NIR spectroscopy system.[1123] (a) Scheme. (b) Functional image, changes in blood volume (total hemoglobin) [see Eqs. (6.9) and (6.10)] during visual simulation (a stationary dodecahedron) as detected over the occipital cortex.

6.1.3 Example of a tissue spectroscopy system

A typical experimental system for *in vivo* backscattering spectroscopy and the corresponding spectra for normal and pathological tissues are shown in Fig. 6.5.[47,94,95] Figure 6.5(b) is an example of spectra taken from the colon of one patient. The absorption bands are of oxyhemoglobin (the Soret band and Q bands are clearly evident). The 400–440-nm segment encloses the hemoglobin Soret band, but also encompasses some absorption from compounds such as flavin mononucleotide, beta-carotene, bilirubin, and cytochrome. The 540–580-nm segment covers the hemoglobin Q band, with minor absorption from cytochrome and other components. On the basis of measurement of the spectral differences of normal and pathological tissue, the corresponding spectral signature "identifiers" can be created. Such spectral "identifiers" for *in vivo* medical diagnostics usually use the ratios of the reflection coefficients integrated within selected spectral bands or the measurement of the spectrum slope for the spectral bands selected.

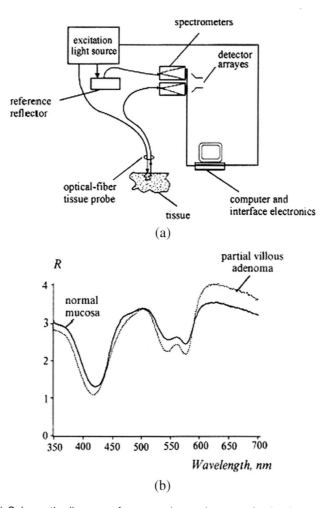

Figure 6.5 (a) Schematic diagram of an experimental system for *in vivo* measurements of spectral reflectance of internal organs.[47] (b) Typical tissue spectra, shown as examples, for two measurements made in the colon of one patient (the spectra have been normalized to the same total integrated signal between 350 and 700 nm); normal mucosa and partial villous adenoma.

A more analytic and quantitative study, provided in Ref. 1124, might yield more insight into the sensitivity of CW reflection spectroscopy. Three different detecting probes were used in the measurements within the spectral range from 400 to 1700 nm. Authors have paid attention to a proper calibration of the probes using a reflection standard (SRS-99-010, LabSphere, North Sutton, UK). For the precise recognition of the absorptions peaks of the tissue, the first derivative of the NIR spectra was computed using the method of Savitzky and Golay. It was shown that CW NIR spectroscopy can detect the presence of lipid in atherosclerotic plaque of the aorta with good sensitivity.

6.2 Time-domain and frequency-domain spectroscopy and tomography of tissues

6.2.1 Time-domain techniques and instruments

One of the designed time-resolved laser systems for *in vivo* measurements of optical properties of the human breast is presented in Fig. 6.6.[288] This system consists of a mode-locked Ti:sapphire laser at a wavelength of 800 nm with a pulse duration of 80 fs and a repetition frequency of 82 MHz. The probe laser beam transilluminates the female breast and the forward-scattered light reaches the detection side of the synchroscan streak camera (S1 photocathode, Hamamatsu C3681). For the enhancement of tissue transmittance, making it more homogeneous and providing stable boundary conditions, the breast was slightly compressed between two transparent plates. Such compression was much less than in a conventional x-ray mammography in order to avoid any influence of changed blood perfusion on the absorption properties. The scattered light is imaged onto the slit of the streak camera with a 1:1 magnification. The dimensions of the slit are 50 μm × 6 mm and the numerical aperture of the camera optics is 0.22. To provide a temporal reference, the reference laser beam was optically delayed and imaged on the streak camera slit; a trigger beam synchronized the streak camera. The working principles of a streak camera and other instrumentation used in time-resolved techniques are described in detail in Refs. 301, 302, 303.

The temporal profile of the light intensity incident on the camera was recorded with a time resolution of about 10 ps and displayed as a spatial profile. A pre-

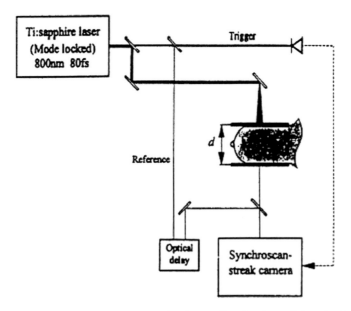

Figure 6.6 Schematic setup for time-resolved transillumination of female breast tissue *in vivo*.[288]

cise shading correction and dark count subtraction were performed for each measurement of the dispersion curve. For *in vivo* measurements, the probe laser beam with a total power of 100 to 150 mW was expanded to a diameter of 10 mm to keep the power density below the maximum permissible exposure of 200 mW/cm^2.

Normalized dispersion curves for three volunteers T_1, T_2, and T_3 and the corresponding results of the theoretical fit according to the diffusion model [see Eq. (1.36)] are shown in Fig. 6.7. The dispersion curves range over a typical period of 6 ns with a mean time of flight of more than 2 ns. Owing to the strong scattering and low absorption, most photons travel ten times the geometrical distance through the compressed breast. The signals T_1 and T_2 ($d = 45$ mm) overcome the background noise for a time of flight of about 510 ps, which is more than twice the minimum time of flight of a ballistic photon (refractive index of tissue 1.4). For a thicker tissue layer T_3 ($d = 59$ mm), this time shifts to 830 ps, which is about three times longer than for ballistic photons.

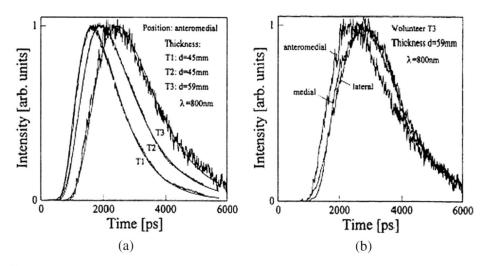

(a)   (b)

Figure 6.7 Normalized *in vivo* dispersion curves of the breasts of three volunteers (T_1, T_2, and T_3; thickness d) and corresponding theoretical fit curves: (a) three breasts in one position; (b) one breast in three positions.[288]

Measurements at different positions of the breast reflect the influence of physiological alterations within different areas of the organ (different types of tissue, different blood volumes, and oxygenation) and serves as a basis for diffuse optical mammography [see Fig. 6.7(b)]. It should be noted that the slightly different boundary conditions and degree of compression, as well as the inhomogeneity of superficial tissue pigmentation, can be critical for obtaining reliable mammograms. Table 2.1 presents the results of *in vitro*, *ex vivo*, and *in vivo* measurements of optical parameters of the human female breast and some other thick tissues carried out by the methods discussed as well as some other optical techniques.

Algorithms for the solving of the inverse problem on determination of μ_a and μ_s' using Eqs. (1.35) and (1.36) can be successfully used not only for tissue spectroscopy but also for tomography. For tomography purposes, we are not able to provide measurements of the absolute concentrations of absorbers and absolute values of the scattering coefficients (although this is desirable), which allow one to implement these algorithms faster. Generally, the imaging is aimed at the detection of pathology or at the localization of lesions. The detection of a lesion is achieved by recording a 2D image with sufficient contrast, while localization needs optical slicing and tomographic reconstruction to obtain the 3D images by which the size, shape, and position of the hidden object can be determined.[287]

Imaging systems usually use 2D or 3D scanning of a narrow laser beam or a translation optical stage with the object attached. Nonscanning systems are more robust and correspondingly fit medical applications much better. Such nonscanning systems use a multichannel fiber-optical arrangement with fixed positions of light sources and detectors, or low-noise, high-sensitivity, and fast CCD cameras with multichannel plate optical amplification. In any case, the measurement procedure is completed by sampling the intensity of each pixel as a function of time to obtain a time-space intensity mapping. The image is numerically reconstructed by attributing to each pixel the intensity measured over the selected integration time.[287]

A multichannel NIR imager/spectrometer based on the time-correlated single-photon counting technique was designed for breast imaging in clinics [see Fig. 6.8(a)].[1125] The instrument uses two NIR wavelengths, 780 and 830 nm, the mean power of each laser diode is about 40 µW, and they pulse at 5 MHz with a pulse width of about 50 ps. A highly sensitive R5600U-50 GaAs photomultiplier has been chosen for breast examination. For enhancement of the contrast of carcinoma images, intravenous administration of Infracyanine 25 (IC25), an NIR contrast agent, was used. Optical absorption changes were calculated using the following relation:

$$\Delta\mu_a = -\frac{2}{c\Delta t^2} \int_{t_1}^{t_2} \ln\frac{J_2(r,t)}{J_1(r,t)} dt, \qquad (6.11)$$

where c is the speed of light in the medium; Δt is the time resolution of the pulse-height analysis (PHA, Hamamatsu Inc.) of the multichannel analyzer (MCA, Hamamatsu Inc.); J_1 and J_2 are the photon current measurement pre- and post-IC25 injection, respectively, and t_1, t_2 are the width of the J_1 time-resolved curve. Equation (6.11) gives accurate values of $\Delta\mu_a$ for small absorption changes.

The relative displacement of the light sources and detectors, and the positions of the projection plane and pathology (carcinoma, black sphere) are shown in Fig. 6.8(b). The calculations of the absorption coefficient differences along the straight lines connecting those source-detector pairs in space that have comparable separations [shown as diamonds in Fig. 6.8(b)] allow for estimation of the IC25 distribution in tissue. Tests with patients were done simultaneously with the standard MR imaging (MRI) examination protocol.

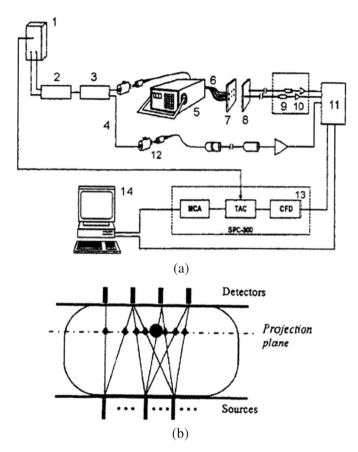

(a)

(b)

Figure 6.8 Multichannel, time-correlated single-photon counting NIR imager/spectrometer: (a) 1, two laser diodes; 2, a wavelength coupler; 3, 19:1 signal splitter; 4, the reference branch; 5, 1 × 24 optical DiCon fiber-optics switch; 6, graded index, 10-m-long optical fibers; 7, compression plates; 8, 8-step index, 10-m-long fiber bundles; 9, PMT; 10, amplification unit; 11, router; 12, attenuator; 13, SPC-300 photon counting system using an SRT-8 8-channel multiplexer; 14, Intel Pentium PC; CFD, constant-fraction discriminator; MCA, multichannel analyzer; TAC, time-to-amplitude converter. (b) Relative displacement of the light sources and detectors, and the positions of the projection plane and pathology (carcinoma, black sphere).[1125]

Figure 6.9 illustrates the possibilities of time-resolved optical diffusion mammography in comparison with MRI study. Measurements were done for a patient (70-year-old Caucasian) diagnosed with an infiltrating ductal carcinoma approximately 10 mm in diameter. For the presented image, six sources and eight detectors were employed. A good correspondence between the MRI and the NIR images is easily seen. In the 780-nm light image, two objects are resolved, either due to measurement noise or due to actual physiology of the tissue. Owing to more absorption (12%) of IC25 at 780 nm than at 830 nm, expected differences should be enhanced more.

A much more comprehensive time-resolved optical tomography system employing 32 channels and designed for imaging of the neonatal brain and the hu-

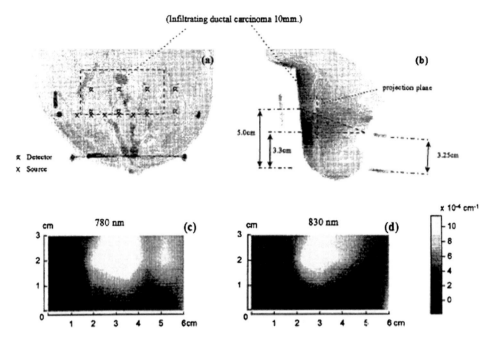

Figure 6.9 MRI and NIR image coregistration. (a) Saggital fast spin echo (FSE) MRI slice from a 70-yr-old patient with infiltrating ductal carcinoma. (b) Spin echo (SE) MRI axial image of the same patient. (c) NIR projection image at 780 nm. (d) NIR projection image at 830 nm.[1125]

man breast is shown in Fig. 6.10.[1126] This is the multichannel optoelectronic near-infrared system for time-resolved image reconstruction (MONSTIR). Light from a pulsed high-power picosecond laser source is switched sequentially into one of 32 fibers that are attached to the surface of an object under study. The detection system is used to record the temporal distribution of light exiting the tissue at certain positions around the object with a temporal resolution of about 80 ps and a rate of photon counting up to a few 10^5 per second per channel. This is accomplished by utilizing 32 fully simplex ultrafast photon-counting detectors. The scattered photons are collected by 32 low-dispersion, large-diameter (2.5-mm) fiber bundles that are coupled to 32-stepper motor-driven variable optical attenuators (VOAs). Because of the large dynamic range of light intensities around the object, the VOAs are required to ensure that the detectors are not saturated or damaged and that the system operates within the single-photon counting mode. Light transmitted via VOA is collected by a short 3.0-mm diameter single polymer fiber and then is transmitted via a visible blocking filter to the photocathodes of four ultrafast eight-anode multichannel plate-photomultiplier tubes (MCP-PMTs). The resulting electronic pulse is preamplified and converted into a logic pulse, and a histogram of the photon flight times is recorded and transfered to the control computer. A dedicated image reconstruction software package, TOAST (time-resolved optical absorption and scattering tomography, http://www.medphys.ucl.ac.uk/toast/index.html),

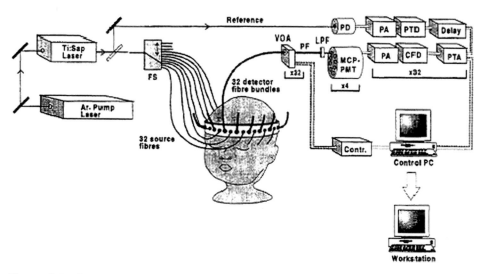

Figure 6.10 Schematic diagram of the MONSTIR imaging system: FS, fiber switch; VOA, variable optical attenuator; PF, polymer fiber; LPF, long-pass filter; MCP-PMT, multichannel plate-photomultiplier tube; PA, preamplifier; CFD, constant fraction discriminator; PTA, picosecond time analyzer; PD, photodiode; PTD, picotiming discriminator.[1126]

is used for the reconstruction of the tomographic images of the absorption and scattering profiles.

A portable three-wavelength NIR time-resolved spectroscopic (TRS) system (TRS-10, Hamamatsu Photonics K.K., Japan) is available on the market.[1127] In the TRS system, a time-correlated single-photon-counting technique is used for detection. The system is controlled by a computer through a digital I/O interface consisting of a three-wavelength (761, 795, and 835 nm) picosecond (about 100 ps) pulsed light source, a photon-counting head for single-photon detection, and signal-processing circuits for time-resolved measurement. The average power of the light source is at least 150 μW at each wavelength at a repetition rate of 5 MHz. The instrumental response of the TRS system, which included a 3-m length light source fiber (graded index type single fiber with a core diameter of 200 μm) and a 3-m length light detector fiber (a bundle fiber of 3-mm diameter), was around 150 ps full width at half maximum (FWHM) at each wavelength. This system was used for estimating the absorption and reduced scattering coefficients of the head in a piglet hypoxia model.[1127] Measurements of absolute values of the absorption coefficient at three wavelengths enable estimation of the hemoglobin concentration and its oxygen saturation in the head.

6.2.2 Frequency-domain techniques and instruments

Considerable progress in the investigation of tissues and molecules of biological importance with the use of the modulation technique provided the foundation for the development and commercial production of spectrometers of a new type (for example, ISS Fluorescence & Analytical Instrumentation). A typical scheme of the frequency-domain spectrometer for tissue study is shown in Fig. 6.11.[1–4,6,301–303,311,312] Such systems for phase measurements use a heterodyning principle (two photomultipliers with heterodyning in one of the first dynodes) to transfer a measuring signal to a low-frequency range (100 Hz for the shown system), where phase measurement can be done much more precisely.

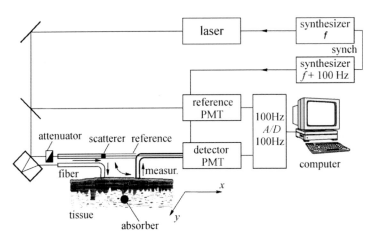

Figure 6.11 A typical scheme of a frequency-domain light-scattering spectrometer or photon-density wave imaging system (if 2D scanning of irradiating and receiving fibers is provided).[1,3]

We should also mention models of compact and comparatively cheap devices—modulation spectrometers—for noninvasive quantitative determination of oxygen saturation of blood hemoglobin, monitoring of optical parameters of tissues, and localization of absorbing or scattering inhomogeneities inside a tissue. Such spectrometers include diode lasers as radiation sources at one or two wavelengths and a photomultiplier with heterodyning in one of the first dynodes or a fast semiconductor photodetector with a high-frequency amplifier.[303,325,326] Specifically, NIM Incorporated produces a PMD 3000b two-wavelength spectrometer (with $\lambda = 760$ and 810 nm and a fixed modulation frequency of 200 MHz) for noninvasive quantitative determination of oxygen saturation of hemoglobin.[1128] Carl Zeiss Jena is the manufacturer of a more sophisticated and universal system, which operates at the wavelength of 685 nm with two fixed modulation frequencies equal to 110 and 220 MHz. This system also includes a computer-controlled optical table, which ensures the regime of tissue transillumination.[325,326] A much simpler and more universal research system has been developed and manufactured at Saratov University.[1129] This system includes quantum-well lasers ($\lambda = 790$ and 840 nm), which ensure a highly efficient low-noise modulation of laser radiation within the range of 100–1000 MHz; a set of optical fibers; and a computer-controlled optical table, which allows one to implement different geometric schemes for an experiment. The detection unit employs an avalanche photodiode with a high-frequency amplifier (20 dB). The total dynamic range of the detection unit along with a spectrum analyzer or a network analyzer is 70 dB.

From the point of view of medical devices, the requirements of a phase-measuring system are very high (better than 0.03 deg in a 2-Hz bandwidth) and close to that imposed to multifrequency, multiwavelength optical-fiber communication systems [time division multiplex (TDM) and wavelength division multiplex (WDM)].[4] Communication systems work at much higher modulation frequencies than medical ones and are well developed in their usage of digital equipment, and have a high degree of multiplexing. The last two features should be very useful for the designing of a new generation of medical equipment. While requirements for medical systems are currently quite modest (three wavelengths and two modulation frequencies), the appearance of the first generation of optical tomographs with a spatial resolution of about 1 cm^{-3} increases the need for multiplexing up to 16/32 channels. In the near future, for providing of a resolution much less than 1 cm^{-3}, the use of 10^3 source-detection combinations is expected.[4]

Phase systems are divided into homodyne and heterodyne groups, which means that, respectively, they do not and they do convert the radio frequency (RF) prior to phase measurements. Heterodyne systems have been termed "cross-correlation" or phase-delay measurement devices (PDMDs). These devices are intended to measure tissue optical properties (μ_a and μ'_s) to an accuracy of 5% and hemoglobin saturation to an accuracy of 3% in the 40–80% range, requiring phase and amplitude precision as follows[4]:

- Phase and amplitude noise in a 2-Hz bandwidth should be less than 0.03 deg and 0.1% of the total signal at a carrier frequency of 50–200 MHz.
- Source-to-detector attenuation may be more than 100 dB, with radio RF coupling causing less than 0.03 drg phase error.
- Amplitude-phase cross talk should be limited—a signal attenuation of 10 dB should not cause more than a 0.03 deg phase error.
- Multifrequency operation should not cause more than 0.03 deg phase interchannel cross talk (at 50 dB attenuation).
- Optical multiplexing employing light sources of different wavelengths should cause less than 0.03 deg phase interchannel cross talk.
- Bandwidth signal output should be variable from 0.2 to 2 Hz, or in special cases of brain study, 40 Hz.
- Sufficient information from multiple RF or multiwavelength operation should be available.

Four types of PDMDs adapted to study tissue optical characteristics study are presented in Fig. 6.12.[4] Two are homodyne [(a) and (b)] and two are heterodyne [(c) and (d)]. The system in (a) uses an in-phase quadrature (IQ) demodulator;

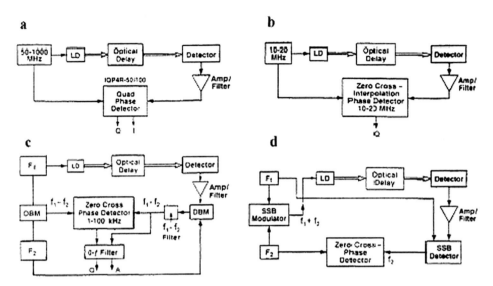

Figure 6.12 Four types of optical propagation delay measurement devices for tissue study: (a) and (b) homodyne systems; (a) has IQ demodulation, (b) a zero-crossing phase detector. (c) and (d) heterodyne systems; (c) is amplitude modulated and (d) is single sideband (SSB) with F_1 as the RF oscillator and F_2 as the local oscillator (audio); the upper sideband $f_1 + f_2$ is used.[4] LD, laser diode; DBM, double-balanced mixer.

(b) uses a zero cross-phase detector. The system in (c) uses amplitude modulation at two close RF, f_1 and f_2, and (d) uses single sideband (SSB) modulation; f_1 is a RF and f_2 is an audio frequency; both systems with zero cross-phase detectors.

A number of phase-measurement systems are described in Refs. 1 and 4. The amplitude measurements are relatively simple, but sometimes they do not provide the accuracy needed because, for example, of the influence of stray light. Phase measurements are amplitude independent and can be carried out with acceptable accuracy. Moreover, multiwavelength phase measurements alone are sufficient to estimate such important quantities as hemoglobin concentration and its degree of oxygenation. Simultaneous amplitude and phase measurements are used for the determination of absolute values of absorption coefficients.

The basic form of a homodyne system with an IQ demodulator is presented in Fig. 6.12(a). Determining the phase shift path (the phase difference between the reference oscillator and the signal pathway) involves a laser diode, an optical detector, an amplifier, and a narrowband filter. The working principle of an IQ demodulator is shown in Fig. 6.13. It includes a 90-deg splitter (hybrid), two double-balanced mixers (DBMs), and a 0-deg splitter. In the demodulator, the carrier (as reference signal) is recovered from an incoming modulated signal and fed to the 90-deg hybrid and the modulated signal (as the signal under test) is fed to the 0-deg hybrids.

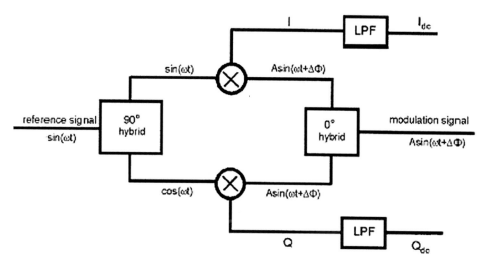

Figure 6.13 Diagram of IQ demodulator as a phase and amplitude detector;[1130] LPF, low-pass filter.

Functioning as a multiplier, the in-phase mixer produces an output of[4,1130]

$$I(t) = \sin(\omega t)2A\sin(\omega t + \Delta\Phi) = A\cos\Delta\Phi - A\cos(2\omega t + \Delta\Phi), \quad (6.12)$$

where $\sin(\omega t)$ is the carrier signal, $2A\sin(\omega t + \Delta\Phi)$ is the modulated signal, and the phase delay $\Delta\Phi$ is caused by the scattering medium.

The quadrature mixer produces an output of

$$Q(t) = \cos(\omega t)2A\sin(\omega t + \Delta\Phi) = A\sin\Delta\Phi + A\sin(2\omega t + \Delta\Phi), \quad (6.13)$$

where $A\sin\Delta\Phi$ and $A\cos\Delta\Phi$ are dc signals that carry the information on amplitude (A) and phase ($\Delta\Phi$) caused by light interaction with the scattering medium. $A\sin(2\omega t + \Delta\Phi)$ and $A\cos(2\omega t + \Delta\Phi)$ are high-frequency components that are blocked by using low-pass filters (LPFs); therefore, after filtration, such signals as I_{dc} and Q_{dc} are registered. The phase and amplitude caused by a medium can be found from the equations

$$\Delta\Phi = \tan^{-1}\left(\frac{Q_{dc}}{I_{dc}}\right), \quad A = \left\{Q_{dc}^2 + I_{dc}^2\right\}^{1/2}. \quad (6.14)$$

For backscattering geometry, such as that presented in Fig. 6.2, the analytical expressions for the phase shift $\Delta\Phi$ and modulation amplitude A in the diffusion approximation are defined as follows:[4,325,326]

$$\Delta\Phi = r_{sd}\left\{\frac{[(\mu_a c)^2 + \omega^2]^{1/2} - \mu_a c}{D}\right\}^{1/2} + \Delta\Phi_0, \quad (6.15)$$

$$A = \left(\frac{A_0}{4\pi D r_{sd}}\right) \exp\left(-r_{sd}\left\{\frac{[(\mu_a c)^2 + \omega^2]^{1/2} + \mu_a c}{2D}\right\}^{1/2}\right), \qquad (6.16)$$

where r_{sd} is the source-detector separation, $\Delta\Phi_0$ is the initial phase due to the instrumental response, A_0 is the initial amplitude due to the instrumental response, $D \approx c/(3\mu_s')$, and c is the speed of light in the medium.

For relatively small modulation frequencies, when $\omega < \mu_a c$, the phase shift is a linear function of frequency,

$$\Delta\Phi = \left(\frac{r_{sd}\omega}{2}\right)(D\mu_a c)^{1/2} + \Delta\Phi_0 \approx \left(\frac{r_{sd}\omega}{2c}\right)\left(\frac{3\mu_s'}{\mu_a}\right)^{1/2} + \Delta\Phi_0. \qquad (6.17)$$

For relatively large frequencies, when $\omega > \mu_a c$ ($\omega/2\pi \leq 500$ MHz),

$$\Delta\Phi = r_{sd}(\omega/2D)^{1/2} + \Delta\Phi_0 \approx r_{sd}(3\omega\mu_s'/2c)^{1/2} + \Delta\Phi_0. \qquad (6.18)$$

A_0 and $\Delta\Phi_0$ can be calibrated by using a standard model (phantom) with known μ_s' and μ_a. Then, after calibration of the experimental setup, optical parameters of the tissue under study can be calculated from the measured amplitude and phase shift on the basis of Eqs. (6.10) and (6.11) using the following iteration formulas:

$$\mu_a = [r_{sd}^4 \omega^2 - 4D^2(\Delta\Phi - \Delta\Phi_0)^4]/4cD(\Delta\Phi - \Delta\Phi_0)^2 r_{sd}^2,$$

$$\mu_s' = c/3D - \mu_a, \qquad (6.19)$$

$$D = -r_{sd}^2 \omega/2(\Delta\Phi - \Delta\Phi_0)[\ln(A/A_0) + \ln(4\pi D r_{sd})].$$

Therefore, the homodyne system measures the phase difference between the reference oscillator and the signal pathway. The analogue IQ detector (see Fig. 6.13) allows one to reach an accuracy of $\sim$0.2 deg in phase and $\sim$0.5 dB in amplitude with carrier frequencies of 140 MHz.[4,1130]

The heterodyne principle is characteristic for many communication systems. Since the error of the phase measurements decreases when oscillator frequency slows down and the bandwidth of a detector is constant, instruments designed for a low-frequency range may give higher accuracy. The nonlinear mixing of two signals with different frequencies, f_1 and f_2, gives signals with the sum and difference frequencies, one of which, namely, $(f_1 - f_2)$, is selected, amplified, filtered, and coupled to a phase detector as a reference signal [see Fig. 6.12(c)]. Propagation of the modulation signal f_1 through the optical system, the biological tissue, the optical detector, and the amplifier/filter leads to phase and amplitude changes. The signal on an intermediate frequency [$(f_1 - f_2)$, 1–100 kHz] is obtained from a second mixer and serves as the measuring signal for the phase detector. The intermediate frequency should be high enough to avoid a $1/f$ noise problem and low enough to exclude high-frequency errors in zero-crossing phase detection. The drawback of the heterodyne system is that the oscillators F_1 and F_2 must have a phase coherence equal to the required system accuracy.

The SSB system [see Fig. 6.12(d)] provides both efficient light modulation and efficient signal detection. It has important advantages: (1) the carrier modulation and the laser diode modulation are present only when the local oscillator (F_2) activates the sideband selected (thus, convenient control of RF light modulation is available); (2) the local oscillator frequency can be in the convenient audio range; and (3) all of the RF power is in a single narrow band of frequencies set by the low-frequency oscillation.

As an example, let us consider in more detail the functioning of a two-wavelength heterodyne IQ detection system, presented in Fig. 6.14.[4,1130] The two-wavelength NIR systems are usually used to detect the hemoglobin saturation of living tissue. Two RF signal sources are used and operate at slightly different frequencies, namely, 140.00 and 140.01 MHz, which provide the driving signals for two laser diodes with different wavelengths. The two laser beams are combined and directed simultaneously with the fiber coupler to the tissue under study. The optical signals collected from the tissue are fiber coupled to the PMT (or a number of PMTs). After passing an amplifier, the two-wavelength optical signals go into each IQ demodulator at the same time (the signal differentiation is due to different RF frequencies).

If channel 1 is characterized by the RF signal $\sin(\omega_1 t)$ and the detected signal is characterized by $2A_1 \sin(\omega_1 t + \Delta\Phi_1)$, and channel 2 by $\sin(\omega_2 t)$ and $2A_2 \sin(\omega_2 t + \Delta\Phi_2)$, then the IQ signals for each channel can be expressed as

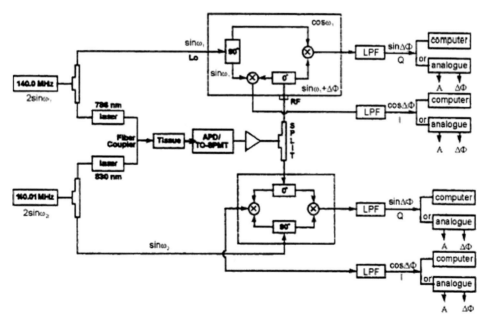

Figure 6.14 A two-wavelength phase-modulated spectroscopy system operating at 140 MHz and using analog IQ demodulation (see text for details).[4]

the following:[1130]

$$I_1(t) = [2A_1 \sin(\omega_1 t + \Delta\Phi_1) + 2A_2 \sin(\omega_2 t + \Delta\Phi_2)] \sin(\omega_1 t)$$
$$= A_1 \cos\Delta\Phi_1 - A_1 \cos(2\omega_1 t + \Delta\Phi_1) + A_2 \cos[(\omega_1 - \omega_2)t + \Delta\Phi_2]$$
$$- A_2 \cos[(\omega_1 + \omega_2)t + \Delta\Phi_2],$$

$$Q_1(t) = [2A_1 \sin(\omega_1 t + \Delta\Phi_1) + 2A_2 \sin(\omega_2 t + \Delta\Phi_2)] \cos(\omega_1 t)$$
$$= A_1 \sin\Delta\Phi_1 + A_1 \sin(2\omega_1 t + \Delta\Phi_1) + A_2 \sin[(\omega_1 - \omega_2)t + \Delta\Phi_2]$$
$$+ A_2 \sin[(\omega_1 + \omega_2)t + \Delta\Phi_2], \tag{6.20}$$

$$I_2(t) = [2A_1 \sin(\omega_1 t + \Delta\Phi_1) + 2A_2 \sin(\omega_2 t + \Delta\Phi_2)] \sin(\omega_2 t)$$
$$= A_2 \cos\Delta\Phi_2 - A_2 \cos(2\omega_2 t + \Delta\Phi_2) + A_1 \cos[(\omega_1 - \omega_2)t + \Delta\Phi_1]$$
$$- A_1 \cos[(\omega_1 + \omega_2)t + \Delta\Phi_1],$$

$$Q_2(t) = [2A_1 \sin(\omega_1 t + \Delta\Phi_1) + 2A_2 \sin(\omega_2 t + \Delta\Phi_2)] \cos(\omega_2 t)$$
$$= A_2 \sin\Delta\Phi_2 + A_2 \sin(2\omega_2 t + \Delta\Phi_2) + A_1 \sin[(\omega_1 - \omega_2)t + \Delta\Phi_1]$$
$$+ A_1 \sin[(\omega_1 + \omega_2)t + \Delta\Phi_1].$$

This system uses two low-pass filters (LPF): one is a dc 1.9-MHz band to reject the high-frequency components ($2\omega_1$, $2\omega_2$, $\omega_1 + \omega_2$) in each channel; another is a dc 10-kHz band to block the low-frequency component ($\omega_1 - \omega_2 = 10$ kHz). According to Eqs. (6.20), such filtration allows one to separate the combined signal into two signals for each wavelength, and each channel itself contains only I and Q signals [see underlined terms in Eqs. (6.20)]. However, it was shown experimentally that the third-order mixing effects influence the low-frequency cross-correlation between channels. The interchannel cross talk for a phase is less than 1.4 deg/dB, and for amplitude is less than 3.8 mV/dB (phase or amplitude changes in one channel caused by changes of amplitude in another one).

For more effective separation of signals, a fast Fourier transform analysis can be used. A much simpler solution is to use time-share control of the system. The computer-controlled time share ensures that at any one time only one wavelength optical signal may pass through the whole system. In this way, the interchannel cross talk can be reduced for phase up to 0.1 deg/dB and for amplitudes up to 0.5 mV/dB.[4]

6.2.3 Phased-array technique

In the NIR region, the wavelengths of diffusive photon-density waves in tissues are equal to 5–14 cm for modulation frequencies from 500 to 100 MHz [see Eqs. (1.47) and (1.48)]. This means that there is low resolution of imaging with the usual source and detection combination in spite of the high accuracy of phase and amplitude measurements. The photon-density waves interference method described for the first time in Ref. 331 (phase and amplitude cancellation method, or phased-array method) is very promising for the improvement of the spatial resolution of the modulation technique.[4,53,342]

The concept of this method is illustrated in Fig. 6.15. It is based on the use of either duplicate sources and a single detector or duplicate detectors and a single source so that the amplitude and phase characteristics can be nulled and the system becomes a differential. If equal amplitude at the 0- and 180-deg phases are used as sources, appropriate positioning of the detector can lead to null in the amplitude signal and a crossover between the 0- and 180-deg phase shifts, i.e., at 90 deg:

$$A \sin(\omega t + 0) + A \sin(\omega t + 180) = 2A \cos(90) \sin(\omega t + 90), \qquad (6.21)$$

where ω is the light modulation frequency.

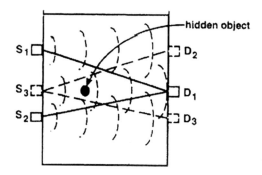

Figure 6.15 The geometry of the amplitude and phase cancellation technique; two sources (S_1 and S_2) and a single detector (D_1), or two detectors (D_2 and D_3) and a single source (S_3).[1133]

In a heterogeneous medium, the apparent amplitude's null and the phase's crossover may be displaced from the geometric midline (see Fig. 6.16). This method is extremely sensitive to perturbation by an absorber or scatterer. A spatial resolution of about 1 mm for the inspection of an absorbing inhomogeneity has been achieved and the similar resolution is expected for the scattering inhomogeneity. Another good feature of the technique is that at the null condition, the measuring system is relatively insensitive to amplitude fluctuations common to both light sources. On the other hand, inhomogeneities, which affect a large tissue volume common to the two optical paths, cannot be detected. The amplitude signal is less useful in imaging since the indication of position is ambiguous (see Fig. 6.16). Although this can be accounted for by further encoding, the phase signal is robust and a phase noise less than 0.1 deg (signal-to-noise ratio is more than 400) for a 1-Hz bandwidth can be obtained.[4]

The phase modulation system requires, for optimal results, SSB measuring technology [see Fig. 6.12(d)]. Nine sources and four detectors are used in the 50-MHz single-wavelength (780 nm) phase-array imaging system presented in Fig. 6.17.[1131] The number of sources and detectors can readily be increased;

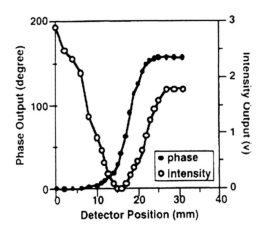

Figure 6.16 Demonstration of the existence of amplitude null and phase crossover for a phased-array measuring system used to study an adult human brain. The detector is scanned between two sources placed at 4 cm apart and excited by 0-deg and 180-deg phase-shifted RF signals at 200 MHz.[1131]

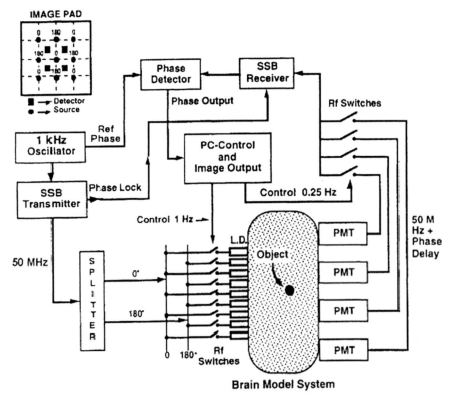

Figure 6.17 Single-wavelength (780 nm), 50-MHz phased-array, single sideband (SSB) phase modulation imaging system.[1131]

furthermore, the number of source and detector combinations can be increased simply by moving the source detector pad in two dimensions with respect to its original position by half the minimal distance between source and detector, equal to 2.5 cm. The image pad dimensions are 9×4 cm (see the upper left part of Fig. 6.17).

A local oscillator at 1 kHz modulates a 50 MHz SSB transmitter, the RF output of which is connected to a 0/180-deg phase splitter/inverter and then to nine switches appropriate to the nine light sources, which are sequenced at 0- and 180-deg phases by 1-Hz switches. The four PMTs are sequentially connected to the 50-MHz SSB receiver (0.5 mV sensitivity), and the audio output at 1 kHz is coupled to a zero-crossing phase detector to give the phase signals in sequence. The phase detector is coupled via a controlling computer to a computer for image computation and display (not shown). The transmitter and receiver are phase locked by RF coupling. The phase noise of the system is less than 0.1 deg (1-Hz bandwidth). A complete set of data from 16 source and detector combinations is obtained every 16 s.

Figure 6.18 illustrates a single-wavelength (780 nm), 50-MHz phased-array image test of neurovascular coupling in human brain. The image represents the increment in phase shift caused by touching the contralateral finger. Calibration with models verifies that this signal is due to an increase in hemoglobin concentration.[1131]

A more universal and comprehensive phased-array imaging system that can be used for testing the brain function of neonates is described in Ref. 1132. Single-wavelength laser diode light sources were replaced by a set of two laser diodes (750 and 830 nm, total of 18 lasers) with a 20-mW power source. The wavelength of 780 nm was shifted to 750 nm because at 750 nm the signal gain is more than

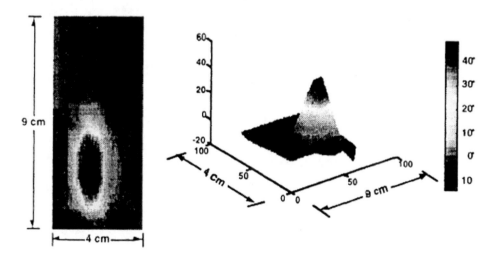

Figure 6.18 The image test for an adult human brain (blonde hair) using the 50-MHz phased-array system presented in Fig. 6.17. Parietal stimulation for 48 s by touching a contralateral finger was provided.[1131,1133]

double, owing to the respective extinction coefficients. Detection of the optical signals was provided by four PMTs (TO8, Hamamatsu). Two independent phasemeters and two SSB radio transmitters/receivers were used with 50- and 52-MHz frequencies. The size of the optical probe is slightly larger (10×5 cm) because two lasers are located in a point, but the source-detection separation was the same, 2.5 cm (see Fig. 6.17). A dual-wavelength phased-array imaging system can be used for testing the brain function of neonates and its relationship with some neurological disorders by monitoring metabolic activity, which is indicated by oxygen concentration or glucose intake to the brain cells.

Another dual-wavelength imaging system (750 and 830 nm) that uses a simple amplitude-cancellation technique (see Fig. 6.16) was used to image a human breast.[1133] The optical probe of the imager consists of 9 laser diode light sources and 21 silicon photodetectors. The imager sequences through all sources and detectors in a millisecond and gives high-quality breast tumor images every 8 s. As an example, in Fig. 6.19, four *in vivo* images of diseased and healthy breasts are presented. Since the difference image between the right and left breast is obtained, many of the background signals are eliminated and strong signals congruent with the expected position of the tumor are displayed (no evidence of the nipple is presented and two shapes for blood volume and for deoxygenation are clear).

The detection limit in the localizing of macroinhomogeneities hidden in a highly scattering tissue using phased-array imaging systems is discussed in Ref. 304.

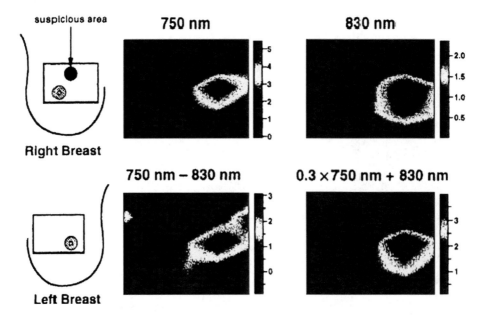

Figure 6.19 Four *in vivo* images of a diseased (with tumor) right breast with reference to the contralateral breast (left, healthy) at 750 nm and 830 nm, and two calculated images: 750–830 nm (deoxygenation image) and 0.3×750 nm + 830 nm (blood volume image).[1133]

6.2.4 *In vivo* measurements, detection limits, and examples of clinical study

Let us consider briefly a few spectroscopy and imaging frequency-domain systems that demonstrate the achievements in the field of optical *in vivo* diagnostics and that have been applied for clinical studies. It was shown previously that for the accurate evaluation of the absolute absorption and reduced scattering coefficients for a single source-detection position, frequency-dependent measurements of the amplitude and phase of photon-density waves should be provided. Therefore, to obtain quantitative measurements of the absolute optical parameters of various types of tissue, a portable, high-bandwidth (0.3–1000 MHz), multiwavelength (674, 811, 849, and 956 nm) frequency-domain photon migration (FDPM) instrument was designed[306,308] (see Fig. 6.20). The key component of an FDPM system is a network analyzer (8753C, Hewlett Packard) that is used to produce modulation sweep in the range of 0.3–1000 MHz. The RF from the network analyzer is superimposed on the direct current of four different diode lasers using individual bias tees and an RF switch. Four 100-μm diameter gradient-index fibers are used to couple each light source to an 8 × 8 optical multiplexer (GP700, DiCon Instruments). Dynamic phase reference and real-time compensation for source fluctuations were provided by an optical tap, which diverts a portion of the source output (5%) to a 1-GHz PIN diode coupled to the network analyzer channel B.

Light is launched onto the tissue under study using up to eight source fibers corresponding to eight source positions. An avalanche photodetector (APD) (C5658, Hamamatsu) is used to detect the diffuse optical signal. Both the APD and probe

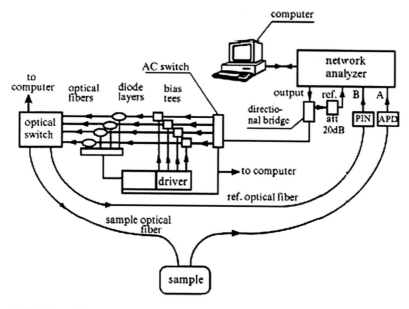

Figure 6.20 A multiwavelength, multifrequency, and multichannel frequency-domain spectrometer.[306,308]

end of the source optical fiber are in direct contact with the patient's skin surface. The optical power coupled into the tissue averages approximately 10–30 mW,
roughly a factor of ten below thermal damage threshold levels for used fibers
and wavelengths. Up to eight separate sources can be directed onto up to eight
unique measurement positions using the 8×8 optical multiplexer. Measurement
time depends on the precision required, the number of sweeps performed, and the
RF/optical switch time. For studies in human subjects, about 0.5 s is used to sweep
over the entire 1-GHz range of modulation frequencies. However, total elapsed
time for four laser diodes (two sweeps per laser), data transfer, display, and source
switching is about 40 s. The source-detector separation used for human subject
measurements was fixed and equal to 1.7, 2.2, or 2.7 cm.

The results of experimental study for three patients using the developed FD
spectrometer are presented in Tables 2.1 and 6.1. Table 6.1 also shows the calculated physiological parameters of a living tissue, such as absolute concentrations of
deoxy- and oxyhemoglobin, total hemoglobin, and water. It was assumed that the
chromophores contributing to the absorption coefficient μ_a in the human subject
are principally oxy- and deoxyhemoglobin, and water. Therefore, the concentration of each component in the tissue is determined from the FDPM measurements
of μ_a at three different wavelengths (674, 811, and 956 nm) in accordance with the
following system of three equations:

$$\varepsilon_{Hb}(\lambda_i)c_{Hb} + \varepsilon_{HbO_2}(\lambda_i)c_{HbO_2} + \varepsilon_{H_2O}(\lambda_i)c_{H_2O} = \mu_a(\lambda_i), \qquad (6.22)$$

where $\varepsilon_{chrom}(\lambda_i)$ is the extinction coefficient in units of $cm^{-1} mol^{-1}$ of a given
chromophore at the wavelength λ_i (674, 811, and 956 nm) defined by the matrix.

$$\begin{bmatrix} 6578300 & 740100 & 0.0748 \\ 1833100 & 2153900 & 0.427 \\ 1500600 & 3048600 & 7.24 \end{bmatrix} \begin{bmatrix} c_{Hb} \\ c_{HbO_2} \\ c_{H_2O} \end{bmatrix} = \begin{bmatrix} \mu_a (674) \\ \mu_a (811) \\ \mu_a (956) \end{bmatrix}. \qquad (6.23)$$

Each column of this matrix contains values of extinction coefficients for each of the
chromophores considered at three chosen wavelengths. The values of an absorption
coefficient at each wavelength were determined from experimental study.

The spectroscopy system discussed can be used as an imaging system as
well. Many FD imaging systems are described in the literature (see Refs. 1, 3,
4, 301–303, and 338). One of them was designed by the University of Pennsylva-
nia and NIM Inc. for regional imaging of brain tissue.[1134] The system can operate
at selectable RFs ranging from 50 to 400 MHz. A dual-wavelength light source
(two laser diodes at 779 and 834 nm), APD photodetection, and SSB modula-
tion/demodulation electronics are the main features of the imager. It was successfully used for a preliminary clinical study, i.e., the positions of the shunt components were defined on the basis of reconstructed images (at a depth of 1.2 cm) of
brain tissue of a patient with hydrocephalus [abnormal increase in the amount of
cerebrospinal fluid (CSF)] who was undergoing surgery to have a shunt replaced.

Table 6.1 Results of *in vivo* measurements of optical and physiological parameters of healthy and diseased tissues of patients (source-detector separation is equal to 2.2 cm, in the brackets given r.m.s. values)[306,308]

Tissue	λ, nm	μ_a, cm^{-1}	C_{Hb}, μM	C_{HbO_2}, μM	C_{Hb+HbO_2}, μM	C_{H_2O}, M
Female breast (56 yr):						
Normal	674	0.04				
	811	0.035	4.96	10.6	15.56	6.39
	849	0.035				
	956	0.085				
Fibroadenoma with ductal hyperplasia	674	0.055				
	811	0.06	5.65	22	27.65	6.02
	849	0.055				
	956	0.12				
Female breast (27 yr):						
Normal	674	0.035				
	811	0.03	4.1	8.13	12.23	9.4
	849	0.038				
	956	0.09				
Fluid-filled cyst	674	0.07				
	811	0.07	8.1	23.6	31.7	11.3
	849	0.08				
	956	0.16				
Multiple subcutaneous large-cell adenocarcinoma(male 62 yr):						
Abdominal:						
Normal tissue	674	0.0589 (0.0036)				
	811	0.0645 (0.0032)	6.22 (0.64)	23.9 (1.9)	30.1 (2.0)	4.09 (2.23)
	849	0.0690 (0.0025)				
	956	0.1110 (0.015)				

Table 6.1 (Continued).

Tissue	λ, nm	μ_a, cm^{-1}	C_{Hb}, μM	C_{HbO_2}, μM	C_{Hb+HbO_2}, μM	C_{H_2O}, M
Tumor	674	0.169 (0.02)	17.4 (3.6)	73.4 (8.3)	90.8 (9.0)	–
	811	0.190 (0.015)				
	849	0.276 (0.03)				
	956	–				
Back:						
Normal tissue	674	0.0883 (0.006)	9.68 (1.04)	33.2 (2.7)	42.9 (2.9)	–
	811	0.0892 (0.005)				
	849	0.0915 (0.0030)				
	956	0.127 (0.03)				
Tumor	674	0.174 (0.02)	19.1 (3.7)	66.0 (7.4)	85.1 (8.2)	–
	811	0.177 (0.013)				
	849	0.190 (0.01)				
	956	0.186 (0.16)				

A very stable and fast scanning and imaging system that uses the diffraction of diffuse photon density waves is described in Ref. 319. The system consists of an RF-modulated (100 MHz), low-power (about 3 mW) diode laser (786 nm). The source light is fiber guided to the tissue. A detection fiber couples the detected diffuse wave to a fast APD. SSB IQ demodulation electronics were used. The dynamic range of the system is about 2500. The source position was fixed, and a single detection fiber was scanned over a square region 9.3 × 9.3 cm; the amplitude and phase of the photon density wave were recorded at each position for a total of 1024 points. To obtain projection images of hidden macroinhomogeneities in a highly scattering tissue, imaging algorithms based on K-space spectral and fast Fourier transform (FFT) analysis were developed and tested clinically. The FFT approach has yielded clinical projection images with processing times much smaller than current collection times. It was shown that boundary effects present important problems. Matching substances might be used to reduce the boundary effects; nevertheless, the boundary effects may be incorporated in the reconstruction algorithm.

A schematic diagram of an FD optical mammography apparatus (LIMA), developed at Carl Zeiss is shown in Fig. 6.21.[325,326] It uses two diode lasers at 690 and 810 nm and the lasers' intensities are sinusoidally modulated at 110.0010 and 110.008 MHz, respectively. The average power is about 10 mW. Both laser beams (2 mm in diameter) are collimated, made collinear, and directed to the object. An optical fiber (5 mm in diameter) located on the opposite side of the breast delivers light to the detector. A PMT with modulated gain at 110 MHz is used as a detector. The differences in frequencies of light and gain modulation are $\Delta f_1 = 1$ kHz (relative to the signal at 690 nm) and $\Delta f_2 = 0.8$ kHz (relative to the signal at 810 nm), and are called cross-correlation frequencies. Appropriate electronic filtering allows separation of signals at these frequencies, i.e., at the two wavelengths.

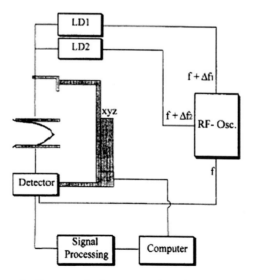

Figure 6.21 Schematic diagram of a frequency-domain mammograph (LIMA).[325,326]

The breast is slightly compressed between two parallel glass plates. The dual-wavelength laser beam and the detector fiber are scanned in tandem along the upper and lower plane, respectively, so that source-detection separation is fixed. The entire compression assembly with the two glass plates can be rotated by 90 deg to allow data to be acquired in craniocaudal and mediolateral projections. The extension of the scanning step (the image pixel size) can be set by software, but it is generally defined by the spatial resolution needed, the total acquisition time, and the signal-to-noise ratio. For this system, a scanning step of 1.5 mm in both directions requires a total acquisition time of about 3 min for a whole mammogram and has noise of about 2 deg for phase and 0.1% for amplitude measurements. The boundary effects were overcome using an appropriate algorithm [$N(x, y)$ function] based on the idea of exploring the phase information in a given pixel (x, y) to obtain an estimate of the breast thickness at that pixel. As a second step, the dependence of the amplitude signal on tissue thickness is modeled using the empirically determined dependence on the thickness in the optically homogeneous case. The LIMA system was clinically tested on 15 patients affected by breast cancer.

Two mammograms, x ray and optical (810 nm), for a female left breast with a tumor are presented in Fig. 6.22. A comparison of these mammograms clearly shows that this optical technique has good contrast and tumor detectability, rather than high spatial resolution, which is intrinsically limited by the diffusive nature of light propagation in tissue. The promise of optical imaging methods lies in high contrast, detectability, and specificity, which provide diagnostic capabilities. Further enhancement in contrast can be achieved by introducing additional light sources, wavelengths, modulation frequencies, and/or multiple detectors (see above discussion). In addition to contrast enhancement, FD and TD methods have the potential to provide an *in situ* optical biopsy by measuring localized optical properties.[306,308]

One of the first phase-imaging systems for *in vivo* studies was designed at the University of Illinois, Urbana-Champaign.[328] The optical signal at 760 nm from a mode-locked titanium:sapphire laser (Mira 900, Coherent) was modulated at 160 MHz. Heterodyne mixing at the dynode chain of the PMT produces a cross-correlation signal (1.25 kHz) carrying the same phase and amplitude information as the original signal. The imaging system provides subsecond data integration times per pixel (10^4 total pixels, 8×8 cm grid in a gradation of 101 steps of 0.8 mm each), resulting in a total measurement time of about 10 min. To partially compensate for limits on the detector's dynamic range and reduce the influence of boundary effects, the human hand under investigation was immersed in a highly scattering aqueous solution of Liposyn III (20%) (an intravenous fat emulsion) with the scattering and absorption properties approximately matched to those of the hand by diluting the emulsion with water and serial additions of black India ink.

An FD tissue spectrometer described in Ref. 305 uses an arc lamp as a light source that is intensity modulated at 135 MHz by a Pockel's cell and the heterodyne mixing at the dynode chain of the PMT with a cross-correlation signal at

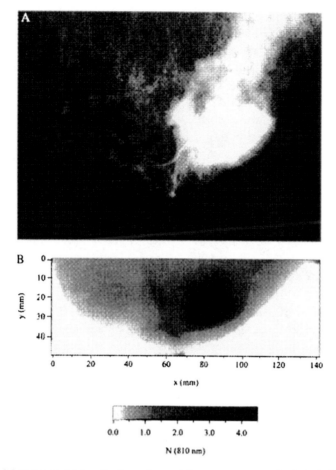

Figure 6.22 (a) X-ray and (b) optical craniocaudal mammograms of a female left breast with a tumor (55-yr-old Caucasian woman with an invasive ductal breast cancer—lateral lower quadrant; the major tumor is 3.0 cm in diameter). The x-ray and optical images cannot be compared point by point because the degree of compression and the compression geometry are different in the x-ray and optical measurements.[325,326]

100 Hz. For typical signal levels, the noise in this system is dominated by photon shot noise. The effects of the noise can be minimized by using a phased-array configuration with two detectors. For best results, signals from the two detectors should be equalized so that noise in the weaker signal is not dominant. The system can provide, at best, about 4% uncertainty in μ_a and μ_s' if the signals at the two detectors are equalized. Increasing the modulation depth and frequency allows random errors to be further reduced up to about 1%.

Systematic errors caused by finite tissue volumes and curved surfaces can be much larger than random errors induced by shot noise. As discussed earlier, these systematic errors can presumably be reduced if appropriate scattering and absorbing immersion surrounding a substance is applied or enough information about the tissue geometry is available to justify the use of a more corrected algorithm.[325,326,328] *In vivo* measurements made on the femoral biceps muscle of rabbits show that it is difficult to achieve shot noise limits in practice. The rms values for μ_a and μ_s' are typically 20% for a 15-s measurement time because the shot noise contribution is estimated to be about 8% in μ_a and 4% in μ_s'. This means that other sources of variation (tissue blood content or oxygenation, tissue inhomogeneity when scanning, finite source and detection size, uncertainty in their relative positions, etc.) with time were more important than the inherent instrument noise in determining the precision of the μ_a and μ_s' estimates.

The results of measurements of absorption and reduced scattering coefficients through the forehead on 30 adult volunteers using a multidistance FD NIR spectrometer (Imagent, ISS, Champaign, IL) were reported.[1135] The spectrometer employs laser diodes modulated at the frequency of 110 MHz and PMTs whose gain is modulated at a slightly offset frequency of 110.005 MHz to heterodyne the high frequency down to the frequency of 5 kHz. In studies described in Ref. 1135, 33 laser diodes (16 at 758 nm and 16 at 830 nm) and four PMTs were used. The laser diodes were multiplexed so that two lasers with the same wavelength and at the same location were on simultaneously. The light from the lasers was guided by optical fibers with a core diameter of 400 μm to the tissue surface and the photons reemitted from the tissue were collected simultaneously by the fiber bundles with a diameter of 5.6 mm, placed several centimeters apart from the source fibers. The collected light was carried to the PMTs and then the signals from the PMTs were digitally processed to yield the average intensity, modulation amplitude, and phase difference. These data were used for accurate estimation of the absolute absorption and reduced scattering coefficients of the adult brain. It was found that the adult head can be reasonably described by a two-layer model and the nonlinear regression for this model can be used to accurately retrieve the absolute absorption and reduced scattering coefficients of both layers if the thickness of the scalp/skull is known. For example, optical coefficients of the brain were estimated at 830 nm as $\mu_a = 0.145 \pm 0.005$ cm^{-1} and $\mu_s' = 4.1 \pm 0.1$ cm^{-1}. The hemoglobin concentration and oxygen saturation of the adult brain were also calculated with sufficiently good accuracy to provide monitoring of cerebral oxygen saturation and hemodynamics in order to assess cerebral health related to tissue oxygen perfusion.

A portable, multiwavelength, FD, NIR spectroscopy instrument similar to that shown in Fig. 6.14 was used for investigation of the optical properties of the brain in 23 neonates *in vivo*.[1136] It was found that the absorption coefficients of the infant forehead are lower than the values reported for adults and, being averaged for 23 infants, were equal to $\mu_a = 0.078 \pm 0.014$ cm^{-1} at 788 nm and $\mu_a = 0.089 \pm 0.019$ cm^{-1} at 832 nm. A large intersubject variation in μ_s' was also demonstrated, $\mu_s' = 9.16 \pm 1.22$ cm^{-1} at 788 nm and $\mu_s' = 8.42 \pm 1.23$ cm^{-1} at 832 nm. Physiological parameters derived from the absorption coefficients at two wavelengths were determined as the following: the mean total hemoglobin concentration was 39.7 ± 9.8 μM and the mean cerebral blood oxygen saturation was $58.7 \pm 11.2\%$. Therefore, it was shown that the bedside FD, NIR spectroscopy could provide quantitative optical measurement of the infant brain.

6.3 Light-scattering spectroscopy

Novel techniques capable of identifing and characterizing pathological changes in human tissues at the cellular and subcellular levels and based on light scattering were recently described.[47,61,94,95,129,130,150,170,180,450,452,620,731–735,798,811,1137] Light-scattering spectroscopy (LSS) provides structural and functional information of a tissue. This information, in turn, can be used to diagnose and monitor disease. One important application of biomedical spectroscopy is the noninvasive early detection of cancer in human epithelium.[180,452,620,732–734,1137] The enlarging, crowding, and hyperchromaticity of epithelium cell nuclei are the common features to all types of precancerous and early cancerous conditions. LSS can be used for detection of early cancerous changes and other diseases in a variety of organs such as esophagus, colon, uterine cervix, oral cavity, lungs, and urinary bladder.[452,620,733,734,1137] Eye lens cataract and other ophthalmic diseases can also be diagnosed using LSS.[811]

Cells and tissues have complex structures with very a broad range of the scatterers' sizes: from a few nanometers, the size of a macromolecule, to 7–10 μm, the size of a nucleus, and to 20–50 μm, the size of a cell itself. Most subcellular organelles are not uniform and have complex shapes and structures; nevertheless, they can be referred to as scattering "particles" (see Chapter 1). A great variety of cell organelle structures are small compared to the wavelength. Light scattering by such particles is known as Rayleigh scattering and is characterized by a broad angular distribution and a scattering cross-sectional dependence on the particle's linear dimension a as a^6 and on light wavelength λ as λ^{-4}. When the particle is not small enough, the coupled dipole theory or another approach such as the Rayleigh-Gans approximation (RGA) can be used. The RGA is particularly applicable to particles with sizes comparable to the wavelength and may be useful to study light scattering by small organelles such as mitochondria, lysosomes, etc. For RGA, the scattering in the forward direction prevails, and the total scattering intensity increases with the increase of the particle relative refractive index m as $(m - 1)^2$ and with its size as a^6.

The scattering by a particle with dimensions much larger than the wavelength, such as a cell nucleus, can be described within the framework of van de Hulst approximation that enables obtaining scattering amplitudes in the near-forward direction [see Eq. (2.25)]. For large particles, the scattered intensity is highly forward directed and the width of the first scattering lobe is about λ/a; the larger the particle, the stronger and narrower the first lobe. The intensity of the forward scattering, exhibits oscillations with the wavelength change. The origin of these oscillations is interference between the light ray passing through the center of the particle and one not interacting with it. The frequency of these oscillations is proportional to $a(m - 1)$, so it increases with the particle size and refractive index. The intensity of scattered light also peaks in the near-backward direction, but this peak is significantly smaller than the forward-scattering peak.

These results agree well with the rigorous scattering theory developed for the spherical particles (Mie theory).[148] To discriminate cell structure peculiarities, originated by the pathology, the difference in light scattering can be used. Structures with large dimensions and high refractive index produce a scattered field that peaks in the forward and near-backward directions in contrast to smaller and more optically "soft" structures, which scatter light more uniformly. Perelman et al.[150,180,620,732] studied elastic light scattering from densely packed layers of normal and T84 tumor human intestinal cells affixed to glass slides in a buffer solution (see Fig. 1.2). The diameters of the normal cell nuclei ranged from 5 to 7 μm, and those of the tumor cells from 7 to 16 μm. The reflectance from the samples exhibits distinct spectral features. The predictions of Mie theory were fit to the observed spectra. The fitting procedure used three parameters, average size of the nucleus, standard deviation in size (a Gaussian size distribution was assumed), and relative refractive index. The solid line of Fig. 6.23 is the distribution extracted from the data, and the dashed line shows the corresponding size distributions measured by light microscopy. The extracted and measured distributions for both normal and T84 cell samples were in good agreement, indicating the validity of the physical picture and the accuracy of the method of extracting information.

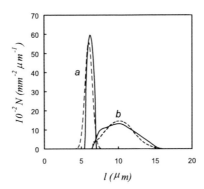

Figure 6.23 Nuclear size distributions of the samples presented in Fig. 1.2. (a) Normal intestinal cells; (b) T84 cells. In each case, the solid line is the distribution extracted from the data using Mie theory, and the dashed line is the distribution measured using light microscopy (from Ref. 180).

In tissues, the photons returned after a single scattering in the backward or near-backward directions produce a so-called single-scattering component. The photons returned after multiple scattering events produce diffuse reflectance. The spectra of both single-scattering and diffusive signals contain valuable information about tissue properties. However, the type of information is different. The single-scattering component is sensitive to the morphology of the upper tissue layer, which in the case of any mucosal tissue almost always includes or is limited by the epithelium. Its spectroscopic features are related to the microarchitecture of the epithelial cells, and the sizes, shapes, and refractive indices of their organelles, inclusions, and suborganellar components and inhomogeneities. Thus, analysis of this component might be useful in diagnosing diseases limited to the epithelium, such as preinvasive stages of epithelial cancers, dysplasias, and carcinomas *in situ* (CIS).[180,620,732,733,1137] The diffusive component contains information about tissue scatterers and absorbers as well, and its diagnostical possibilities and instrumentation are discussed early in this chapter.

The single-scattering component is more important in diagnosing the initial stages of epithelial precancerous lesions, while the diffusive component carries valuable information about more advanced stages of the disease. However, single-scattering events cannot be directly observed in *in vivo* tissues, because only a small portion of the light incident on the tissue is directly backscattered.

Several methods to distinguish single scattering have been proposed. Field-based light-scattering spectroscopy[735] and spectroscopic optical coherence tomography (OCT)[142] were developed for performing cross-sectional tomographic and spectroscopic imaging. In these extensions of conventional OCT, information on the spectral content of backscattered light is obtained by detection and processing of the interferometric OCT signal. These methods allow the spectrum of backscattered light to be measured either for several discreet wavelengths[734] or over the entire available optical bandwidth from 650 to 1000 nm simultaneously in a single measurement.[142]

A much simpler polarization-sensitive technique based on the fact that initially polarized light loses its polarization when traversing a turbid tissue is also available.[150] The conventional spatially resolved backscattering technique with enough small source-detector separation can be used as well.[180] In that case, the single-scattering component (2–5%) should be subtracted from the total reflectance spectra, which can be done using the diffusion approximation-based model by fitting to the coarse features of the diffusive component.

Zonios et al. studied the capability of diffuse reflectance spectroscopy to diagnose colonic precancerous lesions and adenomatous polyps *in vivo*.[735] Figure 6.24 shows typical diffuse reflectance spectra from one adenomatous polyp site and one normal mucosa site. Significant spectral differences are readily observed, particularly in the short-wavelength region of the spectrum, where the hemoglobin absorption valley around 420 nm stands out as the prominent spectral feature. This valley is much more prominent in the spectrum of the adenomatous polyp. This feature, as well as more prominent dips around 542 and 577 nm, which are characteristic of hemoglobin absorption as well, are all indicative of the increased hemoglobin presence in the adenomatous tissue.

Apparently, the differences between these spectra are due to changes in the scattering and absorption properties of the tissues. Both the absorption dips and the slopes of the spectra are sensitive functions of the absorption and scattering coefficients, providing a natural way to introduce an inverse algorithm that is sensitive to such features. The authors quantified the absorption and scattering properties using the diffusion-based model discussed in Section 1.1.2. The equation[735] analog to Eq. (1.27) was fit to the data using the Levenberg-Marquardt minimization method. Thus, the total hemoglobin concentration c_{Hb} and hemoglobin oxygen saturation

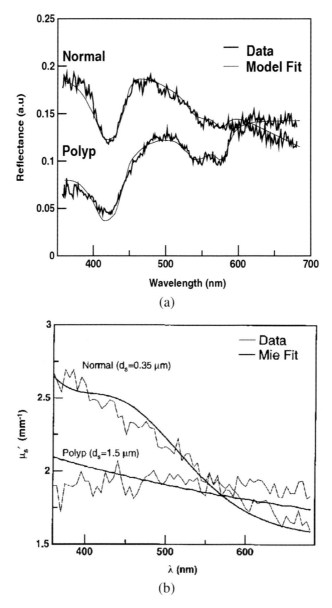

Figure 6.24 Diffuse reflectance analysis: (a) measured reflectance spectra (noisy lines) and modeled fits (smooth lines); (b) scattering spectra obtained from the reflectance measurements (noisy curves) and corresponding Mie theory spectra (smooth curves). The effective scatterer sizes d_s are indicated (from Refs. 732 and 735).

α were obtained. Also, the optimal reduced scattering coefficient $\mu_s'(\lambda)$ was found for each wavelength λ, ranging from 360 to 685 nm. It was found that $\mu_s'(\lambda)$ has a spectral dependence that resembles a straight line declining with wavelength λ. The slope of $\mu_s'(\lambda)$ decreases with an increasing effective size of the scatterers, d_s [Fig. 6.24(b)]. This allows the effective scatterer size to be determined from known $\mu_s'(\lambda)$. The model fits shown in Fig. 6.24(a) are in very good agreement with the experimental data.

The promise of LSS to diagnose dysplasia and CIS was tested in *in vivo* human studies in four different organs and in three different types of epithelium: columnar epithelia of the colon and Barrett's esophagus, transitional epithelium of the urinary bladder, and stratified squamous epithelium of the oral cavity.[733] All clinical studies were performed during routine endoscopic screening or surveillance procedures. In all of the studies, an optical fiber probe delivered white light from a xenon arc lamp to the tissue surface and collected the returned light. The probe tip was brought into gentle contact with the tissue to be studied. Immediately after the measurement, a biopsy was taken from the same tissue site. The spectrum of the reflected light was analyzed and the nuclear size distribution determined. Both dysplasia and CIS have a higher percentage of enlarged nuclei and, on average, a higher population density, which can be used as the basis for spectroscopic tissue diagnosis.

7

Polarization-Sensitive Techniques

In this chapter, polarization-sensitive techniques for imaging and functional diagnosis of biological tissue are considered. Methods based on the polarization discrimination of a probe-polarized light that is scattered by, or transmitted through, a tissue or a cell structure are described. The advantages of polarization methods for tissue imaging and functional diagnostics are discussed. It is shown that polarization-spectral selection of scattered radiation used with the polarization-fluorescence method significantly improves the diagnostic potential of the method.

7.1 Polarization imaging

7.1.1 Transillumination polarization technique

The polarization discrimination of light transmitted through a multiply scattering medium may provide high-quality images of inhomogeneities embedded in the scattering medium. Principles of transillumination polarization diaphanography of a heterogeneous scattering object are described in the literature.[1138] This technique makes it possible to locate and to image absorbing objects hidden in a strongly scattering medium. The method uses modulation of the polarization azimuth of a linearly polarized laser beam and lock-in detection of polarization properties of light transmitted through the object. The scattering sample was probed by an Ar-ion laser beam. The orientation of the polarization plane of the probe beam was modulated by a Pockel's cell as follows: during the first half-period of the modulating signal it was not changed, and during the second half-period it was rotated by 90 deg. Transmitted (depolarized) and forward-scattered (polarized) components of the probe light were collimated by two diaphragms and divided into two channels by a polarizing beamsplitter. It was found that in comparison with conventional diaphanography, polarization diaphanography allows one to get shadow images of a hidden object in a scattering medium that is characterized by up to approximately 30 scattering events on average.[1138]

A comparison of polarization and conventional transillumination imaging was carried out in Refs. 353–355 and 1139. The absorbing inhomogeneity, such as an absorbing plate placed in a scattering slab, was probed by a linearly polarized laser beam (Fig. 7.1). The shadow images were reconstructed from the profiles of the intensity and the degree of polarization P of the transmitted light (Fig. 7.2). Note that the dependencies of the degree of linear polarization on the edge position exhibit an increase in P in the vicinity of the edge. The explanation of this peculiarity is similar to that proposed by Jacques et al.[383] for the polarization-sensitive detection of backscattered light (see Section 7.1.2) and is connected with that near the absorber edge; the degree of polarization may be approximately doubled in value (for a highly scattering media) because no $I_\perp$ photons are scattered into the shadow-edge pixels by the shadow region, while $I_\parallel$ photons are directly scattered into these pixels. Better quality of the shadow images of the object is obtained with the use of the polarization imaging technique in comparison with conventional transillumination.

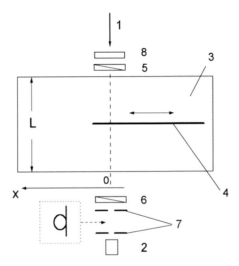

Figure 7.1 Scheme of the experimental setup for transillumination polarization imaging.[1139] 1, Linearly polarized beam of He:Ne laser (633 nm); 2, detector; 3, glass tank filled by scattering medium (diluted milk); 4, absorbing half-plane; 5, polarizer; 6, analyzer; 7, collimating diaphragms or light-collecting optical fiber; 8, chopper.

7.1.2 Backscattering polarization imaging

The principle of polarization discrimination of multiply scattered light has been fruitfully explored by many research groups in morphological analysis and visualization of subsurface layers in strongly scattering tissues.[129,135,136,138,376,378,379,382,383,1139–1143,1147] One of the most popular approaches to polarization imaging in heterogeneous tissues is based on using linearly polarized light to irradiate the object (the chosen area of the tissue surface) and to reject the backscattered light with the same polarization state (copolarized radiation) by the imaging system. Typically, such polarization discrimination is achieved simply by placing a polarizer between the imaging lens and the object. The optical axis of the polarizer is oriented perpendicularly to the polarization plane of the incident light. Thus, only the cross-polarized component of the scattered light contributes to the formation of the object image. Despite its simplicity, this technique has been demonstrated to be an adequately effective tool for functional diagnostics and for the imaging of

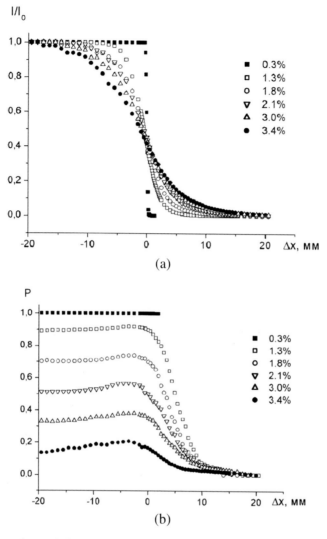

Figure 7.2 Experimental dependencies of (a) the normalized intensity and (b) degree of linear polarization of transmitted light on the absorbing half-plane edge position at different concentrations of the background scattering medium (diluted milk).[1139]

subcutaneous tissue layers. Moreover, the separate imaging of an object with copolarized and cross-polarized light permits separation of the structural features of the shallow tissue layers (such as skin wrinkles, the papillary net, etc.) and the deep layers (such as the capillaries in derma). The elegant simplicity of this approach has stimulated its widespread application in both laboratory and clinical medical diagnostics.

A typical scheme of instrumentation for polarization imaging using the approach discussed above is presented in Fig. 7.3. In the imaging system developed by Demos et al.,[1141] a dye laser with Nd:YAG laser pumping is used as the illumination source. The probe beam diameter is 10 cm and the average intensity is

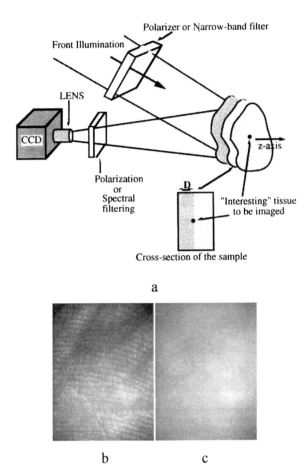

Figure 7.3 (a) An instrument for selective polarization or spectral imaging of subsurface tissue layers; (b) copolarized and (c) cross-polarized images of the human palm at 580-nm polarized laser light illumination.[1141]

approximately equal to 5 mW/cm^2. A cooled CCD camera with a 50-mm focal length lens is used to detect back-reflected light and to capture the image. The first polarizer, placed after the beam expander, is used to ensure illumination with linearly polarized light. A second polarizer is positioned in front of the CCD camera with its polarization orientation perpendicular or parallel to that of the illumination.

The efficiency of selective polarization imaging is illustrated in Fig. 7.3, where the copolarized and cross-polarized images of a human palm are presented. Figure 7.3(b) illustrates surface imaging (copolarized), where the superficial skin papillary pattern is clearly seen. Figure 7.3(c) illustrates subsurface skin imaging (cross-polarized image).

A similar camera system, but one that uses an incoherent white light source such as a xenon lamp, is described in Refs. 36, 383, and 1144, where results of a pilot clinical study of various skin pathologies using polarized light are presented. The image-processing algorithm used is based on the evaluation of the degree of polarization, which is then considered to be the imaging parameter. Two images are acquired: one "parallel," I_{par}, and one "perpendicular," I_{per}. These images are algebraically combined to yield a polarization image as

$$PI = \frac{I_{par} - I_{per}}{I_{par} + I_{per}}. \qquad (7.1)$$

It is important to note that in the polarization image, the numerator rejects randomly polarized diffuse reflectance; therefore, PI may be used to monitor tissue birefringence. Normalization by the denominator makes the expression for PI less sensitive to attenuation, which is common to the individual polarization components and is due to tissue absorption, i.e., melanin pigmentation for skin.

The polarization images of pigmented skin sites (freckles, tattoos, and pigmented nevi) and unpigmented skin sites [nonpigmented intradermal nevi, neurofibromas, actinic keratosis, malignant basal cell carcinomas, squamous cell carcinomas, vascular abnormalities (venous lakes), and burn scars] are analyzed to find the differences caused by various skin pathologies (see some examples in Fig. 7.5).[383] Also, the point-spread function of the backscattered polarized light is analyzed for images of a shadow cast from a razor blade onto a forearm skin site. This function describes the behavior of the degree of polarization at the imaging parameter near the shadow edge. It was discovered that near the shadow edge, the degree of polarization approximately doubles in value because no I_{per} photons are superficially scattered into the shadow-edge pixels by the shadow region, while I_{par} photons are directly backscattered from the superficial layer of these pixels. This result suggests that the point-spread function in skin for cross talk between pixels of the polarization image has a half width at half maximum (HWHM) of about 390 μm.

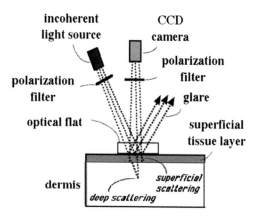

Figure 7.4 A prototype of a polarization camera for skin examination.[383] The incident light is linearly polarized parallel to the scattering plane. An optical flat enforces a uniform skin-glass interface for the reflection of glare away from the camera. The polarized scattered light and the diffusely scattered light reach the camera after passing through a linear polarizer that can be oriented parallel with or perpendicular to the scattering plane.

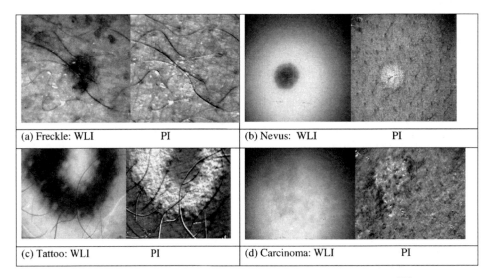

Figure 7.5 Comparison of white light (WLI) versus polarization (PI) images.[383] (a) A freckle; the polarization image removes the melanin from a freckle. (b) A benign pigmented nevus; the polarization image removes the melanin and shows apparent scatter, the drop of polarized light reflectance from epidermis lining the hair follicles is seen. (c) Tattoo; the polarization image lightens the "blackness" of the tattoo, specular reflectance of polarized light off the carbon particles yields a strong image. (d) Malignant basal cell carcinoma; the white light image underestimates the extent of the skin cancer.

Comparative analysis of polarization images of normal and diseased human skin has shown the ability of the above approach to emphasize image contrast based on light scattering in the superficial layers of the skin. The polarization images can visualize disruption of the normal texture of the papillary and upper reticular layers caused by skin pathology. Polarization imaging can be considered as an adequately effective tool for identifying skin cancer margins and for guiding surgical excision of skin cancer [see Fig. 7.5(d)]. Various modalities of polarization imaging are also considered in Ref. 1145.

The evaluation of the quality of the polarization images is based on the presentation of multiply scattered light as a superposition of partial contributions characterized by different values of the optical paths s in the scattering medium.[135,1143] The statistical properties of the ensemble of partial contributions are described by the probability density function of the optical paths $\rho(s)$, whereas the statistical moments of the scattered light are represented by the integral transforms of $\rho(s)$ with the properly chosen kernels. The degree of polarization of multiply scattered radiation with initial linear polarization can be approximately represented in the form of the Laplace transform of $\rho(s)$ as

$$P_L = \frac{I_{//} - I_\perp}{I_{//} + I_\perp} \approx \frac{3}{2} \int_0^\infty \exp\left(-\frac{s}{\xi_L}\right)\rho(s)\,ds, \qquad (7.2)$$

where I_{II} and $I_\perp$ are, respectively, the intensities of the copolarized and cross-polarized components of the scattered light. The parameter ξ_L is the depolarization length for linearly polarized light.

By considering the polarization visualization of the absorbing macrohetero-geneity, along with the degree of polarization of backscattered light as the visualization parameter, the contrast of the polarization image can be defined as[135]

$$V_P = \frac{P_L^{in} - P_L^{back}}{P_L^{in} + P_L^{back}}, \qquad (7.3)$$

where P_L^{in} is the degree of residual linear polarization of the backscattered light detected in the region of the localization of the heterogeneity and P_L^{back} is the analogous quantity determined far from the region of localization. The contrast V_P of the reconstructed polarization image can be represented as a function of the scattering layer thickness l, the depth of inhomogeneity position h, the transport mean free path l_t of the scattering medium, the scattering anisotropy factor g, and the depolarization length ξ_L.[135] The probability density function of the optical paths $\rho(s)$ can be obtained by a Monte Carlo simulation.

To compare the efficiency of the various polarization imaging modalities, the experimental setup shown in Fig. 7.6 was used.[1139] The total normalized intensity (the intensity of the copolarized and cross-polarized components) or the degree of residual linear polarization of the backscattered light were taken as the visualization parameters. A scattering medium (a water-milk emulsion) in a rectangular glass tank ($18 \times 26 \times 26$ cm) was used as a tissue model. The side and rear walls of the tank were blackened. The absorbing object (a rectangular plate with blackened rough surfaces) was positioned in the central part of the tank at different distances h (1 to 4 cm) from the transparent front wall. The white light probe was linearly polarized perpendicular to the plane of incidence. To avoid specular reflection from the front wall of the tank, the illuminating beam was directed at an angle of 30 deg relative to the normal to the wall.

The capture of the object images for each of three chromatic coordinates (R, G, and B) was done using a color CCD camera (Panasonic NV-RX70EN) and a Miro DC20 frame grabber (MiroVideo, Germany). The color 8-bit images of the object were captured with 647×485 resolution with the use of copolarized and cross-polarized backscattered light. The brightness distributions for each of the R, G, and B image components along an arbitrarily chosen line of the image [Fig. 7.6(b)] are applied to reconstruct the images of the absorbing heterogeneity with different visualization parameters [Fig. 7.6(c)]. In the absence of a scattering medium and, hence, the backscattered radiation, the image contrast is equal to zero. An increase in the milk concentration results first in a sharp increase in the image contrast up to the maximum value, with the subsequent monotonic decrease caused by the increase of scattering multiplicity.

Comparison of the experimental data and the MC simulations[135,1139,1143] allows one to conclude that maximal contrast in the polarization image is obtained at

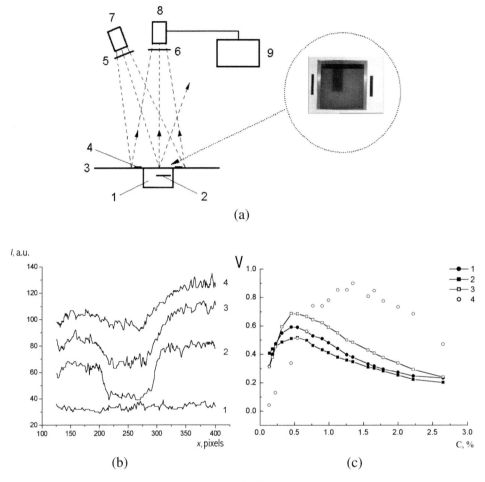

(a)

(b)   (c)

Figure 7.6 Polarization imaging experiment.[1139] (a) Schematic diagram: 1, a cell with a scattering medium; 2, absorbing plate; 3, white marker; 4, black markers; 5, polarizer; 6, analyzer; 7, a halogen lamp; 8, CCD camera; and 9, PC. (b) Distributions of backscattered radiation intensity along an arbitrarily chosen line of the image for different volume concentrations of milk emulsion: 1, 0%; 2, 0.66%; 3, 1.96%; and 4, 5.51% (R-component of the color image). (c) Dependencies of the polarization image contrast on the volume concentration of the milk emulsion when the normalized intensity of the (1) unpolarized light, (2) copolarized components, (3) cross-polarized components, and (4) the degree of polarization of backscattered radiation are used as the visualization parameter.

the depth of an inhomogeneity position on the order of $(0.25-0.6)$ ξ_L (depending on the degree of residual polarization in the backscattered background component detected outside the region of the inhomogeneity localization). In particular, this conclusion agrees with data on polarization imaging of skin, which points to the efficiency of polarization imaging for epidermis and upper layers of papillary derma $(100-150$ $\mu m)$.[1144]

7.2 Polarized reflectance spectroscopy of tissues

7.2.1 In-depth polarization spectroscopy

Imaging and monitoring of the morphological and functional state of biological tissues may be provided on the basis of spectral analysis of the polarization properties of the backscattering light.[1146] Tissue probing by a linear polarized white light and measuring of the spectral response of the copolarized and cross-polarized components of the backscattered light allow one not only to quantify chromophore tissue content, but also to estimate in-depth chromophore distribution.

In the visible wavelength range, skin may have a reduced scattering coefficient $\mu_s' \sim 30\text{--}90\ \text{cm}^{-1}$ and absorption coefficient $\mu_a \sim 0.2\text{--}5\ \text{cm}^{-1}$ (see Table 2.1); therefore, the expected transport mean free path of a photon $l_t = (\mu_a + \mu_s')^{-1}$ [see Eq. (1.22)] is in the range 100–300 μm. Because of the dominating of scatterers with sizes, characterized by a diffraction parameter $ka > 1$ [see Eqs. (1.99) and (1.100) and Figs. 1.27 and 1.28], the depolarization length ξ_L in skin is comparable with the transport scattering length l_t, and exceeds the skin epidermis thickness. On the other hand, such absorbers as melanin in epidermis and hemoglobin in dermis must increase the degree of the residual polarization of the backscattered light in the spectral ranges that correspond to the absorbing bands of the dominating chromophores. Moreover, these chromophores are placed at different depths; thus, their localization may be estimated owing to the characteristic absorbing bands on the differential polarization spectra.[135,1146]

Figure 7.7 shows the experimental setup for the backscattering polarization spectral measurements. Light from a white light source (halogen lamp of 200 W) is guided to the object by a fiber bundle with an attached wideband linear polarization filter. The diameter of the irradiated skin surface is approximately 8 mm. The backscattered light from the skin is collected by another fiber bundle to the input of which a polarized filter is attached. This filter is variable and may change its orientation, being parallel with or perpendicular to the optical axis of the first polarization filter. Fiber bundles are used to exclude some polarization sensitivity that may take place at the use of monofibers. To exclude specular reflection from

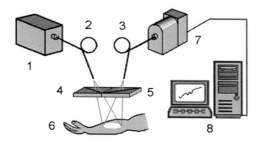

Figure 7.7 Scheme of the experimental setup for the backscattering polarization spectral measurements:[1146] 1, a white light source (halogen lamp, 200 W); 2 and 3, fiber bundles; 4 and 5, polarization filters; 6, object under study; 7, photodiode-array grating spectrometer; 8, PC.

the skin surface, the detecting fiber bundle was placed at an angle of $\sim$20 deg with respect to the normal to the skin surface. A distal end of the fiber was connected with the spectrometer.

This instrument is able to measure reflectance spectra at both parallel and perpendicular orientations of filters $R_{II}(\lambda)$ and $R_{\perp}(\lambda)$. From these spectra, the differential residual polarization spectra $\Delta R^r(\lambda)$ or residual polarization degree spectra $P_L^r(\lambda)$ are calculated as

$$\Delta R^r(\lambda) = R_{II}(\lambda) - R_{\perp}(\lambda), \tag{7.4}$$

$$P_L^r(\lambda) = \frac{R_{II}(\lambda) - R_{\perp}(\lambda)}{R_{II}(\lambda) + R_{\perp}(\lambda)}. \tag{7.5}$$

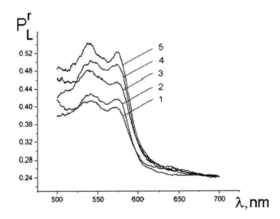

Figure 7.8 *In vivo* measured residual polarization degree spectra for a volunteer with UV-induced erythema of different degrees (erythema index, EI): 1, EI = 157; 2, EI = 223; 3, EI = 249; 4, EI = 275; and 5, EI = 290.[1146] Erythema index was measured using the erythema-melanin meter described in Ref. 1147.

Two examples of *in vivo* human skin studies using polarization spectroscopy are presented in Figs. 7.8 and 7.9.[1146] Figure 7.8 demonstrates spectral distributions of the residual polarization degree $P_L^r(\lambda)$ of the backscattered light for different values of the index of erythema induced by UV light. For a higher erythema index (EI) or skin redness (increased blood volume), $P_L^r(\lambda)$ is improved within the absorption Q-bands of the blood. This happens owing to the reduction of multiplicity of scattering for the photons with wavelengths that fall down to the absorption bands because of more intensive absorption of such photons.

Figure 7.9 presents differential polarization spectra, which are also sensitive to the absorption properties of skin that was controlled using a tape stripping technology. Less epidermal thickness corresponds to a higher blood volume within the measuring volume (higher hemoglobin absorption). Therefore, in spite of the

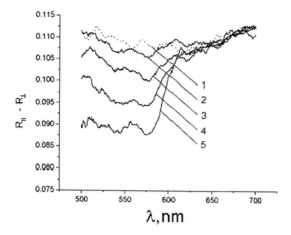

Figure 7.9 *In vivo* measured differential residual polarization spectra for a volunteer at epidermal stripping of different thicknesses:[1146] 1, normal skin; 2, thickness of the removed skin layer is 40 μm; 3, 50 μm; 4, 60 μm; and 5, 70 μm.

$[R_{II}(\lambda) - R_{\perp}(\lambda)]$ value reduction within the blood absorption bands (545 and 575 nm) due to the attenuation of both polarization components, its difference from the values measured far from the blood absorption bands, i.e., in the range 650–750 nm, is significant to providing in-depth profiling of epidermal thickness and blood vessels in skin.

As a criteria of epidermal thickness, the following parameters can be used:[1146]

$$V_{545} = \frac{\Delta R^r_{(650-700)} - \Delta R^r_{545}}{\Delta R^r_{(650-700)} + \Delta R^r_{545}} \quad \text{or} \quad V_{575} = \frac{\Delta R^r_{(650-700)} - \Delta R^r_{575}}{\Delta R^r_{(650-700)} + \Delta R^r_{575}}. \quad (7.6)$$

In these equations, indices 545 and 575 denote the differential residual polarization backscattering coefficients at the wavelengths of the hemoglobin absorption bands' centers ($\lambda = 545$–575 nm) and (650–700) at the wavelength range, where hemoglobin absorption is small ($\lambda = 650$–700 nm).

This method is still simple and, owing to the spectral information received, may provide information about living tissue that is more valuable than the nonspectral polarization methods described in Section 7.1.2. The experimental system is shown in Fig. 7.10. Monochrome images of the skin are captured by a video system, VS-CTT 60-075 (Videoscan Ltd., Russia). To get smoother white light irradiation of the skin surface and to avoid specular light detection, four halogen lamps of 50 W were positioned with their irradiation directed from four different sides at an angle ∼30 deg with respect to the normal to the skin surface. The output of each light source was filtered by identical linear polarized filters, and a rotatable polarization filter-analyzer was placed in front of the monochrome CCD camera. Figure 7.11 shows an example of skin burn lesion polarization-spectral imaging at the wavelength of hemoglobin absorption (∼550 nm). It is well seen that the maximal image contrast of ∼0.49 is provided for the polarization degree as a visualization parameter.

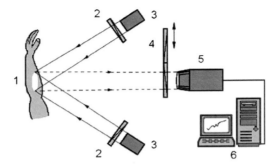

Figure 7.10 Experimental setup for polarization-spectral imaging of *in vivo* tissues:[1146] 1, tissue; 2, polarization filters; 4, polarization and interferential filters; 3, light sources; 5, monochrom CCD camera; 6, PC.

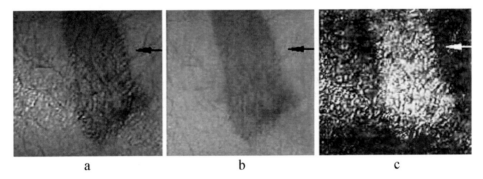

Figure 7.11 Polarization-spectral ($\lambda = 550$ nm) images of a skin burn lesion of the volunteer:[1146] (a) a copolarized image; (b) a crossed-polarized image; (c) degree of polarization image, P_L^r.

7.2.2 Superficial epithelial layer polarization spectroscopy

One of the promising approaches to early cancer diagnosis is based on the analysis of a single scattered component of light perturbed by tissue structure (see Section 6.3). The wavelength dependence of the intensity of the light elastically scattered by the tissue structure appears sensitive to changes in tissue morphology that are typical of precancerous lesions. In particular, it has been established that specific features of malignant cells, such as increased nuclear size, increased nuclear/cytoplasmic ratio, pleomorphism, etc., are markedly manifested in the elastic light scattering spectra of probed tissue. A specific fine periodic structure in the wavelength of backscattered light has been observed for mucosal tissue.[180] This oscillatory component of light scattering spectra is attributable to a single scattering from surface epithelial cell nuclei and can be interpreted within the framework of Mie theory. Analysis of the amplitude and frequency of the intensity spectrum's fine structure allows one to estimate the density and size distributions of these nuclei. However, the extraction of a single-scattered component from the masking multiple scattering background is a problem. Also, as it was shown in Section 7.2.1, absorption of stroma related to hemoglobin distorts the single-scattering spectrum of the epithelial cells. Both of these factors should be taken into account when interpreting the measured spectral dependencies of the backscattered light.

The negative effects of a diffuse background and hemoglobin absorption can be significantly reduced by the application of a polarization discrimination technique in the form of the illumination of the probed tissue with linearly polarized light followed by the separate detection of the elastic scattered light at the parallel and perpendicular polarization states (i.e., the copolarized and cross-polarized components of the backscattered light).[150,163] This approach, called polarized elastic light scattering spectroscopy or polarized reflectance spectroscopy (PRS), will potentially provide a quantitative estimate not only of the size distributions of cell nuclei, but also of the relative refractive index of the nucleus. These potentialities, which have been demonstrated in a series of experimental works with tissue phantoms and *in vivo* epithelial tissues,[150,163,166,180] allow one to classify the PRS technique as a new step in the development of noninvasive optical devices for real-time diagnostics of tissue morphology and, consequently, for improved early detection of precancers *in vivo*. An important step in the further development of the PRS method will be the design of portable and flexible instrumentation applicable to *in situ* tissue diagnostics. In particular, fiber-optic probes are expected to "bridge the gap between benchtop studies and clinical applications of polarized reflectance spectroscopy."[1148]

7.3 Polarization microscopy

Polarized light microscopy has been used in biomedicine for more than a century to study optically anisotropic biological structures that may be difficult, or even impossible, to observe using a conventional light microscope. A number of commercial microscopes are available on the market, and numerous investigations of biological objects have been made using polarization microscopy. However, modern approaches in polarization microscopy have the potential to enable one to acquire new and more detailed information about biological cells and tissue structures. At present, it is possible to detect optical path differences of even less than 0.1 nm.[168,388,754,1149–1151] Such sensitivity as well as the capability to examine scattering samples are due to recent achievements in video, interferential, and multispectral polarization microscopy. Full Mueller matrix measurements and other combined techniques, such as polarization/confocal and polarization/OCT microscopy, promise new capabilities for polarization microscopy including *in vivo* measurements.

In addition to that discussed in Sections 1.4, 3.3, and 5.7.1, in this section, we will discuss only a few of the recent studies and novel techniques that have the potential to examine the anisotropic properties of scattering samples. One of the examples is the multispectral imaging micropolarimeter (MIM), which can detect the birefringence of the peripapillary retinal nerve fiber layer (RNFL) in glaucoma diagnosis.[168,388] The optical scheme of the MIM is presented in Fig. 7.12.

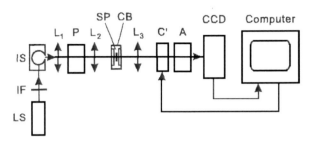

Figure 7.12 Optical scheme of the multispectral imaging micropolarimeter used in transmission mode.[388] LS, light source; IF, interference filter; IS, integrating sphere; P, linear polarizer; SP, specimen; CB, chamber; L_1, L_2, and L_3, lenses; C', linear retarder; A, linear analyzer; CCD, charge-coupled device.

Light from a tungsten-halogen lamp, followed by an interference filter (band of 10 nm), provides monochromatic illumination to an integrating sphere (IS). Lens L_1 ($F = 56$ mm) collimates the beam incident onto a polarizer (P). The use of an integrating sphere assures that the output intensity of the polarizer varies less than 0.2% as it rotates 360 deg. Lens L_2 ($F = 40.5$ mm, NA $= 0.13$) focuses the image of the exit aperture of the integrating sphere onto a specimen (SP) in a chamber (CB) with a flat entrance and exit windows. Lens L_3 ($F = 60$ mm, NA $= 0.07$) focuses the specimen image onto a cooled CCD camera that provides a pixel size of about 4 µm on a specimen in an aqueous medium (magnification ≈ 5.8). Although the lenses are achromatic, the wide spectral range (440–830 nm) requires only small changes in the detection optics' position (moving the lens L_3 and CCD together within a 0.5-mm range) to adjust the focus for each wavelength. A liquid crystal linear retarder (C'), followed by a linear analyzer (A), is used to measure the output Stokes vector of the specimen. Both polarizer and analyzer are Glan-Taylor polarization prisms. The azimuth and retardance of the retarder are set for a few discrete values, and the azimuth of the analyzer is always fixed at 45 deg. Each setting of the retarder (respectively, azimuth and retardance)—(1) 0, 90 deg; (2) 0, 200 deg; (3) 22.5, 207 deg; (4) −22.5, 207 deg)—is characterized by a 1×4 measurement vector. The four retarder/analyzer settings together are characterized by a 4×4 matrix **D**, with each row corresponding to one measured vector.

A Stokes vector $\overline{S}$ can be calculated as

$$\overline{S} = D^{-1}\overline{R}, \tag{7.7}$$

where D^{-1} is the inverse of the measurement matrix and $\overline{R}$ is a 4×1 response vector corresponding to the four retarder/analyzer settings.[816] To evaluate the linear retardance of a specimen, the Mueller matrix should be found from the measurements of the incident $\overline{S}_{inc}$ and the output $\overline{S}$ Stokes vectors (see Section 1.4) as

$$\overline{S} = K\mathrm{M}(\rho, \delta)\overline{S}_{inc}, \tag{7.8}$$

where the factor K accounts for the losses of intensity in transmission and ρ and δ are, respectively, the azimuth and retardance of the specimen. This expression includes four equations for the three unknowns, K, ρ, and δ. In most cases, it is useful to overdetermine the system of equations in Eq. (7.8) by using more than one $\overline{S}_{inc}$.

The retardance and azimuth of a living and fixed rat's RNFLs were measured over a wide spectral range.[388] It was found that the RNFL behaves as a linear retarder and that the retardance is approximately constant in a wavelength range from 440 to 830 nm. The average birefringence measured for a few unfixed rat RNFLs, with an average thickness of 13.9 ± 0.4 µm, is 0.23 ± 0.01 (nm/µm) $\equiv$ 2.3×10^{-4}. The influence of the polarization properties of the retina on the measured RNFLs' anisotropic properties was found. Images presented in Fig. 7.13 illustrate the importance of correcting for the polarization properties of the retina and for the distributions of retardance and azimuth within the sample.

 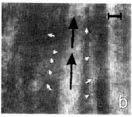

Figure 7.13 Estimated retardances (arrows' lengths) and slow axes (arrows' directions) of the bundle and gap areas of rat RNFLs.[388] The images are at the wavelength 440 nm. Images sizes: (a) 222 × 199 µm; (b) 187 × 177 µm. Nerve fiber bundles appear as brighter bands. Each arrow starts in the center of the area measured. The calibration bar is 1 nm of retardance. (a) The white arrows represent measurements that are not corrected for retinal polarization ability, and the black arrows are corrected ones. (b) The black arrows are corrected bundle retardances; the small white arrows in the gaps show the variation of residual retardances, also after correction.

Another technique, which is related to quantitative polarized light microscopy, is based on a video microscopy technique that is applied to measure variations in the orientations of the collagenous fibers arranged in lamellae within eye corneal tissue.[1149] The lamellar structure of the cornea and sclera is very visible in Figs. 3.2, 3.3, and 3.4. Within a lamella, the fibrils are parallel, but the fibrils of adjacent lamellae do not, in general, run in the same direction. They may have a relative orientation at any angle between 0 and 180 deg.

As was discussed in Section 3.1.1, in corneal stroma, the fibrils have a diameter of 25–39 nm while the mean diameter of scleral fibrils is equal to 100 nm. Therefore, the individual fibrils cannot be resolved with light microscopy; but due to the intrinsic birefringence of collagen fibrils and its dependence on the angle of their orientation from lamella to lamella, they can be recognized. Along its fiber axis, collagen is highly birefringent, so those lamellae that are cut parallel to the fiber axes [$\theta = 0$ deg, see Fig. 7.14(a)] appear brighter under polarized light when the polarizer and analyzer are crossed and the length of the tissue section is oriented at 45 deg to the polarizer/analyzer axis [see Fig. 7.14(b)]. Collagen is not birefringent perpendicular to its fiber axis [θ = 90 deg, see Fig. 7.14(a)], so those lamellae that are cut perpendicular to their fibril direction appear completely dark in this section [see Fig. 7.14(b)]. The variation of intensity along the transect X–Y across the cornea section [see Fig. 7.14(b)] is caused by the different angular orientation of the particular lamella (totally, about 15 lamellae are seen) and presented in Fig. 7.14(c).

Because of the regular arrangement of the lamellae, the angle θ is all that is necessary to define the three-dimensional orientation of the fibrils in sections of normal cornea. Nevertheless, to find this angle distribution for a specific tissue section, the lamellar birefringence of form that contributes about 67% to the total birefringence should be accounted for.[1149] For sections of disrupted pathological cornea and for sections of sclera and limbus (the region where the cornea and sclera fuse), the situation is more complicated because the lamellae have a much less ordered "wavy" arrangement [see Fig. 3.5(b) for sclera].

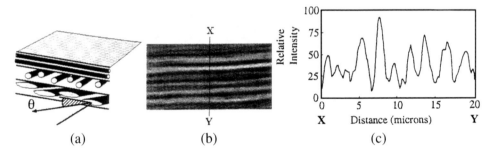

(a)   (b)   (c)

Figure 7.14 Polarization microscopy of a collagenous tissue structure.[1149] (a) Schematic diagram of a tissue section containing three lamellae (the number of fibrils is greatly reduced and the relative fibril diameter greatly exaggerated; an actual electronic micrograph is shown in Fig. 3.2); angle θ is the angle of inclination of the fibrils relative to the plane of sectioning. (b) Digital photomicrograph of a section through part of a rabbit cornea viewed under polarized light (×500). (c) The variation in intensity along the transect X–Y across the cornea section of photomicrograph (b).

Many tissues possess very complex patterns of alignment of the structure-forming elements. Some polarization microscopic methods that show promise for generating alignment maps for such tissues have been developed (see, for example, Refs. 1152–1154). The method presented in Refs. 1152 and 1153, as well as the method used in Ref. 1149, is a useful tool in cases where the tissue structure along the direction of the probing light propagation can be considered to be a uniform one. In Refs. 1152 and 1153, a microscopic polarimetry method for generating fiber alignment maps, which can be used for characterization of the structure of fibrous tissues, tissue phantoms, and other fibrillar materials, is considered. This method is based on probing the sample with elliptically polarized light from a rotated quarter-wave plate and an effective circular analyzer. Nonlinear regression techniques are implemented for estimating the optical parameters of the optic train and the sample. The processing of the sequence of images obtained with different mutual orientations of the rotated quarter-wave plate and the analyzer, which is based on fast harmonic analysis, permits the recovery of an alignment direction map and a retardation map. These maps describe a spatial distribution of a sample's local linear birefringence and, therefore, can be used for morphological analysis of tissues with an expressed structural anisotropy that have linear birefringence as their dominant optical property. The potential of this method for accurately generating alignment maps for samples that act as linear retarders to within a few degrees of retardation has been demonstrated in experiments with an *in vitro* sample of a porcine heart valve leaflet.

The method proposed in Ref. 1154 can also be used in more general situations when the orientation of the structure-forming elements (for instance, the collagen fibers) varies along the probing direction. Such variation should be taken into account when thick (>50 μm) tissue layers (e.g., dermis) are analyzed. In the method discussed,[1154] a standard polarization microscope arranged with a CCD camera can be used. The measurements are usually carried out with wide-spectral-range color filters and without a quarter-wave plate. The following expression describes the dependence of the detected signal at any detection point on the angles of orientation of the polarizer (ϑ) and the analyzer (ϑ') of the microscope:

$$i_C \approx B_0 + B_1 \cos\eta + B_2 \cos\varsigma + B_3 \sin\eta + B_4 \sin\varsigma,$$
$$\eta = 2(\vartheta - \vartheta'), \quad \varsigma = 2(\vartheta + \vartheta'), \tag{7.9}$$

where B_i ($i = 0, 1, 2, 3,$ and 4) are the coefficients, which depend on the local optical properties of the sample in the probed region and the spectral properties of incident light. As was shown in Ref. 1154, the measured values of B_i are capable of providing important information about the sample structure. They can also be used for characterization of specific features of light propagation in the sample; in particular, they allow us to recognize the so-called adiabatic regime of light propagation in the studied medium. This means that in the adiabatic regime, the orientation of the local optical axis changes smoothly in the probed region of the tissue. Note that for fibrous tissues, the direction of the local optical axis typically coincides with the local preferred direction of fiber orientation. If the adiabatic regime is realized, then the angles υ and ϕ, which are calculated from the obtained values of B_i as $\upsilon = (1/4)\arctan(B_2/B_4)$ and $\phi = (1/2)\arctan(B_1/B_3)$, provide information about the structure of the sample. The angle ϕ is equal to the angle between the azimuthal projections of the local optical axes of the medium at the upper and lower boundaries of the sample, and the angle υ defines the orientation of the bisector of the angle between these projections. Analysis of the experimentally obtained "B_i-maps," as well as the spatial distributions of υ and ϕ, can be proposed as an effective tool for tissue structure characterization.[1154] In particular, the B_0-map (this coefficient characterizes the local transmittance of the sample for nonpolarized light), the υ-map, and the ϕ-map for the *in vitro* sample of human epidermis (stratum corneum) are presented in Ref. 1154.

The study of collagen structure and function is important for understanding a wide range of pathophysiological conditions, including aging. One of the prospective laser techniques, which can provide *in vivo* microscopic monitoring of collagen structure, is polarized second-harmonic generation (SHG) microscopy (see Sections 4.7 and 5.75).[941,944] The backscattered SHG signal induced by a 100-fs titanium:sapphire laser, with a mean wavelength of 800 nm, a maximum energy of 10 nJ, and a pulse repetition rate of 82 MHz, was measured by a polarized SHG scanning confocal microscope.[944] The microscope objective has a transverse resolution of about 1.5 µm and an axial resolution of about 10 µm. It should be noted that inside the scattering media, both numbers increase. The maximum intensity in the sample was about 4×10^{11} W/cm^2. To avoid sample damage, a continuous scanning technique was used. A systematic analysis of type I collagen in a rat-tail tendon fascicle was conducted using this microscope. Type I collagen from the fascicles provides one of the strongest SHG signals of all of the various tissues analyzed by the authors in Ref. 944. They hypothesize that such high SHG efficiency is due to the highly ordered architecture of collagen. The polarization properties of collagen are also defined by its ordering. It was shown experimentally that the second-harmonic signal intensity varies by about a factor of two across a single cross section of the rat-tail tendon fascicle.[944] The signal intensity depends both on the collagen organization and the backscattering efficiency. To characterize collagen structure, both intensity- and polarization-dependent SHG signals should be detected. Actually, axial and transverse scans for different linear polarization angles of the input beam show that SHG in the rat-tail tendon depends strongly on the polarization of the input laser beam. In contrast to SHG signal intensity, the functional form of the polarization dependence does not change significantly over a single cross section of the sample, and it is not affected by the backscattering efficiency.

The measured data were in good agreement with an analytical model developed for a SHG signal at linear polarized excitation and were used to determine the fibril orientation and the ratio between the only two nonzero, independent elements in the second-order nonlinear susceptibility tensor, $\gamma \approx -(0.7$–$0.8)$.[944] The small range of values observed for γ in a tendon fascicle suggests that there is structural homogeneity. This parameter might, therefore, be useful in characterizing different collagen structures noninvasively.

The main problem encountered in the *in situ* microscopy of tissues is multiple scattering, which randomizes the direction, coherence, and polarization state of incident light. A number of optical-gating methods have been proposed to filtrate ballistic and least-scattering photons, which carry information about the object structure. One of these is the polarization-gating method and its modifications, which are described in this chapter and Sections 1.4, 3.3, and 5.7.1. The fundamental limitation of all optical-gating methods, including the polarization one, is the fact that only a small number of ballistic and least-scattering photons take part in the formation of an object image. Therefore, polarization-gating techniques in combination with image reconstruction methods can be useful for improving the image resolution in the case of a highly scattering object.[1155,1156] Both reflection-mode and transmission-mode polarization-gating scanning microscopes have been analyzed.[1155,1156]

An optical immersion technique, based on matching the refractive index of the tissue scatterers and the surrounding ground (interstitial) medium, allows one to essentially control the scattering properties of a tissue (see Chapter 5). Usually, the refractive index of the ground medium is controlled. This is accomplished by impregnating the tissue with a biocompatible agent, such as glucose, glycerol, propylene glycol, or x-ray contrasting agents. Due to the fact that the refractive index of the applied agent is higher than that of the tissue ground substance, which is close to the index of water, the refractive index of the ground increases and scattering decreases. Most of the applied agents are hyperosmotic; therefore, they can produce a temporal and local dehydration of the tissue that also leads to an increase in the refractive index of the interstitial space. Figures from 5.35 to 5.38 show experimental results on the temporal transmittance of linear polarized light through tissue sections measured by a white-light video-digital polarization microscope or polarization spectrometer on the application of an immersion agent (x-ray contrasting agent, trazograph-60, or glycerol).[409,410,442,946,1033,1065]

Reduction of the scattering at optical immersion makes it possible to detect the polarization anisotropy of a tissue more easily and to separate the effects of light scattering and intrinsic birefringence on tissue polarization properties. It is also possible to study birefringence of form with optical immersion, but when the immersion is strong and the refractive index of tissue birefringent structure is close to the index of the ground media, the birefringence of form may be too small to be detected [see Fig. 5.36 (the tendency of polarization degree to decay at a later time of tissue impregnation by the OCA) and Fig. 5.38(b) (the complete match of refractive index at the edge region causes tissue to lose not only scattering, but also the birefringence)]. The dynamics of tissue optical clearing and the manifestation of tissue anisotropy at the reduction of scattering are characteristic features, which correlate with clinical data.[343,409,410] Figure 5.10 illustrates the reversibility of the optical immersion effect in the controlling of the polarization properties of a turbid tissue.[343] Practically all healthy connective and vascular tissues show the strong or weak optical anisotropy typical of either uniaxial or biaxial crystals.[409,410] Pathological tissues show isotropic optical properties.

Polarization microscopy is also helpful for investigating individual cells; in particular, for evaluating the amount of glycated hemoglobin in erythrocytes that could be an early diagnostic marker of hyperglycemia in diabetic patients.[754] Hemoglobin glycation causes changes in the cell's refraction index. By using polarizing-interference microscopy, it is possible to measure the light refractive index in an individual erythrocyte. The refractive index of hemoglobin or a red blood cell, containing about 95% hemoglobin, varies approximately linearly with a change in glucose concentration—it saturates only under strong hyperglycemic conditions.[752,753] A Nomarsky polarizing-interference microscope, MPI-5 (Poland), was used for measurements of light phase retardation.[754] Using a Wollaston prism mounted on an object, the erythrocyte images for ordinary and extraordinary light beams were completely separated. In the thickest erythrocyte region, the first interference maxima were visually adjusted to the eye-sensitive purple color for ordinary and extraordinary images by shifting the second Wollaston prism placed in the rear focus of the object. For each erythrocyte measured, the Wollaston prism displacement rendered a second value. From the whole interference bandwidth h and the measured Wollaston prism displacement $2d$, the phase retardation Φ and the refractive index n were calculated for each erythrocyte as[754]

$$n = n_v + \frac{\Phi}{t} = n_v + \frac{d\lambda}{ht}, \tag{7.10}$$

where $n_v = 1.5133 \pm 0.0001$ is the refractive index of the embedding media, t is the thickness of the erythrocyte, and $\lambda = 550$ nm. Separate measurements of the erythrocyte thickness using two embedded media with different refractive indices gives $t = 0.89$ μm. Using this value, the refractive index is calculated with a standard deviation of ± 0.0005.

A robust z-polarized confocal microscope employing only one or two binary phase plates with a polarizer has been suggested by Huse et al.[1157] The major advantage of the microscope having a significant longitudinal field component is that it is then possible to image the z-polarized features in randomly oriented agglomerations of molecules of biomedical interest.

7.4 Digital photoelasticity measurements

Photoelasticity is an established experimental technique that has been applied to study the biomechanics of hard tissues like bone and tooth.[1158,1159] The photoelastic measuring technique is based on the stress-induced optical birefringence effect, which for plane stress analysis is described by the following stress-optic law:[1158,1159]

$$\sigma_1 - \sigma_2 = \frac{\theta}{2\pi} \frac{f_\sigma}{h} = \frac{N f_\sigma}{h}, \tag{7.11}$$

where $(\sigma_1 - \sigma_2)$ is the difference in the in-plane principle stress, θ is the resultant optical phase generated due to stress-induced birefringence in the sample, f_σ is the material fringe value, and h is the thickness of the specimen. Since the values of f_σ and h are constants for the mechanical stresses, recording the optical phase (θ) or fringe order ($N = \theta/2\pi$) at every point of interest on the fringe pattern allows for analysis of the stress distribution.[1158,1159]

As an example, we will consider the results of photomechanical studies of postendodontically rehabilitated teeth, using a conventional circular polariscope and an image processing system, which were the basis for the digital phase shift photoelastic technique described in Refs. 1158 and 1159. A special loading device that applies loads along the long axis (0 deg) and 60-deg lingual to the long axis of the tooth was employed. Using the polariscope, four phase-stepped images were obtained for the sample at each load by rotating the analyzer at 0-, 45-, 90-, and 135-deg angles with respect to the polarizer. The fringe patterns obtained were acquired using a CCD camera, and stored and processed by a computer. The four images were evaluated using a phase-stepping algorithm to obtain a wrapped phase map.[1158] Phase unwrapping was done on selected lines to make the fringe modulation continuous and to get information on the nature of the stress distribution.

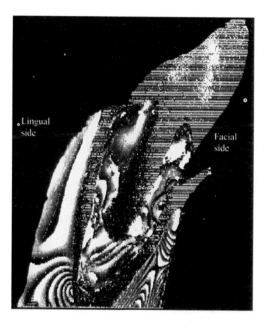

Figure 7.15 Phase-wrapped image obtained from four-phase shifted images in a rehabilitated tooth model, loaded at 125 N, 60 deg lingual to the long axis of the tooth.[1159]

Figure 7.15 shows a phase-wrapped image of the rehabilitated tooth model, loaded at 125 N at an angle of 60 deg in the direction of the long axis of the tooth. It was found that there is a significant (up to threefold) increase in the magnitude of the stress within the rehabilitated tooth model in comparison with the model of the intact tooth. Increased bending stress is identified in the cervical region and in the middle region of the root. This results in higher compressive stress in the cervical region (facial side) and higher tensile stress in the midregion (lingual side). The designed digital phase-shift photoelastic technique is of importance for the investigation of hard tissue elasticity distributions; here, for instance, it highlighted the behavior of a postcore rehabilitated tooth to functional forces.

7.5 Fluorescence polarization measurements

Fluorescence polarization measurements are used to estimate various parameters of the fluorophore environment.[573] They therefore have a potential role to play in biomedical diagnosis; in particular, in discriminating between normal and malignant tissues.[1160–1163] At polarized light excitation, the emission from a fluorophore in a nonscattering media becomes depolarized because of the random orientation of the fluorophore molecules and the angular displacement between the absorption and emission dipoles of the molecules.[573] These intrinsic molecular processes that result in additional angular displacement of the emission dipoles are sensitive to the local environment of the fluorophore. As was already shown in preceding chapters, light depolarization in tissues is determined by multiple scattering; therefore, both excitation and emission radiations should be depolarized in scattering media.[573,1160–1163] Polarization state transformation in scattering media depends on the optical parameters of the medium: the absorption coefficient μ_a, the scattering coefficient μ_s, and the scattering anisotropy factor g. Because of the different structural and functional properties of normal and malignant tissues, the contribution of multiple scattering to depolarization may be different for these tissues. The reduced (transport) scattering coefficient μ_s', or the transport mean free path (MFP) l_t, in particular, determines the characteristic depolarization depth for different tissues. Thus, fluorescence polarization measurements may be sensitive to tissue structural or functional changes, which are caused, for instance, by tissue malignancy at the molecular level (the sensitivity of excited molecules to the environmental molecules) or at the macrostructural level (the sensitivity of propagating radiation to tissue scattering properties).

Mohanty et al.[1162] have considered a fluorophore located at a distance z from the surface of a turbid medium. The homogeneous distribution of the fluorophores and the validity of the diffusion approximation for light transport in a scattering medium were assumed. The average number of scattering events experienced by the excitation light before it reached the fluorophore, and by the emitted light before it exited the medium, are described, respectively, as

$$N_1(z) = z \times \mu_s^{\text{ex}}, \tag{7.12}$$

and

$$N_2(z) = z \times \mu_s^{\text{em}}. \tag{7.13}$$

The fluorescence polarization ability is characterized by polarization anisotropy r, which is a dimensionless quantity independent of the total fluorescence intensity of the object,[573]

$$r = \frac{I_\parallel - I_\perp}{I_\parallel + 2I_\perp}. \tag{7.14}$$

It is defined as the ratio of the polarized component to the total intensity and is connected with the light polarization value P,

$$r = \frac{2P}{3 - P}. \tag{7.15}$$

The polarization, measured as

$$P = \frac{I_{\parallel} - I_{\perp}}{I_{\parallel} + I_{\perp}}, \qquad (7.16)$$

is an appropriate parameter for describing a light source when a light ray is directed along a particular axis. The polarization of this light is defined as the fraction of light that is linearly polarized. In contrast, the radiation emitted by a fluorophore is symmetrically distributed around this axis, and the total intensity is not given by $I_{\parallel} + I_{\perp}$, but rather by $I_{\parallel} + 2I_{\perp}$ (see Section 10.4 of Ref. 573).

Assuming that each scattering event reduces the fluorescence polarization anisotropy r by a factor of A ($A = 0$–1), the anisotropy of fluorescence that is due to a fluorophore embedded at a depth z can be written as

$$r(z) = r_0 \times A^{[N_1(z)+N_2(z)]}, \qquad (7.17)$$

where r_0 is the value of the fluorescence anisotropy without any scattering. For a homogeneous distribution of fluorophores in a tissue of thickness d, the observed value of the fluorescence anisotropy is defined by each ith tissue layer as

$$r_{\text{obs}} = \sum_i (I_i^f r_i) / \sum_i I_i^f, \qquad (7.18)$$

where I_i^f is the contribution to the observed fluorescence intensity from the ith layer of thickness dz at a depth z, and r_i is the value of the fluorescence anisotropy for this layer.

For the broad-beam illumination of a flat tissue surface, the propagation of excitation (ex) light beyond a few MFPs [MFP $\equiv l_{\text{ph}} = \mu_t^{-1}$, $\mu_t = \mu_a + \mu_s$, see Eq. (1.8)] is well described by one-dimensional diffusion theory. In this approximation, and taking into account $\mu_t^{\text{ex}} \gg \mu_{\text{eff}}^{\text{ex}}$ [see Eq. (1.18)], which is valid for many tissues, the excitation intensity reaching depth z is expressed as

$$I(z) \cong C_{\text{ex}} \exp(-\mu_{\text{eff}}^{\text{ex}} z), \qquad (7.19)$$

where C_{ex} is proportional to the excitation intensity and is the function of the tissue optical parameters at the wavelength of the excitation light.

The fluorescence from the fluorophores, embedded at depth z from the tissue surface, reaching the same surface will therefore be

$$I^f(z) \approx \left[C_{\text{ex}} \exp(-\mu_{\text{eff}}^{\text{ex}} z) \right] \left\{ \varphi \left[C_{\text{em}} \exp(-\mu_{\text{eff}}^{\text{em}} z) \right] \right\}, \qquad (7.20)$$

where C_{em} and $\mu_{\text{eff}}^{\text{ex}}$ for the emission wavelength are defined similarly as C_{ex} and $\mu_{\text{eff}}^{\text{ex}}$ for the excitation wavelength, and φ is the fluorescence yield.

By substituting the values I_i^f from Eq. (7.20) and r_i from Eq. (7.17) into Eq. (7.18), the observed fluorescence anisotropy is expressed as

$$r_{obs} = r_0 \frac{\int_0^d \exp(-\mu_{eff}^{tot} z) \times A^{[N_1(z) + N_2(z)]} dz}{\int_0^d \exp(-\mu_{eff}^{tot} z) dz}, \tag{7.21}$$

where

$$\mu_{eff}^{tot} = \mu_{eff}^{ex} + \mu_{eff}^{em}, \tag{7.22}$$

$$\mu_{s}^{tot} = \mu_{s}^{ex} + \mu_{s}^{em}. \tag{7.23}$$

Integration of Eq. (7.21) gives

$$r_{obs} = r_0 \frac{\mu_{eff}^{tot}}{\mu_{eff}^{tot} - \ln(A) \times (\mu_{s}^{tot})} \frac{1 - \exp(-\mu_{eff}^{tot} d) \times (A)^{\mu_{s}^{tot} d}}{1 - \exp(-\mu_{eff}^{tot} d)}. \tag{7.24}$$

Fluorescence anisotropy measurements are usually provided by commercially available spectrometers, the sensitivity of which is different for two orthogonal polarization states. Therefore, all measured fluorescence spectra should be corrected for the system response as[573]

$$r = \frac{I_{\parallel} - G I_{\perp}}{I_{\parallel} + 2 G I_{\perp}}, \tag{7.25}$$

where G is the ratio of the sensitivity of the instrument to the vertically and the horizontally polarized light.

Typical G-corrected polarized fluorescence spectra at 340-nm excitation from malignant and normal breast tissue with thickness ≈ 2 mm are shown in Fig. 7.16.[1162] Collagen, elastin, coenzymes (NADH/NADPH), and flavins contribute to these spectra and the spectra received at a longer wavelength of 460 nm.[1162,1163] The contribution of the NADH dominates with excitation at 340 nm, and different forms of flavins dominate with excitation at 460 nm. In Fig. 7.16, a blue shift in the polarized fluorescence spectra maximum is clearly seen in the malignant, as compared to the normal, tissue. A similar shift of 5–10 nm was also observed for 460-nm excited fluorescence. This shift is associated with the accumulation of positively charged ions in the intracellular environment of the malignant cell.[1160] Some differences, in particular, a spectral shift of the maximum, between the parallel and cross-polarized fluorescence spectra observed for rather thick tissue layers (≈ 2 mm) may be associated with wavelength-dependent scattering and the absorption properties of the tissue.

The mean fluorescence anisotropy values for normal and malignant human breast samples of tissue varying from 10 μm to 2 mm in thickness, determined

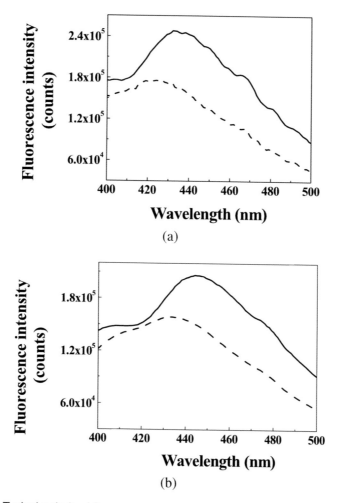

Figure 7.16 Typical polarized fluorescence spectra at 340-nm excitation of human breast tissue samples of 2-mm thickness.[1162] Solid curves, spectra with excitation and emission polarizers oriented vertically ($I_{\parallel}$); dashed curves, spectra with crossed excitation and emission polarizers ($I_{\perp}$). (a) Malignant tissue; (b) normal tissue.

with 440-nm emission and 340-nm excitation, are presented in Fig. 7.17.[1162] The theoretical fit to experimental data using Eq. (7.24) and the parameter of single-scattering anisotropy reduction $A = 0.7$ gives the following data for the anisotropy and optical parameters: $r_0 = 0.34$, $\mu_s^{tot} = 590$ cm^{-1}, $\mu_{eff}^{tot} = 53.5$ cm^{-1} for malignant tissue, and $r_0 \approx 0.25$, $\mu_s^{tot} = 470$ cm^{-1}, $\mu_{eff}^{tot} = 34.5$ cm^{-1} for normal tissue. The anisotropy values are higher for malignant tissues as compared to normal for very thin tissue sections, where $d \leq 30$ μm. By contrast, in thicker sections, the malignant tissue shows smaller fluorescence anisotropy than the normal tissue.

The fact that fluorescence anisotropy varies with tissue thickness is associated with the manifestation of various mechanisms of fluorescence depolarization that are caused by energy transfer and rotational diffusion in the fluorophores and by

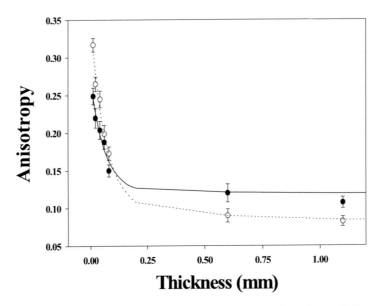

Figure 7.17 Polarized fluorescence anisotropy measured at 440 nm for excitation at 340 nm for malignant (open circles) and normal (filled circles) human breast tissues as a function of tissue thickness.[1162] The error bars represent the standard deviation. The solid and dashed curves show theoretical fits for normal and malignant tissues, respectively [Eq. (7.24)].

the scattering of excitation and emission light. Energy transfer and/or rotational diffusion of the fluorophores dominate in thin tissue sections, and these processes are faster in normal tissues than in malignant ones. In thicker sections, light scattering dominates with more contribution to depolarization during light transport within the malignant tissues.

As was already mentioned in the beginning of this section, the light scattering anisotropy factor g and, correspondingly, the reduced scattering coefficient μ_s' or the transport MFP l_t determine the characteristic depolarization depth in a scattering medium. Parameter A, characterizing the reduction of the fluorescence anisotropy per scattering event in the described model, depends on the value of the g-factor.[1162] The theoretical analysis done by the authors of Ref. 1162 has shown that, for an anisotropy parameter g ranging between 0.7 and 0.9, the value for A varies between 0.7 and 0.8. These results suggest that fluorescence anisotropy measurements may be used for discriminating malignant sites from normal ones and may be especially useful for epithelial cancer diagnostics where superficial tissue layers are typically examined.[1163]

7.6 Conclusion

As it follows from the presented analysis, polarization-sensitive methods are promising tools for optical medical diagnostics and imaging, especially for *in vivo* and *in situ* morphological analysis of living tissue. Polarization discrimination of scattered probe light, which may be easily integrated in traditional optical diagnostical techniques such as diffuse reflectance spectroscopy and imaging, offers a possibility for improving the diagnostic potential of these techniques. Another novel contribution to optical medical diagnostics should emerge from the morphological study of tissues with expressed structural anisotropy. Typically, almost all of the polarization-sensitive techniques that we considered in this chapter can be realized with inexpensive commercially available instrumentation. Neither do they require sophisticated data processing algorithms. In other words, these methods are completely suitable for widespread implementation in clinical diagnostic practice. Fluorescence polarization measurements that can provide additional information at the molecular level may be useful for discriminating malignant sites from normal ones.

8

Coherence-Domain Methods and Instruments for Biomedical Diagnostics and Imaging

In this chapter, we discuss coherent optical methods that hold much promise for applications in biomedicine, such as photon-correlation and diffusion wave spectroscopies; speckle interferometry; full-field speckle imaging; coherent topography and tomography; phase, confocal, and Doppler microscopy; as well as interferential measurements of retinal visual acuity and blood sedimentation.

8.1 Photon-correlation spectroscopy of transparent tissues and cell flows

8.1.1 Introduction

The physical fundamentals of photon-correlation spectroscopy were discussed in Chapter 4. The description of the principles and characteristics of the main modifications of homodyne and heterodyne photon-correlation spectrometers, the laser Doppler anemometers (LDAs), differential LDA schemes, and laser Doppler microscopes (LDMs) can be found in Refs. 5, 6, 22, 76–79, 82, 343, 825–827, 829, 830, 833, 838, 842, 848, and 849. A review of medical applications, mainly limited to the investigation of eye tissues (crystalline lens, cataract diagnosis), hemodynamics in isolated vessels (vessels of eye fundus or any other vessels) with the use of fiber optic catheters, and blood microcirculation in tissues, is provided in Refs. 5, 6, 22, 67, 76–79, 82, 83, 343, 825–827, 829, 830, 833, 835, 838–842, 848–850, 853, 859, and 1164–1201. In this section, we will discuss the photon-correlation technique in application to early cataract diagnostics and to measurement of blood and lymph flow in microvessels.

8.1.2 Cataract diagnostics

The photon-correlation spectroscopy or quasi-elastic light scattering (QELS) technique was originally developed to study small colloidal particles in fluids.[1170] Three decades ago, Tanaka and Benedek[1171,1172] proposed to use this technique to study the onset of cataract in the ocular lens; however, it did not find a wide-scale commercial acceptance in ophthalmology. Owing to innovations since then in the field of optoelectronics, QELS is now emerging as a potential ophthalmic tool, making the study of virtually every tissue and fluid comprising the eye possible.[849] The ability of QELS in the early detection of the molecular morphology has the potential to help develop new drugs to combat not just the diseases of the eye, such as cataract, but to diagnose and study those of the body, such as diabetes and possibly Alzheimer's, as was recently claimed by Ansari.[849,1169]

The coherent fiber-optic photon-correlation spectrometers for study of cataractogenesis and potentially useful for early diagnosis of cataract were designed about a decade ago.[850,1173] The instrument described in Ref. 850 includes two optical fibers. The first, a single-mode fiber, transmits a Gaussian beam of an He:Ne or diode laser to an object. The second, multimode or single-mode fiber is employed to collect backscattered radiation at a certain angle and to transmit this radiation to a photodetector [see Fig. 8.1(a)]. The power of the He:Ne laser radiation (at 633 nm) is on the order of 1 mW. The size of the laser beam on the crystalline lens is about 150 μm. Scattered radiation is detected at angles of 155 deg (detector 1) and 143 deg (detector 2). Figure 8.1(b) shows typical autocorrelation functions measured for bovine crystalline lens under conditions of temperature-induced cataract (reversible cold cataract). The results of the solution of the relevant inverse problem (determination of the sizes of scatterers in human crystalline lenses as functions of age), taking into account Eqs. (4.28), (4.30), and (4.31), are presented in Fig. 8.1(c). These data demonstrate that the method under consideration is sufficiently sensitive for the monitoring of age changes in the structure of the crystalline lens caused by growth in the sizes of aggregated protein components.

In Ref. 850, a clinical modification of the measuring system for the early diagnosis of cataract is presented. According to the estimates, the expected power density incident on a retina that is sufficient to measure the autocorrelation function of intensity fluctuations within a time interval of about 2 min is no higher than 0.05 mW/mm^2, which is almost three orders of magnitude lower than the threshold of retinal damage.

The fiber-optic QELS probe, shown in Fig. 8.2, combines the unique attributes of small size, low laser power, and high sensitivity.[849,1169] The system is easy to use because it does not require sensitive optical alignment nor vibration isolation devices. A low-power (50–100-μW) light from a semiconductor laser, interfaced with a monomode optical fiber, is tightly focused in a 20-μm diameter focal point in the tissue of interest via a GRIN (gradient index) lens. On the detection side, the scattered light is collected through another GRIN lens and guided onto an avalanche photodiode (APD) detector built into a photon-counting module. APD processed signals are then passed on to a digital correlator for analysis. The probe provides quantitative measurements of the pathologies of cornea, aqueous, lens, vitreous, and the retina. By suitable choice of optical filters, it can be converted into a device for spectral measurements (autofluoresence and Raman spectroscopy) and laser-Doppler flowmetry/velocimetry, providing measurements of oxidative stress and blood flow in the ocular tissues. The device also can easily be integrated into many conventional ophthalmic instruments such as slit lamps, Scheimpflug cameras, videokeratoscopes, and fluorometers.

This compact probe (Fig. 8.2) was used for the monitoring of cataractogenesis in mice *in vivo* by examination of the measured autocorrelation function (AF)

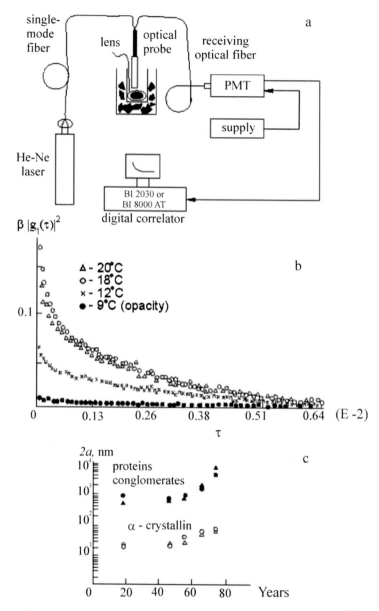

Figure 8.1 Photon-correlation spectrometer for early diagnosis of cataract.[850] (a) Diagram of the spectrometer. (b) Autocorrelation functions for temperature-dependent cataract of bovine eye. (c) Age-dependent changes of scatterers' diameters for the human lens. Two fractions, finely dispersed (α-crystallin, hollow symbols) and coarsely dispersed (protein conglomerates, filled symbols) are excluded from the empirical autocorrelation functions [see Eq. (4.31)]. Triangles and squares represent measurements made with different types of coherent fiber probes.

profiles [see Eqs. (4.28) and (4.30)] at different time lines.[849,1169] As an example, Philly mice were studied. This animal develops cataract spontaneously between day 26 and 33 after birth. The data include a 45-day-old normal mouse of the con-

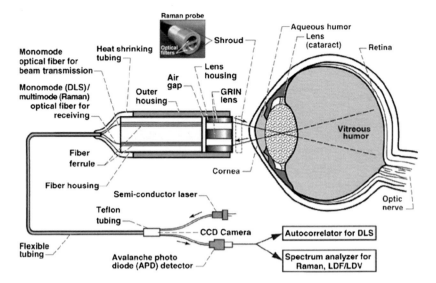

Figure 8.2 Schematic diagram of the sensitive, vibration protected, universal, and easy to use QELS fiber-optic probe.[1169] The probe was originally developed at NASA to conduct fluid physics experiments in the absence of gravity onboard a space shuttle or space station orbiter.

trol FVB/N strain, which does not develop a cataract, and two Philly mice roughly 26–29 days old. Each measurement took 5 s at a laser power of 100 µW. The changing AF slope is an indication of cataractogenesis because the lens crystallins aggregate to form high molecular weight clumps and complexes. The QELS autocorrelation data is converted into particle size distributions using an exponential sampling program and is shown in Fig. 8.3. Although conversion of the QELS data into particle size distributions requires certain assumptions regarding the viscosity of the lens fluid, these size values do indicate a trend as the cataract progress. These measurements suggest that a developing cataract can be monitored quantitatively with reasonable reliability, reproducibility (5–10%), and accuracy.

Besides cataract monitoring, the QELS probe has been proposed and experimentally tested for early, noninvasive, and quantitative detection and monitoring of such disease and abnormalities as vitreopathy, pigmentary glaucoma, diabetic retinopathy, and corneal evaluation of wound healing after laser refractive surgery.[849,1169]

A portable fiber-optic photon-correlation spectrometer based on an He:Ne laser (633-nm), single-mode fibers, a photomultiplier operating in the regime of a photon counting mode, and a 288-channel real-time correlator with a sampling time of 200 ns is described in Ref. 851. This spectrometer allows *in vivo* studies of crystalline lenses of patients. These investigations also confirmed the bimodal character of the distribution of scatterers in the tissue of a human crystalline lens. Specifically, for healthy eyes of patients aged between 39 and 43 (six eyes, three female patients), the finely dispersed fraction has a mean radius of 4.25 ± 1.7 nm, whereas the mean radius of the coarsely dispersed fraction is 497 ± 142 nm. For cataractous

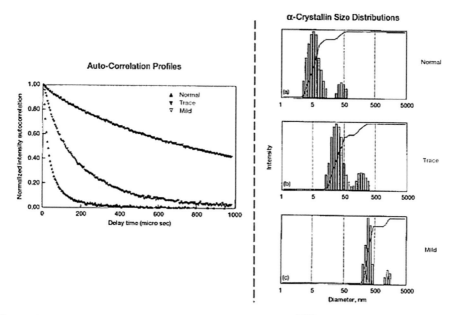

Figure 8.3 *In vivo* cataract measurements in Philly mice.[1169] (a) Autocorrelation profiles. (b) Particle size distribution for the normal eye, control mouse. (c) Particle size distribution for a mouse with trace cataract. (d) Particle size distribution for a mouse with mild cataract.

crystalline lenses, the mean radius of the finely dispersed fraction tends to 160 nm, whereas the mean radius of the coarsely dispersed fraction tends to 1000 nm. The spectrometer permits one to determine the size distribution of species for various localizations of the volume of the measurements. The bimodal size distributions of scatterers measured under conditions when the volume of measurements is shifted along the axis of a cataractous crystalline lens are presented in Fig. 8.4.

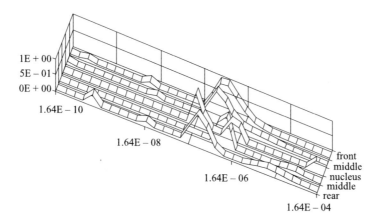

Figure 8.4 Bimodal distribution of the radii of scattering particles for a cataractous crystalline lens (female patient at the age of 76, *in vivo* measurements) with different localizations of the volume of measurements along the axis of the crystalline lens: the front part of the cortical layer, the middle part of the cortical layer, nucleus, the middle part of the rear cortical layer, and the rear part of the cortical layer.[851]

8.1.3 Blood and lymph flow monitoring in microvessels

Parameters of blood or lymph flows in individual vessels can be measured with the use of a technique based on the diffraction of a focused laser beam by moving scatterers (see Section 4.4.2).[77,343,826,833,853,854] A diagram of the relevant speckle microscope is presented in Fig. 8.5. Laser radiation focused into a spot of a small diameter on the order of 4.6λ is projected onto a segment of the microvessel under study. A photodetector whose entrance aperture is much smaller than the mean speckle size registers intensity fluctuations in the scattered light. The intensity fluctuations detected are analyzed with a low-frequency digital spectrum analyzer or converted into a digital signal for subsequent computer analysis. The typical spectra for blood and lymph flows in superficial vessels of rat mesentery averaged over 128 realizations of a random signal are displayed in Fig. 8.6.

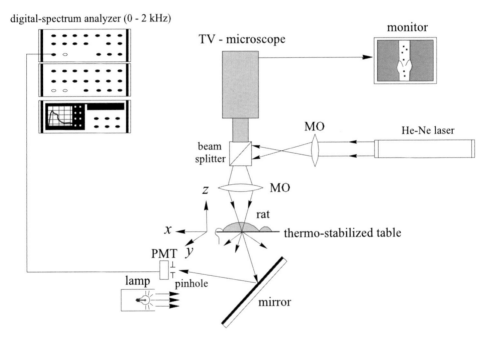

Figure 8.5 Diagram of a speckle microscope for the investigation of blood and lymph flows in microvessels.[853,854]

For a blood microvessel, the spectrum of intensity fluctuations in the scattered field has a nearly Gaussian shape in the low-frequency range. The spectra of intensity fluctuations for lymphatic vessels are rather complicated, which indicates that the motion of lymph in microvessels is much more complex than the motion of blood. For example, lymph may be involved in a characteristic shuttlelike motion, be nonmovable near the vessel wall, or move in a direction opposite to the flow in the central part of a vessel. Such a behavior of lymph is associated with a complex

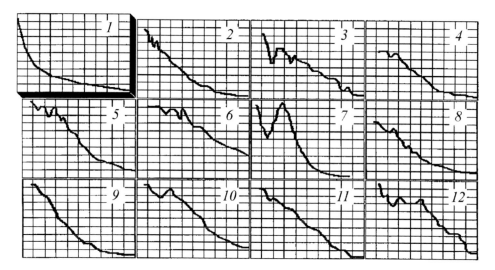

Figure 8.6 Spectra of intensity fluctuations in the scattered light (averaged over 128 realizations) for the diffraction of a focused laser beam on rat mesentery: 1, blood vessel (12 μm in diameter); 2–12, lymphatic vessels of various diameters for different rats. Measurements were performed within the frequency range of 30–1500 Hz. [853,854]

contraction dynamics of smooth-muscle cells in vessel walls, governed by a local rhythm driver (pacemaker) and the behavior of the valves of lymphatic vessels.

To estimate the parameters of blood and lymph flows in microvessels, we introduce the following quantities:[853,854]

$$V_V = \frac{\Delta F}{D_V},$$

$$(8.1)$$

$$\Sigma_V = \frac{\int_0^{\Delta F} |S(f) - G(f)|^4 df}{[\int_0^{\Delta F} |S(f) - G(f)|^2 df]^2 / \Delta F}.$$

Here, ΔF is the width of the averaged spectrum, D_V is the diameter of a microvessel, $S(f)$ is the power spectrum of the intensity fluctuations for the speckle field studied, and $G(f)$ is the spectrum with a Gaussian envelope. The spectra $S(f)$ and $G(f)$ have equal bandwidths and powers. Parameter V_V is directly proportional to the flow velocity, and Σ_V provides the information concerning spatial and temporal variations of the flow rate in the studied area of a vessel.

The characteristics given above have been employed to analyze the influence of a lymphotropic agent (*Staphylococcus* toxin, ST) on the dynamics of lymph flow in mesentery microvessels of experimental animals (rats).[853,854] It was found that even at the fifth minute of ST action, all the vessels studied showed variations in the spectra of intensity fluctuations in scattered light, which indicates that the velocity characteristics of the flow change: 59% of the 17 vessels studied displayed a decrease in the mean velocity V_V of lymph flow by $41 \pm 8\%$ and a growth in the

Σ_V parameter by $33 \pm 7\%$. The remaining 41% of the vessels displayed an increase in the mean rate of lymph flow by $63 \pm 22\%$ and a decrease in the Σ_V parameter by $35 \pm 9\%$. In later stages (between the fifth and twentieth minutes), vasoconstriction progresses and the number of vessels contracting in phase decreases, which leads to changes in lymph dynamics. After the twentieth minute, lymph flow stopped in all the vessels studied.

The optical scheme of the setup providing detection of cell flow direction and velocities in the range from 10 μm/s to 10 mm/s with a temporal resolution up to 50 ms is shown in Fig. 8.7.[833,858] Radiation from a uniphase He:Ne laser (633 nm) is delivered through the illuminator channel and focused by the objective of the microscope into a spot of a diameter of about 2 μm in a plane apart at a distance $z = 100$ μm from the axis of the microvessel. The radius of curvature of the wavefront of the beam illuminating the microvessel is quite small to ensure the acceptable translation length of biospeckles. The measuring volume is formed by the intersection of the diverging laser beam with the microvessel and has the shape of a truncated cone (whose elements have a slope of 10 deg and a mean diameter on the order of 30 μm). The laser radiation scattered by the cell flow is directed with the help of the beamsplitter to the photodetector placed at a distance of 300 mm from the objective plane of the microscope. The diameter of each photodetector is 3 mm, which corresponds to the mean speckle diameter in the observation plane. The distance between the centers of the photodetectors is about 7 mm. Signals from the photodetectors are amplified by the photocurrent transducers and digitized with the help of a two-channel 16-bit analogue-to-digital converter with a sampling frequency of 44.1 kHz. A PC is used to determine the cross-correlation function of the photodetector signals as well as the positions of its peaks.

Depending on the time resolution, the processing of the realization of photodetector signals of duration 60 s takes between 90 and 300 s. A digital video camera combined with a transmission microscope is used for analysis of microvessels' functions *in vivo* in real time: estimate the mean flow velocity and its direction, measure the diameter of a microvessel, and register the appearance of phasic contraction in the investigated lymphatics. Dynamic digital images were processed with specially developed software. The cell velocity was determined as the ratio of the difference in cell coordinates in two consecutive frames to the time interval between the two frames. The mean flow velocity was calculated by averaging the velocities of four to six cells. Dynamic digital microscopy allows one to record cell flow velocity in the range from 25 μm/s to 2–2.5 mm/s with a time resolution of 40 ms.

The described setup was tested for *in vivo* measurements of lymph flow velocity in the mesentery vessels of narcotized white rats. Animals were placed on a thermostabilized stage (37.7°C) of the microscope (see Fig. 8.7) and the mesentery and intestine was kept moist with Ringer's solution at 37°C (pH $\sim$ 7.4). The images of microvessels were evaluated by transmission microscopy and laser speckle-velocimeter simultaneously. Figure 8.8 shows the temporal dependences of the flow velocity in the investigated microlymphatic with mean diameter 170 ± 5 μm

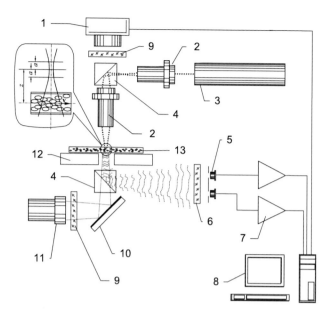

Figure 8.7 Scheme of the experimental setup of a laser speckle velocimeter integrated with dynamic digital microscopy providing measurements of absolute values of cell flow velocity and its direction: 1, digital video camera; 2, microobjective; 3, He:Ne laser (633 nm); 4, beamsplitter; 5, photodiodes; 6, red light filter; 7, photocurrent converters; 8, PC; 9, green light filters; 10, mirror; 11, illuminator; 12, thermally stabilized table; 13, lymph microvessel of mesentery. The inset shows the illumination of a lymphatic vessel by a focused Gaussian laser beam (a is the length of the laser beam waist and z is the separation between the flow axis and the waist plane of the laser beam).[858]

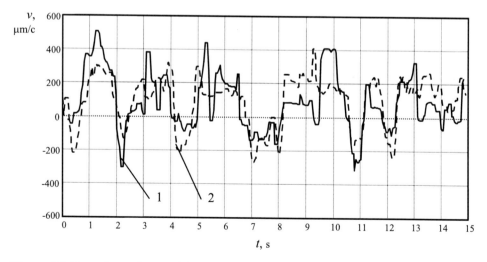

Figure 8.8 Time dependence of the lymph flow velocity in a lymphatic vessel of mean diameter 170 ± 5 μm of white rat mesentery: 1, recorded with a speckle velocimeter; and 2, with dynamic digital microscopy (see Fig. 8.7).[858]

and mean lymph flow velocity 169 ± 4.6 μm/s.[833,858] These dependences were obtained concurrently by laser speckle velocimeter and by processing of the video images. A laser speckle velocimeter allows one to measure the lymphocyte velocity in relative units only. The proportionality coefficient between the data of laser speckle velocimetry and the mean flow velocity, measured by dynamic digital microscopy, was determined from the slope of the line of linear regression between the velocities (measured by these two methods). The correlation coefficient of linear regression was equal to 0.723 for measurements in the lymph vessel and 0.966 for calibration measurements in the glass capillary.

8.2 Diffusion-wave spectroscopy and interferometry: measurement of blood microcirculation

Experimental implementation of diffusion-wave spectroscopy is very simple: a measuring system should irradiate the scattering object under investigation with a light beam produced by a continuous wave laser and measure intensity fluctuations in scattered radiation within a single speckle with the use of a photomultiplier and an electronic correlator. A typical setup employed for model experiments is presented in Fig. 8.9.[80,81,875,1174,1175] Radiation (with a wavelength of 514 nm and a power on the order of 2 W) produced by an argon laser with an intracavity etalon passes through a multimode fiber-optic cable and irradiates the surface of a solid-state bulk sample (finely dispersed TiO_2 powder suspended in resin). The sizes of the sample are $15 \times 15 \times 8$ cm. A spherical cavity 2.5 cm in diameter filled with a 0.2% aqueous suspension of polystyrene spheres 0.296 μm in diameter at the temperature of 25°C is placed at the center of the sample 1.8 cm below its upper surface. The transport MFP lengths of photons for the suspension and the sample are $l_t = 0.15$ and 0.22 cm, respectively. The absorption coefficients of these media are equal to each other, $\mu_a = 0.002$ cm^{-1}. The diffusion coefficient of Brownian motion in suspension is $D_B = 1.5 \times 10^{-8}$ cm^2/s. The single-mode fiber collects light emerging from a certain area of the object and transmits it to the photomultiplier. The output signal of the photomultiplier is fed to a digital autocorrelator, which reconstructs the time-domain autocorrelation function (AF) of the intensity fluctuations. This AF is related to the time-domain autocorrelation function of the field by the Siegert formula [see Eq. (4.28)]. Optical fibers were designed in such a manner as to pick up radiation from any area at the surface of the sample.

Figure 8.10 displays the experimental results for the normalized time-domain AF of the field for three different arrangements of optical fibers connected to a source of radiation and the detector, and compares these experimental data with theoretical predictions. Since the origin of the x–y coordinate frame lies on the surface of the sample above the center of the dynamic cavity, the source of radiation and the detector were placed along the y-axis in such a manner that the coordinate of the source was $y = 1.0$ cm, and the coordinate of the detector was $y = -0.75$ cm. Measurements were performed for $x = 0.0, 1.0,$ and 2.0 cm. The distance between the source and the detector remained constant. The error of these

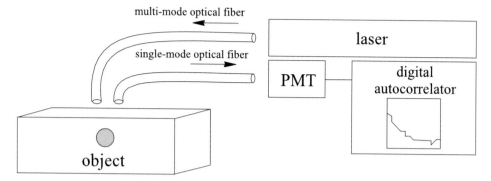

Figure 8.9 Typical experimental setup for diffusion-wave (correlation) spectroscopy of scattering media.[1175]

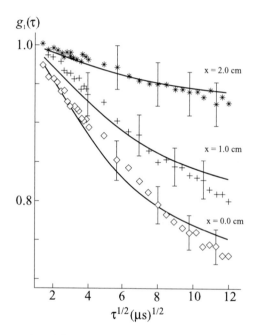

Figure 8.10 Experimental and theoretical normalized autocorrelation functions of intensity fluctuations in light scattered from a TiO_2 sample with a spherical cavity filled with a suspension of polystyrene spheres (see the text for the details).[1175]

measurements was estimated as 3%. The main source of errors was associated with uncertainties in the positioning of the optical fibers. Theoretical curves represent the results of simulations based on the diffusion theory of autocorrelation with allowance for the experimental data.[875,1174,1175] It can be easily seen that the AF decays faster when the source of radiation and the detector are located close to the dynamic sphere, which gives rise to fluctuations in the time domain. It is in this area that most of the detected photons pass through the dynamic volume. Such a behavior of AFs allows one to employ the variation in their slopes (decay rates) as a parameter for the imaging of dynamic inhomogeneities in a medium. This model corresponds to a situation where the microcirculation rate of blood locally increases near by, e.g., a growing tumor. Using a similar approach, one can also model a directed blood flow. For this purpose, a through hole should be drilled in a solid sample at a certain depth, and scattering fluid (e.g., Intralipid) should be circulated through this hole with a definite flow rate.[1174,1175]

To determine the AF of the field on nanosecond and subnanosecond time scales, we should replace an electronic correlator by a Michelson interferometer with a large difference in arm lengths, which should be on the order of 3 m.[80,1176] In this case, the intensity $\langle I(\tau)\rangle$ averaged in time depends on the delay time τ between the interfering fields in the interferometer, the carrier frequency ω of the optical signal, the average intensity I_{ave} of speckles, and the time-domain AF $g_1(\tau)$ of the field as

$$\langle I(\tau)\rangle = I_{ave}\frac{1 + g_1(\tau)\cos(\omega\tau)}{2}. \tag{8.2}$$

A diagram of the experimental setup and the results of model experiments are presented in Fig. 8.11. Radiation of a continuous-wave laser scattered in the forward direction by an object within a single speckle is coupled into a long Michelson interferometer. The difference between arm lengths of this interferometer can be smoothly adjusted by variation of air pressure in the short arm. Effects arising due to a finite correlation length of laser radiation and geometric factors were excluded through the calibration of the measuring system with the use of diluted samples whose AFs do not decay on the time scales studied.

The possibilities of the DWS technique for medical applications have been demonstrated in Ref. 1177. The experimental setup employed in this study is shown in Fig. 8.12(a). The experimental system was based on a titanium:sapphire laser with a power of about 100 mW and a wavelength of 800 nm. Laser radiation was transmitted onto an object through a multimode optical fiber with a core 200 μm in diameter. Radiation was detected within a single speckle with the use of a single-mode fiber 5 μm in diameter. The distance between the optical fibers on the surface of an object remained constant and was equal to 6 mm. The rate of blood flow in the bulk of a tissue from a human forearm was adjusted with the use of a medical tonometer. A digital autocorrelator coupled with a photomultiplier in the regime of a photon counter was used to measure the time-domain AF of intensity fluctuations and the dependence of the AF shape on the pressure P produced by the tonometer. Such dependences for the AF of the field, which is related to the intensity AF by the Siegert formula, are presented in the logarithmic scale in Fig. 8.12(b). These dependencies show a sufficiently high sensitivity of the AF slope to variations in the applied pressure, i.e., changes in the rate of volume blood flow. In accordance with Eqs. (4.30) and (4.43), the normalized AF of field fluctuations can be represented in terms of two components related to the Brownian and directed motion of

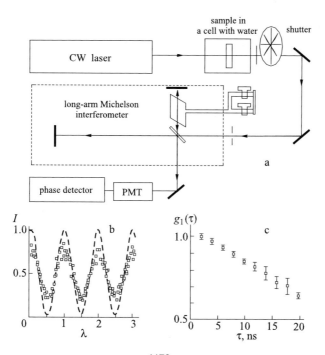

Figure 8.11 Diffusion-wave interferometry:[1176] (a) Experimental setup. (b) Normalized output signal (interference fringes) for (dots) $\tau = 0$ and (squares) 20 ns for two-phase aqueous suspensions of polystyrene spheres 0.0385 and 0.299 μm in diameter. (c) The relevant autocorrelation function of the field $g_1(\tau)$.

scatterers as[1177]

$$g_1(\tau) = \int_0^\infty p(s) \exp\left\{-2\left[\frac{\tau}{\tau_B} + \left(\frac{\tau}{\tau_S}\right)^2\right]\frac{s}{l_t}\right\} ds, \qquad (8.3)$$

where $\tau_B^{-1} \equiv \Gamma_T$ is defined in Eq. (4.30), $\tau_S^{-1} \cong 0.18 G_V |\bar{q}|/l_t$ characterizes the directed flow, and G_V is the gradient of the flow rate. The other quantities involved in Eq. (8.3) are defined in Eqs. (4.26) and (4.43). The above relationship allows one to express the slope of the AF in terms of the diffusion coefficient and the gradient of the directed velocity of scatterers. When shear flow significantly dominates under Brownian motion, a semilogarithmic plot of $g_1(\tau)$ versus $\tau^{1/2}$ gives a straight line with a slope proportional to the velocity of the scattering particles flow.

Figure 8.12(c) displays the measured velocity of blood flow as a function of the applied pressure. If we neglect the Brownian component, this dependence is characterized by the variation in the AF slope [see Eq. (8.3)]. Since measurements were performed at the wavelength close to the isosbestic point (805 nm), changes in the degree of oxygenation of the blood due to the variation in the applied pressure only slightly influence the AF slope (the velocity of blood flow). This circumstance allows us to find the correlation between the velocity of blood flow and variations in the diameter of vessels by means of simultaneous and independent measurements

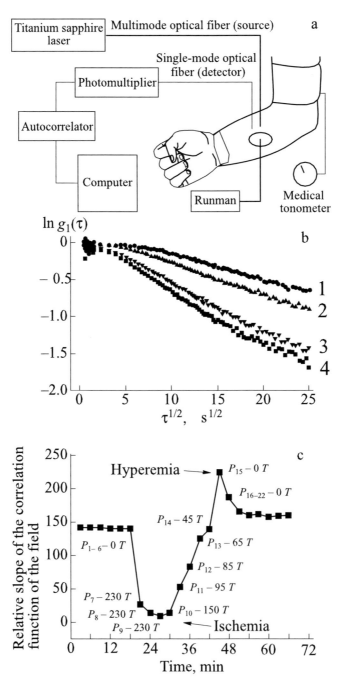

Figure 8.12 *In vivo* measurements of blood flow velocity by means of the DWS technique:[1177] (a) Experimental setup. (b) Experimental AFs of field fluctuations in backward scattering for different pressures applied to an arm (the pressure increases from 1 to 4). (c) Dynamics of the relative slope of the AFs for various pressures applied to an arm, *P* is the tonometer pressure. The arrows indicate the moments of time that correspond to the narrowing (ischemia) and broadening (hyperemia) of the vessels.

of the oxygenation degree and the volume of blood in a tissue with the use of a two-frequency Runman spectrometer (NIM Inc., Philadelphia). These measurements provide a pictorial illustration of the high efficiency of the DWS technique for *in vivo* studies of blood flow in bulk tissues.

We should also note that if the parameters of blood flow remain constant, the measured AFs provide information concerning the static optical parameters of a multiply scattering medium, i.e., l_t or μ_s', μ_a, and g [see Eq. (4.43)]. Indeed, as shown in Ref. 1178, the half-width of the spectrum of time-domain intensity fluctuations under conditions of multiple scattering depends not only on dynamic and geometric parameters of scattering particles, but also on the absorptivity of erythrocytes in blood, which allows us to estimate the degree of blood oxygenation from the results of measurements performed far from the isobestic wavelength.

The hybrid instrument and measuring protocol based on diffuse correlation spectroscopy (blood flow information) and diffuse reflectance spectroscopy (blood oxygenation information) described in Ref. 1179 provide the evaluation of microcirculation and muscle metabolism in patients with vascular diseases. A CW laser (800 nm) with a long coherence length and an avalanche photodiode were used for correlation measurements; source-detector separations ranged from 0.5 to 3 cm and the sampling time was 1.5 s. A complete frame of data, cycling through all source-detector pairs, was acquired in 2.5 s. Ten healthy subjects and one patient with peripheral arterial disease were studied during 3-min arterial cuff occlusions of the arm and leg, and during 1-min plantar flexion exercises. Signals from different layers (cutaneous tissues and muscles) during cuff occlusion were differentiated, revealing strong hemodynamic responses from muscle layers. During exercise in healthy legs, the observed approximately 4.7-fold increase in relative blood flow was significantly lower than the corresponding increase in relative muscle oxygen consumption, which was approximately sevenfold. In the diseased patient, during exercise the magnitudes of both these physiological parameters were ~1/2 of the healthy controls, and the oxygen saturation recovery time was twice that of the controls.

8.3 Blood flow imaging

It can be easily shown that the methods of Doppler flowmetry, which have been extensively developed within the past three decades, are, in general, identical to comparatively new speckle methods (which were proposed in the 1980s) in their applications to the analysis of the parameters of blood microcirculation because these two approaches provide an opportunity to determine the velocity of blood flow at a certain point.[82] The review of Doppler methods for the monitoring of blood microcirculation in tissues is provided in Refs. 5, 22, 112, 826, 827, 831, 833, 838, 841, 842, and 1180–1182 and in several original papers, e.g., Refs. 67, 839, 840, and 1183–1193. Note that the extension of the Doppler method to the investigation of blood microcirculation in thick tissues has stimulated the development of the theory of Doppler signals and the methods of simulations and detection of such signals in multiply scattering media.[247,930,1186–1190] Speckle methods of investigation of blood microcirculation are described in Refs. 3, 5, 76, 82, 83, 112, 821, 827, 829–831, 853, 854, 859, 863, 865, 1164, 1165, and 1194–1204.

The diagnosis of many diseases associated with blood microcirculation disorders requires a monitoring of microcirculation within large areas of a tissue, i.e., imaging of the field of blood flow velocity.[83,112,827,851,861–863,865,1164] Since the methods under study are characterized by a high spatial locality one ensures mechanical scanning or sequential analysis of the intensity fluctuations within a separate pixel of a CCD camera, or implements both mechanical scanning and sequential analysis. Such scanning systems for the imaging of blood circulation have been implemented and developed up to the stage of commercial production.[861,862] One such system is shown in Fig. 8.13.[862] A system for the imaging of blood circulation should be useful for diagnosis and therapy support in cases associated with diseases of the peripheral vascular system and for the medical treatment of wounds and burns. However, mechanical scanning of a laser beam or the necessity to collect and process large data arrays in systems involving CCD cameras prevented designers from creating simple and high-performance imaging systems. Apparently, the only exception was a speckle system based on a 100 × 100-pixel matrix photodetector, which was designed specifically for the analysis of retinal blood flow. The total time of data analysis in this system for a 0.42 × 0.42-mm field was 15 s.[860]

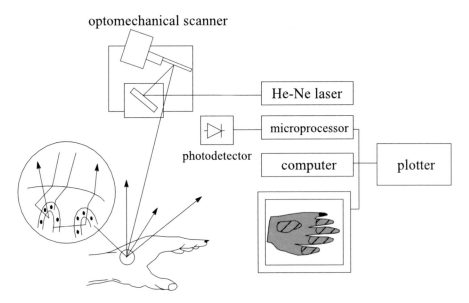

Figure 8.13 Diagram of a scanning Doppler system for the imaging of blood microcirculation in tissues.[862]

The potential exists to build a robust blood-flow imager in tissue with multiple scattering based on focused-laser-beam probing and detection of spatial cross-correlation of the scattered field using a CCD camera.[1196] The method is based on the concept that the region of single scattering within the tissue (volume oc-cupied by a focused laser beam) will produce large correlated areas (speckles) in a transverse dimension, whereas the comparably large halo of multiple scattered photons produced by this beam will give rise to small speckles. Thus, the cross-correlation function of the intensity fluctuations in two spatial points (CCD camera pixels) with the separation Δx larger than the size of the multiscattered speckles will reflect the form of single-scattered AF. The profile of a single-scattering AF will provide information about blood or lymph flow in a volume occupied by a fo-cused laser beam. The attractiveness of this approach is defined by its applicability to the intermediate scattering regimes, whereas QELS provides accurate informa-tion only for a single scattering regime and DWS can be applied only for a case of diffusive photon propagation.

A typical approach to improve the performance of blood flow imagers, partic-ularly to increase the imaging speed, is to parallelize measurements by using 1D or 2D arrays of photodetectors. A new generation of high-speed instruments for full-field blood flow laser Doppler imaging (LDI) was recently developed on the basis of CMOS image sensors.[1191–1193] The LDI system employing an integrating CMOS image sensor delivers high-resolution blood flow images every 0.7–11 s, depending on the number of points in the acquired time-domain signal (32–512 points) and the image resolution (256 × 256 or 512 × 512 pixels). For the in-tegrating imager, a digital CMOS camera based on the VCA1281 monochrome CMOS image sensor from Symagery (Canada) was utilized. This sensor operates in a rolling shutter mode; it has 280 horizontal × 1024 vertical resolution, a 7 × 7-μm pixel size, a 40-MHz sampling rate, and an 8-bit amplitude digital converter (ADC). The sensor has a specified flat spectral response in the range between 500 and 750 nm. The camera was connected to the host PC via a fast LVDS (low-voltage differential signaling) interface providing for a high-speed transfer of the obtained frames.

For the object illumination, a solid-state-diode-pumped laser of 250-mW out-put optical power emitting at 671 nm was used. The laser beam was coupled to a 1.5-mm diameter plastic optical fiber. A GRIN (gradient index) lens of 1.8-mm diameter was placed at the distal end of the fiber. This configuration produced a uniform illumination of the object. The illuminated area was up to a 170-mm di-ameter. The backscattered light was collected with an objective ($f = 6$ mm) with a low f-number (1.2), providing the system with the superior photon collection effi-ciency that becomes critical for short integration times (in the range of a few tens of milliseconds). Typically, the imager head was placed at a distance of 150–250 mm from the investigated tissue surface (see Fig. 8.14).

The specially developed software allows for changing the sensor parameters for control of the data acquisition mode, for acquisition of the data, and for display of the flow-related (perfusion, concentration, speed) maps. A photographic image of the sample and flow-related maps displayed on the monitor are obtained with the same image sensor; therefore, the obtained flow maps can be easily associated with an area of interest on the sample. The signal sampling frequency is inversely

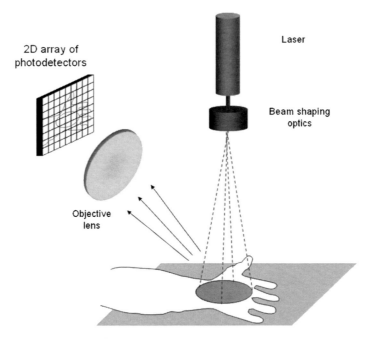

Figure 8.14 Schematic of a high-speed full-field laser Doppler imaging system on the basis of a CMOS image sensor.[1193]

proportional to the time to acquire one subframe. The subframe sampling rate of the sensor depends on its size and the pixel clock frequency. The clock frequency was fixed at 40 MHz for optimum performance speed and quality. The size of the sampled subframe finally defines the signal sampling frequency of the imager. For the 256×4-pixels subframe, the frame sampling frequency was 30 kHz, and at 256×6 pixels it was 20 kHz, at 256×8 pixels it was 14 kHz, etc.

To obtain one flow map over a region of interest (ROI), which was 256×256 or 512×512 pixels, the ROI must be subdivided into smaller regions (e.g., into 32 subframes of 256×8 pixels) and scanned electronically. From 32 to 512, sampled points were obtained for the acquired time-domain signal for each pixel of the subframe; thus, the intensity fluctuation history was recorded for each of the pixels of the ROI.

The signal processing comprises the calculation of the zero moment (M_0) and the first moment (M_1) of the power density spectrum $S(\nu)$ of the intensity fluctuations $I(t)$ for each pixel. The zero moment is related to the average concentration $\langle C \rangle$ of the moving particles in the sampling volume. The first moment (flux or perfusion) is proportional to the root-mean-square (rms) speed of the moving particles, V_{rms}, times the average concentration.[1205] The governing expressions are

$$\text{concentration} = \langle C \rangle \propto M_0 = \int_0^\infty S(\nu)d\nu, \qquad (8.4)$$

$$\text{perfusion} = \langle C \rangle V_{\text{rms}} \propto M_1 = \int_0^\infty \nu S(\nu) d\nu, \qquad (8.5)$$

$$S(\nu) = \left| \int_0^\infty I(t) \exp(-i2\pi\nu t) dt \right|^2. \qquad (8.6)$$

Here, the variable ν is the frequency of the intensity fluctuations induced by the Doppler-shifted photons. To calculate the power density spectrum, an FFT algorithm, optimized for the speed performance and applied to the recorded signal variations at each sampled pixel of the ROI, was used. Noise subtraction was performed on the calculated spectra by setting a threshold level on the amplitude of the spectral components. This filtering is applied to reduce the white noise (e.g., thermal and readout noises) contribution to the signal. Thereafter, the perfusion, concentration, and speed maps were calculated and displayed on a computer monitor. The total imaging time (including data acquisition, processing, and display) depends on the number of samples obtained for each pixel and the ROI size. For the 256×256-pixel ROI, the imaging time is 0.9 s for 64 samples, 1.2 s for 128 samples, 1.7 s for 256 samples, and 2.9 s for 512 samples.

In Fig. 8.15, flow-related maps obtained on finger skin of a healthy person are shown. The images were obtained for the imager settings for the bandwidth from dc to 4000 Hz with 66 Hz resolution; the integration time was 130 µs. The total imaging time was around 5 s. A smoothing filter was applied to the row images: the value of each pixel shown was obtained by averaging the row values of eight neighboring pixels. The flow maps (perfusion, concentration, speed) are false coded with nine colors (not shown). The images clearly show the difference in the speed and concentration distributions measured on the fingers. The lower value for the concentration signal measured on the nail is caused by the higher amount of non-Doppler-shifted photons reemitted from the relatively thick statically scattering nail tissue compared to the thin statically scattering epidermis of the skin. The signal measured on the nail shows a higher speed of the moving blood cells in the undernail tissue. However, it could not be definitely predicted whether this is because the blood speed was really higher under the nail, or because the measured values were obtained due to a multiple scattering influence (see AF presented in Fig. 8.27). This ambiguity is a common problem for all laser Doppler or laser speckle imagers. The black-and-white photographic image of the object of interest is obtained with the same CMOS camera. This image is useful for determining the anatomical boundaries associated with the perfusion regions presented in the blood flow maps.

The imaging time of the this high-speed LDI system approaches the imaging time of laser speckle imaging (LSI) systems,[82,83,112,827,831,863,865–868,1197–1203] which are currently accepted as the fastest.[1200] The LSI systems obtain flow-related information by measuring the contrast of the image speckles (see Section 4.4.3). Effectively, the contrast values measured by LSI are directly proportional to the

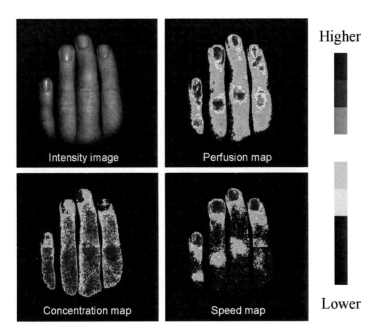

Higher

Lower

Figure 8.15 Flow-related maps obtained with a CMOS integrated imager on finger skin (ROI = 512 × 512 pixels): image of the object (intensity image); perfusion map (lower is 200 a.u. and higher is 700 a.u.); blood concentration map (lower is 140 a.u. and higher is 310 a.u.); flow speed map (lower is 400 a.u. and higher is 1500 a.u.). The imaging area is 11 × 11 cm. The imaging time is 5 s total.[1193]

normalized M_0 value that is measured by laser Doppler with integrating photodetectors. The images shown in Fig. 8.15 demonstrate a difference between the perfusion (M_1) and concentration (M_0) maps. It looks like LDI provides more objective information rather that the LSI method because with the LDI technique, the concentration and speed signals can be measured independently. In LSI, these two signals are typically mixed and it can thus be hard to attribute an exact cause for the changes in the contrast signal.[1204] However, a good correlation ($R^2 = 0.98$) between LDI and LSI measurements of the same area of regional cerebral blood flow (CBF) for different animals (male Wistar rats) was found.[1197] A detailed comparison of the laser-Doppler and speckle contrast methods of blood flow imaging can be found in Ref. 1199. It follows from this analysis that the speckle contrast technique can provide an image of tissue vascular structure with the relative distribution of blood velocity [correspondingly to a nonlinear response, described by Eq. (4.34)], but it does not provide a linear measurement of perfusion in comparison with LDI.

However, the speckle-contrast technique LASCA, whose fundamentals are discussed in Section 4.4.3, is a conceptually simple high-performance technique for blood flow imaging.[82,83,112,821,827,829–831,863,865–868,1194–1201] The measuring system employs a CCD camera, a frame grabber, and dedicated software for the computation of the local contrast of a speckle pattern and conversion of the contrast into a color map (which gives a map of flow velocities). The resulting image represents the contrast of the speckle field averaged in time. However, such an averaging is performed rather quickly (the averaging time is usually 5–30 ms) in order to permit real-time measurements.

Equation (4.34) gives an expression for the speckle contrast in the time-averaged speckle pattern as a function of the exposure time T and the correlation time,

$$\tau_c = \frac{1}{ak_0 v}, \tag{8.7}$$

where v is the mean velocity of the scatterers, k_0 is the lightwave number, and a is a factor that depends on the Lorentzian width and scattering properties of the tissue.[1205] As in LDI, it is theoretically possible to relate the correlation times, τ_c, to the absolute velocities of the red blood cells, but this is difficult to do in practice because the number of moving particles that light interacted with and their orientations are unknown.[1205] However, relative spatial and temporal measurements of velocity can be obtained from the ratios of $2T/\tau_c$, which is proportional to the velocity and defined as the measured velocity.[831,866,867]

The schematic diagram of the experimental setup is shown in Fig. 4.14. A He:Ne laser beam ($\lambda = 633$ nm, 3 mW) was coupled into an 8-mm diameter fiber bundle, which was adjusted to illuminate the area of interest evenly.[831,866,867] The illuminated area was imaged through a zoom stereo microscope (SZ6045TR, Olympus, Japan) onto a CCD camera (PIXELFLY, PCO Computer Optics, Germany) with 480×640 pixels, yielding an image of 0.8 to 7 mm, depending on the magnification; the exposure time T of the CCD was 20 ms. Images were acquired through easily controlled software (PCO Computer Optics, Germany) at 40 Hz.

The raw speckle images were acquired to compute the speckle contrast image. The number of pixels used to compute the local speckle contrast can be selected by the user: lower numbers reduce the validity of the statistics, whereas higher numbers limit the spatial resolution of the technique. To ensure proper sampling of the speckle pattern, the size of a single speckle should be approximately equal to the size of a single pixel in the image, which is equal to the width of the diffraction-limited spot size and is given by $2.44 \lambda f/D$, where λ is the wavelength and f/D is the f-number of the system. In the system, the pixel size was 9.9 μm. With a magnification of unity, the required f/D is 6.4 at a wavelength of 633 nm. Squares of 5×5 pixels were used according to the theoretical studies.[83] The software calculated the speckle contrast for any given square of 5×5 pixels and assigned this value to the central pixel of the square. This process was then repeated to obtain a speckle contrast map. To each pixel in the speckle contrast map, the measured velocity ($2T/\tau_c$) was obtained through Eq. (4.34) that describes the relationship between the correlation time and velocity and therefore measures the velocity map.

To compute the relative blood flows in vessels of interest, first a threshold was set in a region of interest from the measured velocity image and then the vessels of interest were identified by the pixels with values above this threshold. The mean values of the measured velocity in those pixels were computed at each time point.

The relative velocity in the vessel of interest was expressed as the ratio of the measured velocity in the condition of stimuli to that of the control condition.

LSI is a noninvasive full-field optical imaging method with high spatial and temporal resolution, which is a convenient technique in measuring the dynamics of CBF.[831,866,867,1197,1198,1201,1203,1206] In particular, in Ref. 831, the LSI method was used to monitor the dynamics of CBF in several animal models during sciatic stimulation. Stimulation of the sciatic nerve was similar to that used in conventional physiological studies. Blood flow was monitored in the somatosensory cortex in a total of 16 rats under electrical stimulation of the sciatic nerve, and the activated blood flow distribution was obtained at different levels of arteries/veins and at the change of activated areas. One example of the results is shown in Fig. 8.16, in which the brighter areas correspond to the area of increased blood flow. In comparison with LDI, an area of 1 mm^2 ROI in Fig. 8.16(a) was chosen to evaluate its mean velocity (Fig. 8.17): the evoked CBF started to increase at 0.7 ± 0.1 s, peaked at 3.1 ± 0.2 s, and then returned to the baseline level. It is consistent with the conclusions obtained from the LDI technique.[1208,1209] In order to differentiate the response patterns of artery/vein under the same stimulus, six distinct levels of vessels were labeled in Fig. 8.16(a) and their changes of blood flow displayed. The results clearly showed that the response patterns of arteries and veins in the somatosensory cortex were totally different: vein 1 (V-1, $\sim$140 µm in diameter) remained almost unaffected, and arteriole 1 (A-1, $\sim$35 µm in diameter) responded slowly; arteriole 2 (A-2, $\sim$35 µm in diameter) peaked at 3.5 ± 0.5 s after the onset of stimulation and then reached the steady-state plateau; vein 2 (V-2, $\sim$70 µm in diameter) presented a delay and mild response; blood flow in the capillaries (A-3 and V-3, $\sim$ 10 µm in diameter) surged readily and increased significantly. The changes in arteries and veins with different diameters were also measured.[831] The activation pattern of cerebral blood flow was discrete in spatial distribution and highly localized in the evoked cortex with the temporal evolution. This is consistent with the hypothesis of Roy and Sherrington.[1207–1209]

The influence of epidurally applied hyperosmotic glycerol on *in vivo* the resting CBF was also investigated using the LSI technique (see Section 5.81).[831] The skull was removed, and intact *dura mater* was exposed. To study the influence of glycerol on *in vivo* CBF, a small area of *dura mater* was removed. Warm dehydration glycerol was administrated near the exposed area. Velocity images of CBF under the effect of glycerol are shown in Fig. 5.44. When glycerol diffused in brain tissue and influenced CBF under the *dura mater*, the CBF in the exposed area would also change. Figure 5.45 gives the time course of changes in four different vessels.

As described above, LSI is based on the first-order spatial statistics of time-integrated speckle. The main disadvantage of LASCA is the loss of resolution caused by the need to average over a block of pixels to produce the spatial statistics used in the analysis, although it actually has higher resolution than other techniques such as scanning laser Doppler. A modified LSI method utilizing the temporal statistics of time-integrated speckle was recently suggested.[1210] In this method, each pixel in the speckle image can be viewed as the single-point area.

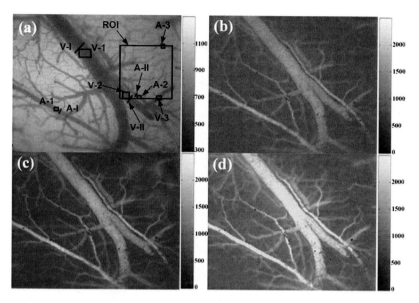

Figure 8.16 Blood flow change in the contralateral somatosensory cortex of rats under unilateral sciatic nerve stimulation.[1207] (a) A vascular topography illuminated with green light (540 ± 20 nm); (b) blood activation map at prestimulus; (c) 1 s and (d) 3 s after the onset of stimulation. The relative blood flow images are shown and converted from the speckle-contrast images, in which the brighter areas correspond to the area of increased blood flow. A-1, A-2, A-3 and V-1, V-2, V-3 represent the arbitrarily selected regions of interest (ROI) for monitoring changes in blood flow. A-I, A-II and V-I, V-II represent the selected loci on the vessel whose diameters are measured in the experiment.

Then, the signal processing consists of calculating the temporal statistics of the intensity of each pixel in the image as

$$N_{i,j} = \frac{\langle I_{i,j,t}^2 \rangle_t - \langle I_{i,j,t} \rangle_t^2}{\langle I_{i,j,t} \rangle_t^2},$$

$$i = 1\text{--}480, \quad j = 1\text{--}640, \quad t = 1 - m, \tag{8.8}$$

where $I_{i,j,t}$ is the instantaneous intensity of the ith and jth pixels at the t frame of raw speckle images, and $\langle I_{i,j,t} \rangle_t$ is the average intensity of the ith and jth pixels over the consecutive m frames. $N_{i,j}$ is inversely proportional to the velocity of the scattering particles. The value $N_{i,j}$ of each pixel in the consecutive m frames $(I_{i,j,t})$ of the raw speckle pattern is computed according to Eq. (8.8). The process is then repeated for the next group of m frames. The results are given as 2D grayscale (65,536 shades) or false-color (65,536 colors) coded maps that describe the spatial variation of the velocity distribution in the area examined.

Other approaches of LASCA technique improvement, in particular noise reduction, based on an active speckle averaging scheme that ensures perfect ensemble averaging are also described.[1199,1202] These approaches can use various methods to generate speckle images in reduced processing time, such as the use of sec-

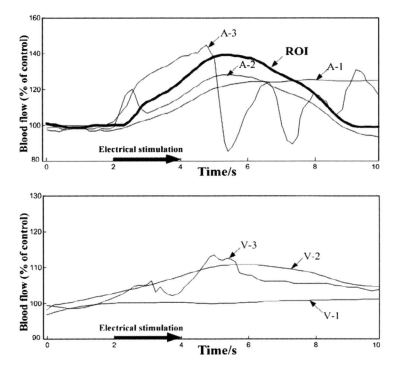

Figure 8.17 The relative change of blood flow in the six areas indicated in Fig. 8.16(a) (divided by the values of the prestimuli).[1207]

ondary low-coherence light sources (illumination with a dispersed laser beam that has passed a rotating diffuser) or vibration techniques.

8.4 Interferometric and speckle-interferometric methods for the measurement of biovibrations

A large number of optical methods have been proposed to date for the monitoring of biovibrations. Specifically, a fiber-optic sensor based on a single-mode x-coupler was successfully employed for the monitoring of heartbeats of patients examined with the use of a magnetic resonance tomograph.[1211] Contactless biovibrometers with an ultrahigh sensitivity based on heterodyne laser interferometers with a high automation degree and well-developed software were described in detail by Khanna et al.[1212,1213] Their investigations have been devoted to the measurement of vibrations of different components in the inner ear of animals. The confocal scheme of the heterodyne interference microscope employed in these studies made it possible to investigate vibrations of various layers of the tissue. This approach provided a record sensitivity with respect to small displacements of objects with low reflectivities (on the order of 10^{-4}–10^{-5}). The sensitivity achieved in these experiments within the range of vibration frequencies from 50 to 2000 Hz was 10^{-11} m. A conceptually similar but much simpler laser system for the investigation of biovibrations was described in Ref. 1214. This system also employs a heterodyne interferometer and is referred to as a laser Doppler vibrometer. This instrument, which operates within a frequency range up to 10 kHz, was used to monitor vibration spectra of a tympanic membrane under various disorders of the inner ear. Holographic analysis of vibrations of a tympanic membrane is also described in the literature (e.g., see Ref. 1215). An optical system for remote monitoring of cardiovibrations is presented in Ref. 864. An optical interferometer with a 633-nm He:Ne laser was utilized to detect micrometer displacements (sensitivity of 366.2 μm/s) of the skin surface.[1216] The detected velocity of skin movement is related to the time derivative of the blood pressure. Motion velocity profiles of the skin surface, near each superficial artery and auscultation point on a chest for the two heart valve sounds, exhibited distinctive profiles. The designed optical cardiovascular vibrometer has the potential to become a simple noninvasive approach to cardiovascular screening.

A laser Doppler technique based on the self-mixing effect in the diode laser[840,841] was used for cardiovascular pulse measurements above the radial artery in the wrist.[1217,1218] The developed self-mixing interferometer was used to measure the skin displacement, which was induced by a cardiovascular pulse. The reconstructed Doppler spectrograms followed the first derivative of the corresponding blood pressure pulse for both normal and abnormal pulse conditions. The correlation coefficient between the shapes of the Doppler spectrograms and the first derivative of the blood pressure pulse for ten investigated volunteers (738 cardiovascular pulsegrams) was found as 0.95 with a standard deviation of 0.05. A self-mixing interferometer was also used to measure the baroreflex effect and the elastic modulus of the arterial wall.

The fact that the interaction of laser radiation with tissues gives rise to the formation of speckle structures did not receive an adequate consideration in biovibrometry. The interference between speckle-modulated (reflected from a tissue) and reference fields has many specific features in this case. In designing biovibrometers, one should take into account these specific features and sometimes even make use of them[76,836,1219–1222] (see Chapter 4). The development of coherent optical contactless biovibrometers for the purpose of medical diagnostics is closely related to the solution of the problem of diffraction of laser beams propagating in nonstationary, randomly nonuniform media.[76] Several speckle techniques have been developed thus far for medical applications.[76,836,1219–1222] A diagnostic probe based on a miniature speckle/electronic-interference system, including a fiber-optic speckle interferometer and a matrix photodetector, was described in Ref. 1221. Sequential frame-by-frame analysis of the distribution of electron speckles makes it possible to image vibrations in three dimensions with a high quality. The probe was designed for the quantitative analysis of vibrations of the tympanic membrane and vocal chords.

The possibility of applying a Michelson speckle interferometer to the investigation of cardiovibrations and the detection of pulse waves was substantiated in Refs. 836, 1219, and 1220. Here, we briefly consider the main results of these studies and present some results on the detection of pulse waves with the use of the speckle technique based on the diffraction of focused laser beams. Medical diagnostics require simple and noise-resistant optical systems. The homodyne speckle interferometer shown in Fig. 8.18 meets these requirements. The output signal of this interferometer reaches its maximum when the speckle fields are matched (see Section 4.2). For focused laser beams, such matching can be easily achieved by the equalization of the arms of the interferometer. The speckle vibrometer can operate in two regimes: with comparatively large vibration amplitudes ($l_0 > \lambda/4$, the regime of fringe counting) and with small vibration amplitudes ($l_0 < \lambda/4$, when the random amplitude of the output signal displays an additional dependence on the initial phase). If the number of speckles within the receiving aperture satisfies the condition $N_{sp} > 4$, then the output signal of the interferometer has Gaussian statistics of the first order, i.e., the amplitude of this signal is characterized by a Rayleigh distribution.[836] The relevant experimental and theoretical dependencies of the averaged amplitude $\langle U \rangle$ and variance σ_U^2 of the output signal on the number of speckles within the aperture of the photodetector show a way to improve the signal-to-noise ratio in homodyne interferometers (see Fig. 8.19). It can be demonstrated that in the case of vibrations with large amplitudes, we have

$$\langle U \rangle \cong d_{av}^2 (N_{sp})^{0.5}, \quad \sigma_U^2 \approx N_{sp}, \tag{8.9}$$

where d_{av} is the mean transverse size of a speckle and $N_{sp} = (2R_a/d_{av})^2$ for a circular aperture with a diameter $2R_a$. In writing Eq. (8.9), we assume that variations in speckle sizes do not change the intensities of the interfering beams.

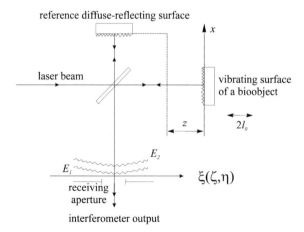

Figure 8.18 Diagram of a homodyne speckle interferometer for the investigation of biovibrations.[836]

Thus, to increase the amplitude of the output signal of a speckle interferometer, one should choose schemes that would ensure the detection of a large number of speckles with a maximum mean size, i.e., employ focused beams and a photodetector with a wide aperture. Considerable longitudinal displacements are usually accompanied by transverse and angular shifts of an object surface. As demonstrated

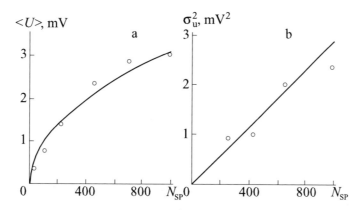

Figure 8.19 Experimental and theoretical dependencies of (a) the averaged amplitude $\langle U \rangle$ of the output signal of a speckle interferometer and (b) its variance σ_U^2 on the number of speckles N_{sp} within the receiving aperture (the rough surface is characterized by a considerable amplitude of vibrations, $l_0 \sim \lambda$; the sizes of speckles remain unchanged, and averaging is performed over 300 realizations of the reference speckle field).[836]

in Ref. 836, these shifts give rise to low-frequency modulation of the output signal of an interferometer. The depth of this modulation varies within a broad range from 0 to 100%, depending on specific realizations of the signal and reference speckle fields. Since it was demonstrated that such a modulation is caused by intensity fluctuations in a group of closely located speckles, one can considerably weaken spurious modulation by blocking the relevant group of speckles with a small opaque screen.

For vibrations with small amplitudes, the output signal of an interferometer has substantially different statistics. The modulus of the output signal in this case has an exponential distribution function with a maximum probable value equal to zero. However, the mean value of the modulus of the signal and its variance are characterized by the same dependencies on N_{sp} as in the case of vibrations with large amplitudes[836] [see Eq. (8.9)].

Taking into account the aforesaid and using Eqs. (4.22) and (4.23), we can represent the output signal of a homodyne interferometer in the following form:

$$U_i(t) = A_i \sin[\phi_i + A_L H(t)], \qquad (8.10)$$

where $H(t)$ is the normalized signal with a variance equal to unity that describes the waveform of vibrations on the surface of a bioobject, A_i and ϕ_i are random quantities determined by the conditions of detection of speckle interferograms and the chosen realization of the surface with an index i, and A_L is the amplitude of vibrations. Equation (8.10) also holds true for a differential interferometer if the quantity A_L is defined as the difference of the vibration amplitudes at two points. A laser differential speckle interferometer with two beams focused onto the surface of an object has been successfully employed for the detection of the human pulse at various points of skin surface in the wrist area.[836] As demonstrated in

Ref. 1222, the investigation of pulse waves through the analysis of phase portraits of the output signal of a differential speckle interferometer holds much promise for cardiodiagnostics. However, an appropriate filtration of the signal should be carried out in order to efficiently eliminate the influence of lateral shifts of skin surface caused by the pulse wave.

Basing on the speckle technology developed, a robust vibrometer for medical applications can be designed[1220] (see Fig. 8.20). However, the simplicity of the instrument is achieved at the cost of a nontrivial description of the response function of the vibrometer. Intensity fluctuations in scattered light in the case of diffraction of a focused Gaussian beam are related to vibrations of an object by some nonlinear random function. Nevertheless, due to regular variations in the speckle field (displacement and decorrelation of speckles) caused by vibrations of the scattering surface, the signal at the output of the photodetector contains spectral components that correspond to vibrations of the surface. The complex motion of the surface gives rise to additional nonlinearities in the response function. For example, a periodic motion of skin surface caused by a pulse wave can be considered as a superposition of at least three displacements: displacement normal to the surface, angular displacement, and transverse displacement (along the surface). For the measuring system shown in Fig. 8.20, normal vibrations do not contribute to the output signal. Comparatively small angular vibrations are responsible for transverse oscillations of speckles in the observation plane (without decorrelation of speckles), whereas the small transverse surface shifts lead to a partial decorrelation of the speckle field. Thus, time-domain intensity fluctuations of the scattered field in the case of periodic vibrations of a surface also involve a periodic component. In such a situation, the nonlinear random operator that relates intensity fluctuations to the displacement of the scattering surface depends on the sizes of the irradiated surface area, conditions of speckle observation, and the specific realization of the surface under study. However, numerical analysis of the diffraction of a focused Gaussian beam from a moving rough surface with Gaussian statistics within the framework of the Kirchhoff approximation shows that when the amplitudes of transverse shifts are less than the surface correlation length (for example, for human skin, $L_c \sim 60$–$80\ \mu m$) and the amplitudes of angular vibrations are less than one degree, statistical and nonlinear properties of the signal do not exert a considerable influence on the detection of the motion law of a rough surface.[1220]

A fiber-optic sensor based on these principles is shown in Fig. 8.20(a). This sensor transforms skin vibrations caused by a pulse wave into the corresponding motion of speckles, which is detected in the observation plane. To standardize the reflective properties of the surface and exclude bulk scattering in tissues, the device employs a thin rubber membrane, which is attached to the surface of the skin in such a manner that it does not perturb the motion of the skin surface. Various thin backscattering films produced by the spraying of a substance on the skin surface can also be employed as standard reflectors. It should be noted that this method also provides a sufficient efficiency in the case of an open skin surface, because the contribution of bulk scattering is suppressed to a considerable extent due to a

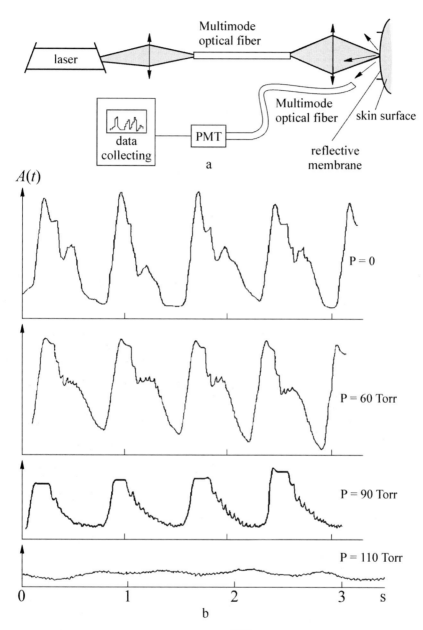

Figure 8.20 Fiber-optic laser speckle vibrometer.[1220] (a) Diagram of the instrument. (b) Pulsograms detected in the wrist area of a young man (24 yr) for various pressures P in the arm cuff of a tonometer.

sharp focusing of radiation onto the skin surface, and the skin surface itself can be satisfactorily described in terms of Gaussian statistics. Figures 8.20(b)–(e) present typical pulsograms measured with the use of a fiber-optic speckle vibrometer in the wrist area of a healthy young man for different external pressures in the forearm area (as before, a medical tonometer was used for these measurements). Obviously,

this device can be considered an advanced prototype of a compact, simple, and reliable remote probe of biovibrations with a high spatial resolution, which can be
assembled with the use of a diode laser and a photodiode integrated with segments
of multimode optical fibers. Such a probe would be useful in cardiology, for a diagnosis of vascular diseases, and in sport medicine (monitoring of self-contractions
of muscles and other tissues).

8.5 Optical speckle topography and tomography of tissues

Methods of speckle topography and tomography are rapidly progressing at
the moment.[76,135,136,138,139,155,343,395,396,566,799,825,827–832,834,835,843–846,1196,1202,
1223–1228,1239] Local statistical and correlation analysis offers much promise as a
method for topographic mapping and structure monitoring of scattering objects.
Local estimates of correlation characteristics [correlation or structure functions
or their parameters, see Eqs. (4.20) and (4.21), and Figs. 4.5 and 4.6] and normalized statistical moments [the contrast V_I and asymmetry coefficient Q_a are
usually employed, see Eqs. (4.5)–(4.7)] are highly sensitive to the structure parameters of an object, such as the correlation length L_c and the standard deviation σ_L of optical altitudes (thicknesses) of inhomogeneities or the relevant
correlation length L_ϕ and the standard deviation σ_ϕ of phase fluctuations of the
boundary field (see Section 4.1) in the case of the diffraction of focused laser
beams.[76,155,343,825,827,829,830,834,835,1224–1233] An object under study can be considered as an irregular system of lenslets with definite statistical characteristics that
display intensity fluctuations similar to those observed when a focused laser beam
is scanned over the surface of an object.[835]

Intensity fluctuations include two components (see Fig. 8.21). The first component is a background with a relatively small and comparatively smooth varying amplitude. The second component is represented by infrequent high-intensity pulses
related to matched inhomogeneities (the distances between the plane of the waist of
the incident laser beam and the object, and between the object and the photodetector are matched with the effective focal length of the inhomogeneity, which ensures
effective reimaging of the waist of the laser beam into the observation plane).

We can classify the inhomogeneities by analyzing the contrast V_I and the asymmetry coefficient Q_a as functions of the distance between the waist plane of the
laser beam and the surface of an object. Using this approach, we can reconstruct
statistical distributions of fluctuations of the refractive index of a medium.[77] The
fact that V_I and Q_a abruptly increase when the ratio of the radius of the laser beam
to the correlation length satisfies the condition $w/L_\phi \sim 1$ is a direct manifestation of the microfocusing effect in the far-field diffraction zone[835,1229] (see Fig.
8.22 for V_I, $\Delta z = \pm 0.4$ mm). The growth in the ratio $w/L_\phi (|z| > 0.4$ mm) is
accompanied by a decrease in the quantities V_I and Q_a, which reach values corresponding to completely developed speckles. These effects can occur in the case of
weakly scattering objects, such as thin tissue layers or cellular monolayers, when
$L_c \sim L_\phi \sim d$ (where d is the thickness of the sample), $L_c \gg \lambda$, and the standard

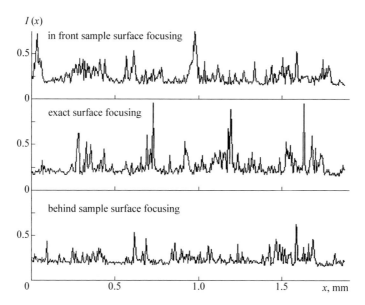

Figure 8.21 Realizations of speckle intensity fluctuations obtained with a focused laser beam scanned over an epidermis sample of psoriatic human skin (epidermal stripping).[343] The upper and lower realizations were obtained with a laser beam focused in front of and behind the sample surface, respectively. The middle realization corresponds to a laser beam focused exactly on the surface.

deviation σ_ϕ related to phase fluctuations of the field is completely determined by fluctuations of the refractive index δn.

Multiple scattering is characteristic of optically thick tissue layers. In this case, the spatial distribution of scattered light has a broad angular spectrum, and depolarization effects play an important role. The spatial distribution of the correlation properties of the scattered field, which is related to the structure of an object, can be considered in a manner similar to diffusion-wave spectroscopy [see Eq. (4.43)] with allowance for the fact that an object has a static structure, and a laser beam (or the object itself) is scanned over the surface of the object at a definite rate. Along with the chosen beam radius and the character of the optical inhomogeneities of the medium, the rate of scanning determines the fluctuations of the scattered field in the time domain. In such a situation, the normalized autocorrelation function of the intensity fluctuations is generally defined by Eq. (4.20) with $\xi \equiv t$ and $\Delta\xi \equiv \tau$. The behavior of the structure function [see Eq. (4.21)] near the zero value of its argument, which corresponds to the highest efficiency of the high-frequency spatial intensity fluctuations, can be conveniently characterized in this case in terms of the exponential factor ν_I as[835,1224]

$$\nu_I = \frac{\ln[D_I(\Delta\tau_2)/D_I(\Delta\tau_1)]}{\ln(|\Delta\tau_2|/|\Delta\tau_1|)}. \tag{8.11}$$

To analyze the polarization properties of speckle fields, we can employ the following time-domain first- and second-order statistical characteristics of the intensity

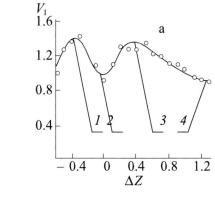

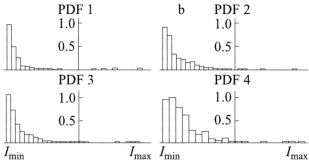

Figure 8.22 The correspondence between (a) the contrast values and (b) the shape of the distribution of speckle intensity fluctuations (PDF) as functions of the position of the waist of a focused laser beam with respect to the sample surface (an epidermal stripping of psoriatic human skin) Δz (mm); $\Delta z = 0$ corresponds to the case where the beam waist lies on the surface of the sample.[835]

fluctuations of scattered light in the paraxial region, which should be measured for two orthogonal linear polarizations (relative to the polarization of the probing beam):[343]

(1) the mean intensity of speckles,

$$\langle I_{sp}\rangle = \langle I_{||}\rangle + \langle I_{\perp}\rangle; \tag{8.12}$$

(2) the cross-correlation function (correlation coefficient) for two polarization states,

$$r_{\perp II}(\tau) = \langle [I_{II}(t) - \langle I_{II}\rangle][I_{\perp}(t+\tau) - \langle I_{\perp}\rangle]\rangle, \tag{8.13}$$

where the indices $(\perp, ||)$ denote combinations of polarization states, and averaging is performed over the trajectory of scanning.

Figure 8.23 presents the optical scheme of a spatial speckle correlometer intended for topography or tomography of comparatively thin samples of tissues.

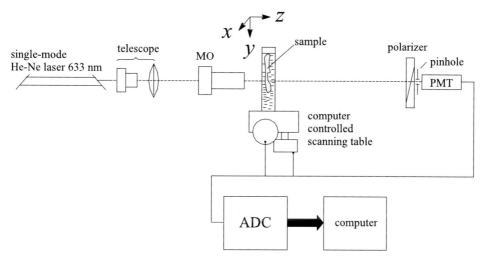

Figure 8.23 Optical scheme of a scanning polarization-sensitive spatial speckle correlometer.[155] PMT, photomultiplier tube; ADC, amplitude-digital convertor; MO, microobjective.

This device employs a focused laser beam about 5 μm in diameter produced by a single-mode uniphase He:Ne laser. Structure patterns of an object were usually reconstructed through the two-dimensional scanning of the object and an appropriate analysis of the statistical and correlation properties of the scattered light. The scanning step on both axes was 5 μm. A photodetector was placed along the direction of the axis of the incident laser beam (scattering exactly in the forward direction). The diameter of the entrance pinhole was about 25 μm, which is much less than the mean diameter of a speckle. The maximum rate of scanning was about 5 mm/s. The electronic units employed made it possible to obtain at least 20 equidistant counts per single step of scanning. An object was usually placed in the waist of the incident laser beam. The position of an object relative to the beam waist and the orientation of a polarization analyzer mounted in front of the photodetector were adjusted manually.

As was shown above, the estimation of the structure parameters of a tissue, such as the characteristic size of local inhomogeneities and spatial fluctuations of the refractive index, generally requires some assumptions concerning the scattering model. One of the simplest models of scattering is the model of a random phase screen with Gaussian statistics of inhomogeneities (see Section 4.1).[157,822,824,835,1230] In many cases, statistical models for living structures are much more complicated. In particular, some of these models are nonlinear and may take into account multiple scattering. In spite of the lack of well-developed models for the structure of many tissues, any empirical information concerning statistical properties of scattered light is useful for the analysis of structure images of tissues. Such information may be also useful for the development of structure models themselves.

Human skin is one of the most natural objects for the application of optical speckle correlometry.[76,834,835,1224,1225] All the structural specific features of skin surface are manifested in the statistical and correlation properties of the speckle field produced in the far-field diffraction zone when skin samples or skin replicas are probed with a focused laser beam. The technology that permits one to obtain thin slices (strippings) of epidermis with the use of medical glues and quartz (or glass, or metal) substrates is very convenient for *in vitro* structure studies of epidermis by means of speckle-correlation optics.[834,835] The thickness of the slices in this case usually ranges from 30 to 50 μm. This technology allows one to obtain from 5 to 7 sequential strippings from the same place. As an example, Fig. 8.21 presents three realizations for the intensity fluctuations obtained by scanning a skin epidermis sample from an area of psoriasis focus of a patient for three different positions of the waist of the laser beam. The thin layers of normal and psoriatic epidermis studied demonstrated that such samples can be described within the framework of a model of single or low-step multiple scattering, because only insignificant depolarization effects were observed in the far-field zone.

The contrast V_I and the asymmetry coefficient Q_a of intensity fluctuations in the far-field zone as functions of the variation in the defocusing parameter Δz for normal and psoriatic epidermis display two maxima near the area of exact focusing ($\Delta z = 0$).[835] Figure 8.24 shows such a dependence for a sample of psoriatic epidermis, along with the relevant probability density functions for the intensity

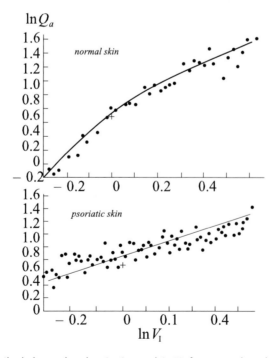

Figure 8.24 Statistical dependencies $\ln Q_a$ and $\ln V_I$ for normal and psoriatic skin samples (strippings of human epidermis); + indicates the values of Q_a and V_I for developed speckles.[1231]

fluctuations. The behavior of the first-order statistical characteristics confirms the validity of the lenslet approach for the description of the scattering properties of epidermis. Normal epidermis (where scatterers have smaller sizes and more uniform distribution) is characterized by a partial overlapping of the maxima mentioned above. The symmetry in the arrangement of V_I and Q_a peaks with respect to the plane of the beam waist allows us to assume that the statistical weights of the negative and positive lenslets in an ensemble of scatterers are equal to each other.

The differences between the curves $V_I(\Delta z)$ and $Q_a(\Delta z)$ for samples of normal and psoriatic epidermis are due to the changes in the structure of a tissue caused by the disease. These changes are associated with the appearance of parakeratotic foci (the structure of cells near such a focus is substantially disordered) and the saturation of the surrounding tissues with interstitial fluids (which decrease the efficiency of scattering, similar to immersion fluids). Later stages of the disease are characterized by the appearance of microspaces filled with air and the scaling in the formation of inhomogeneities, which increases scattering and changes its character.[835,1231] For a small Δz, the quantity V_I is greater than 1, reaching the values of 1.6–1.7 for certain samples. As a rule, samples of normal skin display a higher contrast than samples of psoriatic skin. Parametric dependencies for $\ln Q_a$ and $\ln V_I$ plotted for various Δz illustrate the differences in the statistical properties of samples of normal and psoriatic skin (see Fig. 8.24). For normal skin, the first derivative of the function $\ln Q_a = f(\ln V_I)$ is greater than that for psoriatic skin. This derivative has a negative slope in the case of normal skin and positive slope for pathological skin. For developed speckle fields with $V_I = 1$ and a negative exponential probability density function of intensity fluctuations [see Eq. (4.9)], we have $Q_a = 2$ (these points are indicated by a plus sign in Fig. 8.24).

Thus, the first-order statistics can be used as a simple and efficient criterion for the recognition of structurally specific features of tissue samples. Investigations of special skin replicas demonstrated that this technique is also efficient for the semiquantitative determination of the dryness and fatness of skin.[1227,1232]

Second-order statistical characteristics of intensity fluctuations are also highly sensitive to structural changes in tissues.[834] For normal skin, normalized 1D autocorrelation functions of the intensity fluctuations are characterized by comparatively small values of the correlation length, $L_I \sim 60$–80 μm, and the absence of considerable fluctuations of the correlation coefficient on large scales (see Fig. 4.5). As a psoriatic plaque arises, the AFs display a nearly twofold increase in the correlation length, $L_I \sim 95$–180 μm, and the appearance of large-scale aperiodic oscillations with a comparatively large amplitude. The effective sizes of structure inhomogeneities giving rise to such oscillations correlate with the sizes characteristic of an ensemble of parakeratotic foci. Detailed analysis of oscillatory components of the AF should allow one to estimate the surface density of parakeratotic foci and their mean size.

In the range of high spatial frequencies of about 1 μm^{-1} and higher, the structure function or its exponential factor ν_I [see Eqs. (4.21) and (8.11)] are preferable for the description of intensity fluctuations. Local estimates of the exponential factor permit one to determine the contribution of high-frequency structure components of a tissue and to find their spatial distribution in the form of topograms. Figure 8.25 presents topograms obtained with the use of this technique for samples of normal and psoriatic epidermis.[155,1226,1233] The topograms and the corresponding distributions of the exponential factor (see Fig. 8.26) display structure changes related to different stages of pathology development. The values of ν_I were averaged along the direction of the x-axis with an averaging gate containing no less than 10^3 counts. The epidermis of normal skin with comparatively small-scale inhomogeneities is characterized by a somewhat smaller mean value and a greater variance of the parameter ν_I than the early and middle stages of the formation of a psoriatic plaque (see Figs. 8.25 and 8.26). Large-scale structure features, such as fragments of a skin pattern, are clearly seen in the topogram of normal skin. The middle stage of pathology is characterized by a small variance and a relatively large mean value of the parameter considered. In the later stage of the process, the variance grows, the mean value of the parameter ν_I slightly decreases, and the symmetry of the distribution function lowers. The stages of the formation of a psoriatic plaque, clearly distinguished by means of correlation spectroscopy, are consistent with the results of clinical observations.

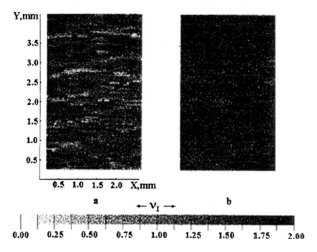

Figure 8.25 Topograms of the exponential factor ν_I obtained for samples of epidermal strippings of human skin: (a) normal skin and (b) psoriatic skin (the middle stage of the disease).[155]

The control of the optical properties of tissues, in particular the possibility of considerably decreasing the scattering coefficient, may become a key point in optical tomography of tissues in the process of searching for small tumors at early stages of their formation (see Chapter 5). Figure 5.8 presents two recorded fragments that correspond to the early and later stages of tissue clearing. The characteristic changes in speckle structures in the far-field zone were visually observed on a screen located in the plane of a photodetector and were recorded in reflected

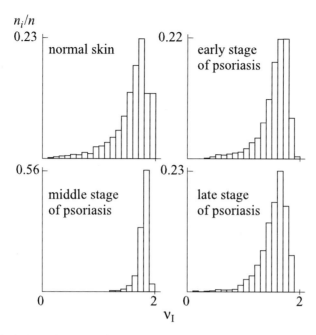

Figure 8.26 Evolution of the distributions of the exponential factor ν_I in the progress of psoriasis:[155] normal skin [corresponding to Fig. 8.25(a)], early stage of psoriasis, middle stage of psoriasis [corresponding to Fig. 8.25(b)], and late stage of psoriasis.

light by means of a CCD camera.[155,172,742] The evolution of typical normalized AFs of intensity fluctuations for sclera in the process of sclera clearing measured with the use of a speckle correlometer (see Fig. 8.23) is shown in Fig. 8.27. Within small time intervals (1–2 min), the time evolution of the shape of the autocorrelation peak, which is associated with the transition of a tissue from one scattering regime to another, can be approximated by an exponential curve; whereas for large

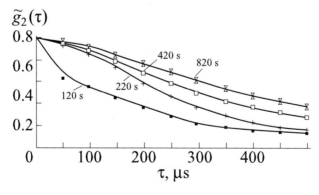

Figure 8.27 Evolution of the normalized autocorrelation function of intensity fluctuations in the speckle field produced by light scattered from a sample of human sclera in the process of scleral enhanced translucence in trazograph-60; the thickness of the sample is 0.6 mm; the measurements were performed with a sample processed in the solution during 120, 220, 420, and 820 s.[155]

time intervals, this process can be approximately described by a Gaussian curve. The initial stage of the process is characterized by the existence of many scales of intensity fluctuations (the AF displays at least three distinguishable values of its slope). At later stages of clearing, the half-width of the AF peak tends to a value of $(0.3–0.4) \times 10^{-3}$ s, which is close to the ratio of the waist radius of the incident beam to the scanning rate, w/v. This effect can be employed as a criterion of the completion of the transition from multiple scattering to single scattering. The time evolution of the exponential factor v_I [see Eq. (8.11) and Fig. 8.28] characterizes the behavior of the high-frequency components of the intensity fluctuations.

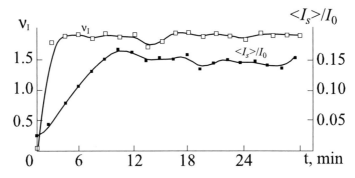

Figure 8.28 Typical time dependencies of the exponential factor v_I and the normalized mean intensity $\langle I_s \rangle / I_0$ measured at the observation point for a scleral sample placed in trazograph-60; $\lambda = 633$ nm.[155]

We should note that the exponential factor and the relative fluctuations of the mean intensity of transmitted light have substantially different rates of response to the action of hyperosmotic optical clearing agents (OCAs): v_I is especially sensitive to the action of OCAs at the early stages of clearing (within $\sim$1.5 min for trazograph-60), whereas transmission reaches its maximum only by the tenth minute. This difference in response rates can be accounted for by the fact that the correlation properties of the field, characterized by transition from uniform to nonuniform speckle distributions, substantially change at the initial stages of matching of the refractive indices. By contrast, maximum collimated transmission is achieved only when the refractive indices are completely matched.

Straightforward modeling of the transport of initially collimated photons with a wavelength of 600 nm through a fibrous tissue consisting of collagen fibers with a mean diameter of 100 nm and a refractive index $n_c = 1.474$ surrounded by a ground substance whose refractive index varies within the range $n_0 = 1.345–1.474$ shows that, even with partial matching of refractive indices, $n_0 = 1.450$, nonscattered ($\sim$67%) and singly scattered ($\sim$24%) photons dominate in transmitted light (see Fig. 5.6).[798] This prediction agrees well with the measured transmission and reflection spectra of sclera and the data on correlation measurements (see Section 5.3).[798,799]

Another specific feature of scleral clearing is the appearance of quasi-periodic oscillations of mean-intensity transmission, which are also manifested in the correlation characteristics. These oscillations have small amplitude and can be clearly seen at later stages of translucence, when the main dynamic process associated with the directed diffusion of the substance from the solution into the tissue and of water from the tissue to the solution is close to its completion. The characteristic oscillation time is $\sim$ 1.5–2.0 min. Apparently a process with such a characteristic time can be attributed to the nonuniformity of the diffusion of substances inside a tissue in space and time. This effect may be accounted for by the multistage character of the diffusion process. At the first stage, the diffusion of the OCA into a tissue and the flow of water out of the tissue partially equalize the refractive indices of the hydrated collagen and the intercollagen substance. Under these conditions, the optical transmission of a tissue grows until the dependence under study saturates. However, at the second stage, a relatively weak process of the interaction of the new ground substance with collagen is manifested. The ground substance somewhat lowers its refractive index through the dehydration of the collagen, whereas the refractive index of collagen increases. The resulting mismatch of the refractive indices slightly decreases the optical transmission. At the next stage, a certain violation of the balance between the pressures of water and the OCA in the solution and tissue gives rise to the diffusion of water from the tissue and the OCA into the tissue, which leads to the more exact equalization of the refractive indices, and transmission grows again. Then, the process described above is repeated, and transmission oscillates in the time domain. Such oscillations are observed within the entire period of time when a hyperosmotic agent acts on a tissue up to 40–60 min in this particular experiment.[155,172,343] Probably, oscillations of a similar nature with the time period of $\sim$2.5–3.5 min were registered for *in vivo* hamster skin at topical application of glycerol as an OCA using an OCT system as a detector.[1059]

As it was shown earlier (see Section 5.7.1), the transition of a scattering object from the regime of multiple scattering to the regime of single scattering should change the polarization properties of scattered radiation, which can be described in terms of the first- and second-order statistical characteristics [see Eqs. (8.12) and (8.13)]. At the early stage of sclera optical clearing, both of the polarization components of transmitted light have approximately equal intensities. However, in the process of clearing, the component polarized along the polarization of the incident beam begins to dominate over the other component (see Fig. 5.10).[343] These experimental data demonstrate the reversibility of the optical clearing process, which is important for living systems, and reveal a high sensitivity of the polarization characteristics to structural changes in a tissue.

Note that translucent sclera features large-scale spatial inhomogeneities of scattering and polarization properties in the form of a domain structure, with domain areas on the order of 0.1–1 mm^2. Such a structure is associated with a spatially nonuniform distribution of the diffusion rate of substances and is clearly manifested in both time-domain realizations of intensity fluctuations for separate polarization components and the behavior of the cross-correlation function for these components (see Fig. 8.29) measured with a laser beam scanned over a sample.[155] It is obvious that, similar to the correlation characteristics of speckle fields, polarization and cross-correlation characteristics can be employed for the imaging of the structure of tissues, as well as for tissue topography and tomography.

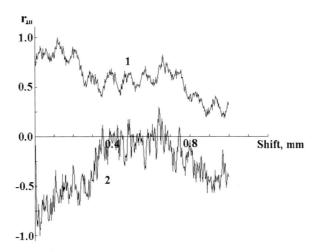

Figure 8.29 Evolution of the cross-correlation coefficient for two orthogonal polarizations of intensity fluctuations in the speckle field produced in the observation plane by a laser beam scanned over a scleral sample processed in trazograph-60 during (1) 200 s and (2) 400 s. The scanning rate was 5 mm/s.[155]

8.6 Methods of coherent microscopy

Modern methods of microscopy are developing toward *in vivo* structural investigations of individual cells without fixation of these cells, with simultaneous monitoring of intracellular dynamic processes caused by the vital activity of the cell. The above-mentioned laser Doppler microscope[5,24,841,1183] and a high-performance phase microscope with an ultrahigh spatial resolution[175,1112,1240–1242] are good examples of such devices. Another tendency in the development of microscopy is *in vivo* layer-by-layer analysis of tissues with a high spatial resolution. Confocal microscopy[1,3,28,76,120,122,614,878–898,1243] is a prominent example of this direction in microscopy (see Sections 4 and 5.7.2). Holographic microscopy[1244–1247] also offers much promise for numerous applications.

A phase microscope described in Refs. 175, 1112, and 1242 is a Linnik-Tolansky interferometer with a computer-controlled piezodriver with a mirror in the reference channel and a dissector (coordinate-sensitive detector) that registers an interference pattern (see Fig. 8.30). In fact, such a microscope is a microprofilometer with a spatial resolution up to 10 nm at a wavelength of 633 nm. The temporal resolution provided by this microscope in the investigation of dynamic processes is on the order of 1 ms. The resolution in height for this device is 0.5 nm,

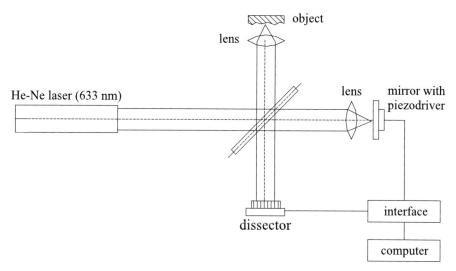

Figure 8.30 Diagram of a phase microscope.[1240]

the imaging area includes from 64×64 up to 256×256 pixels, the total gain is 10^5–10^6, the minimum pixel size is 5 nm, and the time of data processing is 4–20 s. The information concerning the structure of an object is represented in the form of altitude (optical path length) topograms, cross sections, three-dimensional images, etc. When dynamic processes are studied at an arbitrarily chosen point of an object, the results of the investigations are represented in the form of time-domain realizations, Fourier spectra, or histograms. A phase microscope was employed for structural investigations of living and dried fibroblast cells (L 929) and mitochondria extracted from cells of rat liver in the normal and condensed states.[175,1112] The phase images of L 929 cells obtained in these studies are characterized by the mean optical path length of light on the order of 600 nm for a living cell and about 50 nm for a dried cell. These findings indicate the possibility of the efficient monitoring of this type of cell metabolism. Analogous images of mitochondria demonstrate that the optical length of a normal mitochondrion with respect to the environment is about 15 nm. For the condensed state, the optical length is 3 nm. A microscope of this type was also successfully employed for structural investigations of the wall of fungi cells and erythrocytes with a high spatial resolution, for the study of intracellular motility with a high temporal resolution,[1240] and for human carcinoma cells in different physiological states induced by hyperosmotic agents.[1242]

Confocal laser scanning microscopy, the principles of which are discussed in Section 4.4, is a well-developed imaging technique for biomedical investigations.[1,3,28,76,120,122,614,878,879,780–782,883–888,890–898] In *in vivo* morphometry using real-time confocal microscopy of human epidermis, a spatial resolution better than 1 μm and depth profiling up to 150–250 μm, depending on the anatomical site and skin optical characteristics (color, transparency), were shown.[1243] Some of these results on the estimation of the nuclear size and number of keratinocytes for different layers of the living human epidermis are presented in Table 8.1. The resolution

Table 8.1 Estimation of nuclear size and number of keratinocytes in horizontal optical sections of the living epidermis.[1243]

Epidermal nuclei	Diameter, μm	Density, number/mm^2
Stratum granulosum	12–15	1500
Stratum spinosum	9–12	4000
Stratum basale	6–8	7000

achieved in structural studies of tooth dentine is comparable with the resolution characteristic of scanning electron microscopy, and additional subsurface tissue imaging with a resolution of 1 μm for depths up to 30–50 μm is also provided.[882]

Three-dimensional imaging of cells in different layers of corneal epithelium made it possible to reveal the character of mitosis in such cells.[890,891] Figure 8.31 displays a typical scheme of a confocal microscope where scanning is performed along the z-axis (along the light beam) (see Ref. 1, pp. 555–575). Such a microscope is intended for the layer-by-layer analysis of eye structure. Radiation is delivered to the microscope by means of an optical fiber. The microscope is based on two optically conjugate slits. One of these slits is imaged onto an object, whereas the second slit is placed in front of a photodetector. An objective is scanned along the z-axis by a computer-controlled piezodriver. An immersion liquid provides optical matching between the objective and the eye under study.

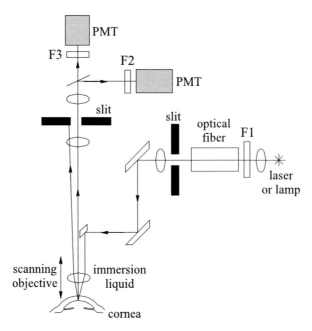

Figure 8.31 Diagram of a confocal microscope with optically conjugate slits and scanning along the z-axis.[1,879] F1, F2 , and F3, filters; PMT, photomultiplier tube.

Figure 8.32 shows a slit-scanning confocal microscope of another type. Such a microscope allows one to obtain images of an object at different depths.[883,884, 888–891] This microscope operates in real time and can be employed for *in vivo* studies of eye tissues. The main element of the microscope is a two-sided mirror, which implements transverse scanning without shifting the axis of the reflected beam (such a scheme was proposed for the first time by Svishchev in 1969 for the investigation of transparent scattering objects, including living nerve tissues). The lowering of the image quality due to the motion of the patient's eyes was excluded with the use of an electronic scheme, which ensured the required scanning frequency and phase synchronization between all the elements of the system. As an example, Fig. 8.33 presents an image of endothelium cells of human cornea obtained with the use of the confocal microscope shown in Fig. 8.32.[888] High-quality layer-by-layer images of cellular structures of eye tissues and skin in the normal and pathological states as well as the dentin structure of human teeth are presented in Refs. 1, 3, 76, 120, 122, 879–882,893, 898, and 1243.

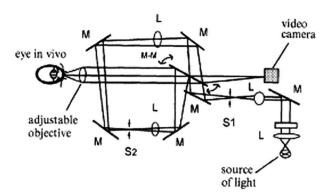

Figure 8.32 Diagram of a slit-scanning confocal microscope operating in real time.[883] M, mirrors; L, lenses; S1 and S2, conjugate slits; and M-M, scanning two-sided mirror.

Depending on the source of light employed, the detecting system, and the type of tissue under investigation, averaging over several expositions (frames) may be necessary to achieve a satisfactory signal-to-noise ratio. For example, for weakly reflecting eye tissues (usually, less than 1%), four to eight frames may be necessary. However, if a highly sensitive video camera in combination with a broadband video tape recorder is employed as a detector, averaging over frames is not necessary, and real-time *in vivo* measurements can be carried out.[883,884] Although confocal microscopes can operate with mercury or xenon arc or halogen lamps, monochromatic laser radiation provides images with a higher quality owing to the absence of chromatic aberrations introduced by the optical system. However, the type of laser should be chosen with allowance for the depth of penetration of the laser radiation into a tissue and the transmission of the optical system of the microscope.

Comparative analysis of confocal and heterodyne scanning microscopes and their applications for the investigation of scattering objects have demonstrated that

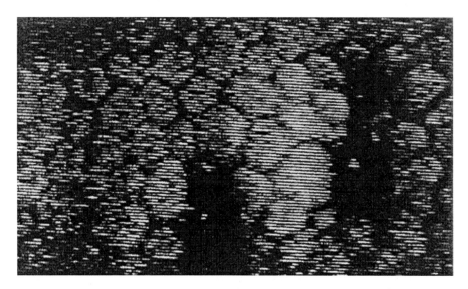

Figure 8.33 Optical map of human cornea *in vivo*. The image shows endothelial cells on the rear surface of cornea. Optical mapping is performed at a depth of 500 μm with respect to the cornea surface. Dark areas in the image correspond to pathological endothelium.[888]

in many cases the limitations of the confocal technique are mainly associated with the small level of the signal rather than with the degradation of an image due to scattered light.[894,895] Indeed, a satisfactory resolution in depth (selection of photons reflected from a definite layer and elimination of the influence of scattered light) can be achieved when a confocal system has conjugate pinholes with a radius (in optical units) of

$$v_p \leq 2, \quad v_p = \frac{2\pi r_p a_1}{\lambda f_1}, \tag{8.14}$$

where r_p is the radius of the pinhole and a_1 and f_1 are the radius and the focal length of a lens that focuses light on the pinhole. Such pinholes do not transmit much light. For example, with $v_p = 2$, a pinhole transmits only 40% of radiation incident on this pinhole within the limits of its aperture. At the same time, the heterodyne (interference) scheme involving a narrowband light source eliminates, to a considerable extent, this restriction due to optical amplification. Eventually, the interference scheme allows one to obtain images of an object with the same signal-to-noise ratio as in confocal microscopy within time intervals shorter than those required in confocal microscopy. Additional advantages of the interference scheme, which stem from the coherence of light and the use of the amplitude response of an object under study, are associated with the appearance of a new mechanism of suppression of scattered light and the possibility of detecting smaller differences in the reflectivities of various tissue layers. It is expected that the integration of the approaches considered above in one microscope and the use of broadband (low-coherence) sources of light (see Chapter 9) may considerably improve the selectivity of the system, which is important for the investigation of nearly uniform tissues.[894,895]

Speckle interferometry using sharply focused laser beams and spatial averaging of optical signals also has advantages in the depth profiling of scattering objects. Using a speckle interferometer (see Fig. 4.7), glue-strippings of human skin attached to metal plates were investigated. Depth profiling of thin tissue layers with a subcellular resolution was obtained (see Fig. 8.34). This method allows one to estimate the thickness of tissue layers and the depth distribution of the refractive index. Appropriate transverse scanning of the object will give tomograms with a spatial resolution of about $3 \times 3 \times 3$ μm. The optical gain of this scheme and the possibility of using powerful lasers may allow one to obtain high values of signal-to-noise ratio.

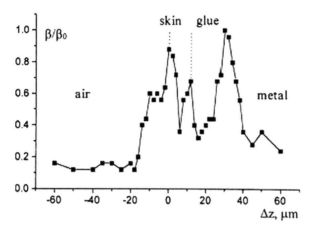

Figure 8.34 Depth profiling for glue-stripped human skin attached to a metal plate (dependence of the normalized modulation factor of the photoelectric signal β/β_0 on the longitudinal displacement of the sample).[837]

Technical developments in the different fields of optical microscopy are increasingly focusing on the *in vivo* and *in situ* imaging of metabolic functions and dysfunctions. The observation of the dynamics of biological processes on a microscale and a nanoscale is required for a more detailed understanding of both cellular physiology and pathology.

8.7 Interferential retinometry and blood sedimentation study

Laser interferential retinometers used for the monitoring of human retinal visual acuity are based on an optical dual-beam interferometer that forms two coherent beams that are focused onto the nodal plane N of the eye and form a spatially modulated laser beam (SMLB) with parallel interferential fringes on the retina (see Figs. 4.9 and 8.35).[5,832,847] The period L of the fringes and their orientation

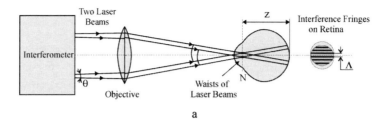

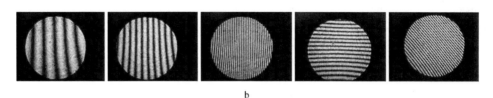

Figure 8.35 (a) Scheme of a laser interferential retinometer and (b) interference fringes of various spacings and orientations on a retina.[832]

depend on corresponding parameters of the incident SMLB[5] as

$$L = \frac{D\lambda}{2l}, \tag{8.15}$$

where D is the mean distance between the eye nodal plane and retina, λ is the wavelength, and $2l$ is the separation between two point light sources formed in the nodal plane.

Normal retinal visual acuity is defined as an angular resolving power of the eye and is characterized by the density of interferential fringes per a degree of the view angle[5] as

$$N_{\text{int}} = \left[\arcsin\left(\frac{\lambda}{2l}\right) \right]^{-1}. \tag{8.16}$$

For the ideal conditions of the front media of the eyes, the fringe pattern contrast at the retina is very high, practically equal to unity because of the high degree of mutual coherence of the laser beams.

The procedure of estimation of retinal visual acuity is simple. First, a fringe pattern with a large period at the patient's retina is formed. The patient lets the doctor know that he/she is able to see a fringe pattern. Then, the period of the fringes is decreased and the patient must indicate his/her ability to see the pattern. This procedure is repeated until the patient is unable to see the pattern. To avoid false patient response, fringes can be rotated on an arbitrary angle [see Fig. 8.35(b)].

For patients with cataractous (turbid) lenses, the scattering of a spatially modulated laser beam by a turbid media prevents the creation of the fringe pattern [see Fig. 8.36(c)]. However, averaging of the interferential pattern may improve the visibility of the initial fringes (see Section 4.3) and, therefore, with this technique some limits may be applicable for retinal acuity estimations in eyes with cataract. The optical scheme presented in Fig. 8.36(a), where the SMLB is moving periodically by a deflector but fringes on the retina are unmovable, provides such an averaging. Some other averaging schemes are also available.[832]

Another example of SMLB application in medicine is the monitoring of the erythrocyte sedimentation rate (ESR). Usually, the test uses the measurement of the distance that erythrocytes have fallen after one hour in a vertical column of anticoagulated blood under the influence of gravity. The test is helpful in the specific diagnosis of several types of cases, including diabetes mellitus and myocardial infarction. The SMLB technique allows one to detect the temporal changes of the scattering properties of a blood suspension at its sedimentation. This method was used to investigate a highly diluted blood, i.e., to monitor the sedimentation of individual and weakly interacting erythrocytes, and was tested in clinical studies.

The experimental setup is presented in Fig. 8.37. A 633-nm He:Ne laser beam was expanded to a diameter of 10 mm; the interferometer created the parallel fringes in the illuminating beam. The piezodeflector was used to make dynamic fringes. The SMLB was positioned in order to incident vertically on a horizontally placed glass vessel with blood samples. The fringe contrast in the course of blood sedimentation was measured. At blood sedimentation, the scattering medium transits from a highly scattering to a low-scattering one due to a high degree of packing

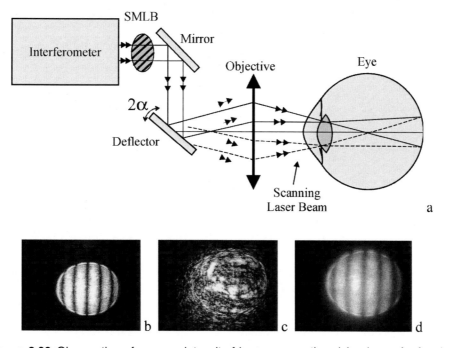

Figure 8.36 Observation of average intensity fringes on a retina: (a) scheme for forming average intensity fringes on a retina; (b) fringes for a clear lens; (c) pattern for a turbid lens; (d) average intensity fringes.[1248]

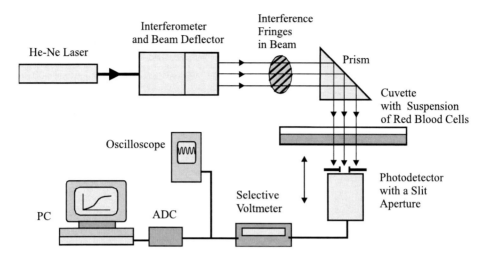

Figure 8.37 Laser system with a spatially modulated laser beam for the study of the dynamic scattering properties of a suspension of red blood cells during their spontaneous aggregation and sedimentation.[1249]

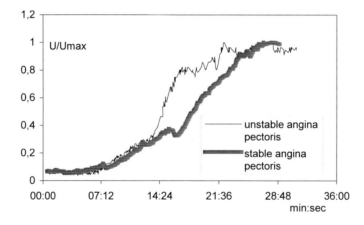

Figure 8.38 Temporal scattering characteristics of erythrocyte suspensions (whole blood diluted by saline as 1:200) at sedimentation process. Blood was taken from the patients with stable and unstable angina pectoris.[1249]

of the fallen erythrocytes; therefore, the contrast improves (see Fig. 8.38). Owing to the different properties of blood samples taken from patients with different stages of disease, the dynamic characteristics of contrast are different; therefore, contrast may serve as a diagnostic parameter.

9

Optical Coherence Tomography and Heterodyning Imaging

9.1 OCT

9.1.1 Introduction

A description of the fundamentals and basic principles of optical coherence tomography (OCT) can be found in Section 4.5. In Section 2.6, OCT is discussed as a method for tissue optical properties' measurements. Methods and data of measured absorption and scattering coefficients and refractive index are presented (see Table 2.1). The enhancement of OCT penetration depth and image contrast owing to the action of hyperosmotic optical clearing agents is given in Section 5.5.5 for skin, in Section 5.6.2 for gastric tissues, and in Section 5.8.2 for blood samples; and in Section 5.9.1, OCT glucose sensing is discussed. In this section, we will briefly give an overview of some typical OCT schemes and illustrate their biomedical applications.

9.1.2 Conventional (time-domain) OCT

A large number of schemes of coherent optical tomography for the investigation of tissues have been described in the literature, with overviews given.[1,3,8,13,17,18,28,45,76,77,84,102,108–111,116,126,127,129,135,136,138,139,141,142,343,717,775,865,901,902,909,931–939,1246,1247] Figure 9.1 presents one of the typical time-domain tomographic schemes based on a superluminescent diode (SLD) ($\lambda = 830$ nm, $\Delta\lambda = 30$ nm) and a single-mode fiber-optic Michelson interferometer.[717,904] The power of IR radiation on tissue surface is about 30 μW. The interference signal at the Doppler frequency, which is determined by the scanning rate of a mirror in the reference arm [see Eq. (4.53)], is proportional to the coefficient of reflection of the nonscattered component from an optical inhomogeneity inside the tissue. One can localize an inhomogeneity in the longitudinal direction by equalizing the lengths of the signal and reference arms of the interferometer within the limits of the coherence length of the light source ($\sim$10 μm) [see Eq. (4.54)]. The transverse resolution of a beam scanning along the surface of a sample is determined by the radius w_0 of the focal spot of the probing radiation (usually $w_0 \leq 20$ μm, which should be consistent with the required length of the probed area in the longitudinal direction, and is determined by the length of the beam waist, $2n\pi w_0^2/\lambda_0$). Figure 5.30 displays OCT tomograms of human skin with psoriatic erythrodermia before and after topical application of glycerol.[717]

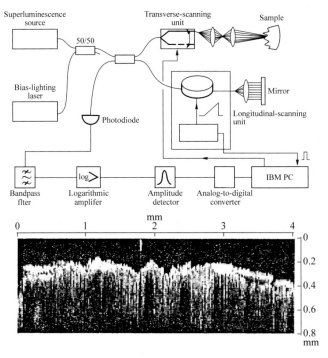

Figure 9.1 Diagram of a fiber-optic coherent optical tomograph.[904]

A scanning velocity of 50 cm/s is required to acquire images with a size of 2.5×4 mm (axial size $\times$ lateral size), resolution of 20×20 μm, and acquisition rate of 1 image/second.[717] The scanning velocity should be maintained constant with an accuracy of at least of 1% to confine the Doppler frequency signal within the detection band. Resonance properties of currently available mechanical scanning systems cannot guarantee constant velocity with the required accuracy throughout the modulation period. For OCT systems developed at the Institute of Applied Physics of the Russian Academy of Sciences, a longitudinal-scanning system is based on a fiber-optical piezoelectric converter (see Fig. 9.1).[717,904] This converter is capable of scanning the path length difference between the interferometer arms at the rate of 50 cm/s and up to 4 mm in depth. Its practically inertia-free response within the range of the amplitudes and modulation frequencies used substantially simplifies the detection of the informative signal at the Doppler frequency.

9.1.3 Two-wavelength fiber OCT

Sometimes multiwavelength images are very helpful in detecting an abnormality within the optically sampled tissue. The sensitivity and recognition range of such systems may be very high because the scattering and absorption properties of normal tissue and pathological inclusions may depend on the probing wavelength in different ways. It is very important to provide the acquiring of OCT images at different wavelengths simultaneously using the same interferometer and focusing system.[1250]

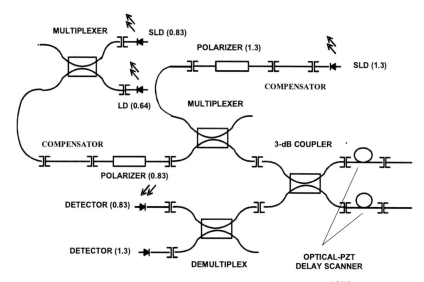

Figure 9.2 Schematic of a two-wavelength OCT.[1250]

The schematic of a two-wavelength fiber OCT system is shown in Fig. 9.2.[717,1250] Two SLDs with central wavelengths of 0.83 μm and 1.3 μm, spectral bandwidths of 25 nm and 50 nm (corresponding axial coherence lengths of 13 μm and 19 μm), and power of 1.5 and 0.5 mW, respectively, were used as light sources. The light from both SLDs was coupled to a Michelson interferometer. The incident radiation was split into two equal parts between the sample and reference arms by a fiber coupler with 3 dB of light separation at both wavelengths. The path length difference between the interferometer arms was modulated by a piezoelectric converter (see Fig. 9.1) providing in-depth scanning up to 3 mm. The most challenging problem of simultaneously compensating the wave dispersion for two different wavelengths in the interferometer arms was solved by inserting into one of the arms of the interferometer an additional piece of fiber whose dispersion properties were quite different from those of the principal fiber. The attained in-depth spatial resolution for the wavelengths 0.83 and 1.3 μm was 15 and 34 μm, respectively.

9.1.4 Ultrahigh resolution fiber OCT

The typical axial resolution of OCT imaging systems with such universally adopted broadband light sources as SLDs or mode-locked lasers varies from 10–15 μm

(SLD) to 4–5 µm (short-pulse laser sources such as organic-dye and Ti:sapphire lasers).[109,116,127,142] To provide significantly higher axial resolution (e.g., on the subcellular level), the broadband light sources covering a few hundred nanometers in the visible and NIR ranges are required. Such an ultrahigh-resolution OCT instrument for *in vivo* imaging is described in Ref. 1251. The optical scheme is presented in Fig. 9.3. A longitudinal resolution on the order of 1 µm was demonstrated for such a system; that is the highest OCT resolution achieved to date. A Kerr-lens mode-locked femtosecond Ti:sapphire laser was used as an illuminating source; it emitted sub-two-cycle pulses corresponding to bandwidths of up to 350 nm, with a center wavelength at 800 nm. Such high performance was achieved with specially designed double-chirped mirrors with a high-reflectivity bandwidth and controlled dispersion response, in combination with low-dispersion calcium fluoride prisms for intracavity dispersion compensation. A pair of fused-silica prisms and razor blades were used to spectrally disperse the laser beam and spectrally shape the laser output. The optical scheme of the low-coherence interferometer was optimized for the ultrabroad bandwidth of the illuminating source. Specially designed lenses with a 10-mm focal length and a numerical aperture of 0.30 in combination with single-mode fibers and special broadband fiber couplers were used. Polarization controllers were also used to exclude the broadening of the shape of the interference envelope due to polarization mismatch. Dispersion was matched by use of variable-thickness fused-silica and BK7 prism in order to reach a uniform group-delay dispersion. The ultrahigh-resolution OCT system was optimized to support optical spectra of up to 260 nm (FWHM), and a 1.5-µm longitudinal resolution in free space, corresponding to 1-µm resolution in tissue, was achieved.

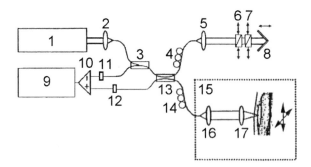

Figure 9.3 Ultrahigh resolution fiber OCT system with a Kerr-lens mode-locked Ti:sapphire laser:[1251] 1, KLM Ti:sapphire laser; 2, 5, 16, and 17, specially designed lenses; 3 and 13, special broadband fiber couplers; 4 and 14, polarization controllers; 6 and 7, dispersion-matching elements; 8, reference mirror; 9, computer-based data processing unit; 10, 11, and 12, dual balanced detector; 15, scanning system.

In vivo subcellular level resolution (1 × 3 µm; longitudinal × transverse) tomograms of an African frog tadpole (*Xenopus laevis*) were obtained to demonstrate the potential of the ultrahigh-resolution OCT instrument described above. The obtained images clearly depict multiple mesanchymal cells of various sizes and nuclear-to-cytoplasmatic ratios, the olfactory tract and intracellular morphology, as well as mitosis of several cells. It should be noted that high lateral resolution throughout the different depths has also been achieved.

9.1.5 Frequency-domain OCT

Frequency-domain, or Fourier-domain, OCT methods are based on backscattering spectral interferometry and, therefore, are also called spectral OCT.[109,116,127,142, 919,934] One of the main advantages of this technique is that it does not require scanning in the depth of a sample. However, for a long time after the first demonstrations, it did not play a significant role among OCT methods. As it was noted in Ref. 934, this probably happened because sufficiently fast and sensitive CCD cameras were not available at the time. At present, it is clear that this technique has a great potential in terms of speed and sensitivity.[109,116,127,142,934,1252,1253] It demonstrates sensitivities that are two to three orders of magnitude greater than its time-domain counterpart.[1253,1254]

One of the first demonstrations of Fourier-domain OCT is described in Ref. 919. Figure 9.4 shows a diagram of the relevant experimental setup based on a Michelson interferometer and a high-resolution spectrometer (a spectral radar device), and the optogram of skin in a human arm measured *in vitro*. Since the spectral radar measures the amplitude of scattering $E(z)$ along the axis from the surface toward the inside of an object during a single exposure of a detector without longitudinal scanning of the beam, the time span of a tomogram recording may be very small. With allowance for the superposition of the object and reference fields on the detector, we can write the intensity of light in the following form:[919]

$$I(k) = |S(k)|^2 \int_0^\infty E(z) \cos(2kz) dz + \cdots, \tag{9.1}$$

where k is the wave vector and $S(k)$ is the spectral distribution of the amplitude of the light source. Performing an inverse Fourier transform, we can find the dependence $a(z)$ of the scattering amplitude on the depth. Higher frequencies in the detected signal correspond to larger depths. We can estimate the maximum depth of probing based on the spectral resolution of the spectrograph. Specifically, for a spectrometer with a resolution of $\Delta\lambda = 0.05$ nm, we have $z_{max} = (1/4n)(\lambda_0^2/\Delta\lambda) \cong 2.4$ mm ($n = 1.5$, $\lambda_0 = 853$ nm). To ensure a signal-to-noise ratio at the level of 10^4, one should employ a highly sensitive CCD or CMOS camera at the output of the spectrometer.

Fourier-domain OCT can be also realized using a swept-laser source—a rapidly tunable laser over a broad optical bandwidth.[1254] In swept-source OCT, instead of CCD or photodiode arrays, a single photodiode in the detection path of the interferometer is employed, which allows the spectral interferometric signal to be encoded with a characteristic heterodyne beat frequency. It was shown that heterodyne de-

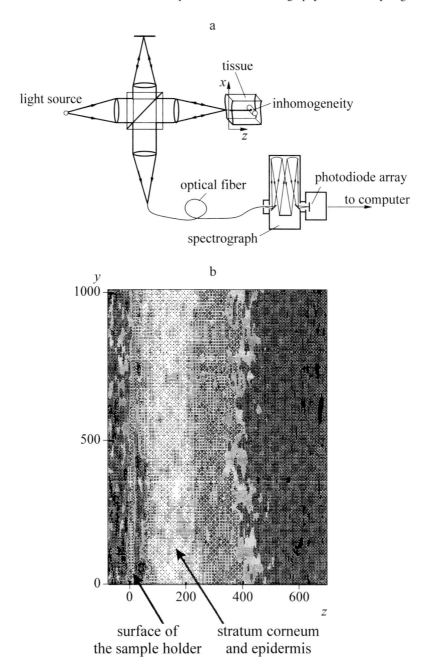

Figure 9.4 (a) Basic scheme of a spectral radar (spectral OCT) device with a source of light with a small coherence length.[919] (b) Optogram of skin in a human arm *in vitro*. The ordinate axis represents the lateral displacement over the arm surface in micrometers. The abscissa axis shows the depth of a tissue in micrometers.

tection in swept-source OCT allows for the resolution of complex conjugate ambiguity and the removal of spectral and autocorrelation artifacts, characteristic of spectral OCT.[1255]

9.1.6 Doppler OCT

Doppler OCT combines the Doppler principle with OCT to obtain high-resolution tomographic images of static and moving constituents in highly scattering biological tissues.[937] When light backscattered from a moving particle interferes with the reference beam, a Doppler frequency shift (f_{Ds}) occurs in the interference fringe,

$$f_{Ds} = \frac{2V_s n \cos \theta}{\lambda_0}, \tag{9.2}$$

where V_s is the velocity of a moving particle, n is the refractive index of the medium that is surrounding the particles, θ is the angle between the particle flow and the sampling beam, and λ_0 is the vacuum center wavelength of the light source. The longitudinal flow velocity (velocity parallel to the probing beam) can be determined at discrete user-specified locations in a turbid sample by measurement of the Doppler shift. The transverse flow velocity can also be determined from the broadening of the spectral bandwidth due to the finite numeric aperture of the probing beam.[937]

Scanning of the reference mirror of the OCT system at velocity v produces a Doppler signal at frequency f_D, described by Eq. (4.53). Blood or lymph flow with velocity V_s produces another Doppler signal, described by Eq. (9.2). Therefore, the signal of the Doppler OCT is proportional to

$$A(t) \cos[2\pi(f_D - f_{Ds})t + \phi(t)], \tag{9.3}$$

where $A(t)$ is the reflectivity and $\phi(t)$ is the phase shift defined by a scatterer position.

A fiber-optic Doppler OCT ($\lambda_0 = 850$ nm, $\Delta\lambda = 25$ nm, $P = 1$ mW) was employed for measurements of the blood flow velocity in a vessel located behind a strongly scattering layer and in a living object (vessel of rat mesentery).[908] This is a new approach to the investigation of directed blood flow in subsurface vessels under a layer of tissue.[908,926–930,937] Electronic data processing allows one to separate the signal that characterizes the amplitude of backward scattering, which is necessary to generate a stationary tomogram of an object from the Doppler signal, which characterizes the velocity of scatterers at a given point of an object. Figure 9.5 presents a structure image of a fragment of rat mesentery with an artery and two veins, images of the blood flow velocity in the artery and veins, and the velocity profile of the total blood flow.

9.1.7 Polarization-sensitive OCT

The specificity of conventional OCT can be improved by providing measurements of the polarization properties of the probing radiation when it propagates through a tissue. This approach was implemented in the polarization-sensitive OCT technique (PS OCT), which is described in detail in numer-

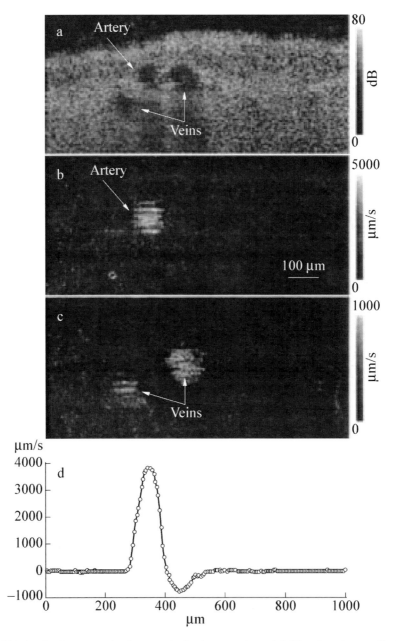

Figure 9.5 *In vivo* images of a fragment of rat mesentery: (a) Structure image (featuring an artery and two veins). (b) Image of the blood flow velocity in the artery. (c) Image of the blood flow velocity in veins. (d) Profile of the blood flow velocity measured at the depth of the artery [indicated with an arrow in Fig. 9.5(a)]. The negative peak on the right-hand side is related to the influence of the blood flow velocity in the vein.[908]

ous original papers,[412–424,913,1256–1258] and a few overview papers and book chapters.[127,135,142,717,936] Advanced PS OCT systems provide tissue imaging using Jones matrix[418] or Mueller matrix elements.[416]

In the majority of studies on PS OCT, the criterion of pathological changes in tissue is a measured decrease in tissue macroscopic birefringence. However, there is difficulty in providing correct measurements of birefringence at depths of more than 300–500 μm. For deeper layers (up to 1.5 mm), a much simpler variant of PS OCT known as cross-polarization OCT (CP OCT) can be employed.[717,913] Light depolarization caused by light scattering and tissue birefringence both lead to the appearance of a cross-polarized component in the backscattered light. Pathological processes are characterized by the changes in the amount of collagen fibers and their spatial organization. Therefore, a comparative analysis of cross-polarization backscattering properties of normal and pathological tissues may be used for early diagnosis of neoplastic processes.

The scheme of an experimental system for measuring conventional OCT and CP OCT images is shown in Fig. 9.6.[717,1256] Using a multiplexer (M), a low-coherence IR radiation from a SLD ($\lambda = 1.3$ μm and $l_c = 21$ μm) is combined with radiation from a red diode laser (RL) used for optical system alignment. One of the eigen polarization modes of a polarization maintaining (PM) 3-dB fiber coupler (FC) is selected by means of a polarization controller (CP). The PM fiber is used to transport radiation with a certain polarization state in both the signal and reference arms of the interferometer. When there is no Faraday rotator (F) in the reference arm, a copolarized component of backscattered radiation is recorded (conventional OCT is realized). The Faraday rotator performs a rotation to an arbitrary polarization state by a specified angle, and the direction of the rotation depends only on the direction of the magnetic field inside the rotator and does not depend on the propagation direction of the radiation. Therefore, in the case of the 45-deg Faraday rotator, the radiation passes through it, and being reflected by a mirror goes back through the rotator and becomes orthogonally polarized. As a result, only the cross-polarized component of the light backscattered by a

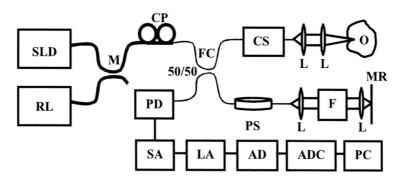

Figure 9.6 Experimental setup for cross-polarization OCT:[717,1256] SLD, superluminescence diode; RL, red diode laser; M, multiplexer; CP, polarization controller; FC, fiber coupler; CS, cross-sectional scanner; O, investigated object; PS, longitudinal piezoscanner; L, lenses; F, Faraday rotator; MR, reference mirror; PD, photodiode; SA, selective amplifier; LA, logarithmic amplifier; AD, amplitude detector; ADC, analog-to-digital converter; PC, personal computer. Bold line corresponds to single-mode fiber; thin line illustrates polarization maintaining fiber.

biological object would interfere with light from the reference arm. The acquisition time of one OCT image is 1 s. For all OCT images, a logarithmic intensity scale was used. The lateral resolution of the system, determined by the diameter of the probing beam in the focus, was chosen close to the axial (in-depth) resolution, which is determined by the coherence length and was 21 μm. Both types of images, conventional and cross-polarized, were obtained from the same tissue site.

Figure 9.7 demonstrates the facilities of the crossed-polarized imaging technique in comparison with a conventional one in the example of the imaging of *ex vivo* human esophagus scar tissue. It is well seen that in contrast to conventional OCT, the crossed-polarized image provides some additional structural information.

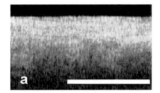 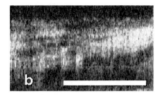

Figure 9.7 (a) Conventional OCT image of *ex vivo* human esophagus scar tissue. (b) The corresponding cross-polarized OCT image. White bar corresponds to 1 mm.[717,1256]

9.1.8 Differential phase-sensitive OCT

Differential phase-sensitive OCT (DPS OCT) provides quantitative dispersion data that are important in predicting the propagation of light through tissues, in photorefractive surgery, and in tissue and blood refractive index measurements.[142] Refractive index variations cause phase variations in the sample beam. One of the DPS OCT schemes is presented in Fig. 9.8.[1113] The probe beam is split by birefringent wedges and collimated by the sample lens. Two orthogonally oriented beams separated by x illuminate the sample. The backscattered beams are combined by the birefringent wedges and separated by the polarizing beamsplitter (Wollaston prism) in the detection arm. From the photodetector signals, three interferograms and their corresponding three images are obtained: two intensity images, and a phase difference image. Experiments have shown that measurements of angstrom/nanometer-scale path length change between the beams $[(\lambda/4\pi)\Delta\varphi]$ in clear and scattering media can be provided.[142,1113] The DPS OCT technique was demonstrated to be suitable for noninvasive, sensitive, and accurate monitoring of analyte concentrations, including glucose (see Section 5.9.1).[1113] Since DPS OCT detects phase contrast in the direction of beam separation, it detects phase gradients caused by transversal variations of the refractive index and/or the phase change on reflection at interfaces.[142]

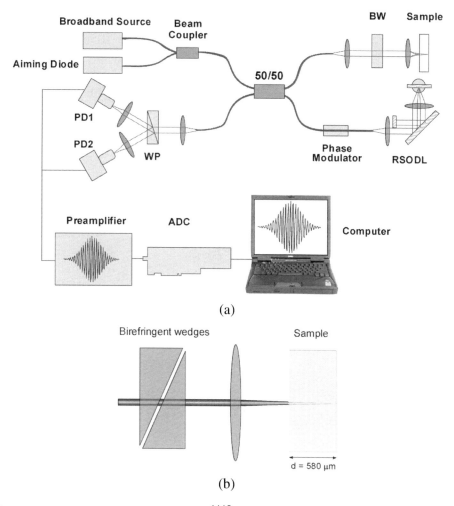

Figure 9.8 Phase-sensitive OCT system.[1113] (a) General scheme: WP, Wollaston prism; RSODL, rapid scanning optical delay line; PD, photodetectors; BW, birefringent wedges; ADC, analog-to-digital converter. (b) Sample arm.

9.1.9 Full-field OCT

Conventional time-domain OCT is a single-point detection technique. It can be used to generate two-dimensional OCT images up to video rates; however, such systems have a limited sensitivity or a limited space-bandwidth product (resolved pixels per dimension).[142] Full-field or parallel OCT uses linear or two-dimensional detector arrays of, respectively, N and N^2 single detectors. The advantage of parallel OCT is that the SNR when using linear or two-dimensional detector arrays can roughly be, respectively, $\sqrt{N}$ and N times larger, compared to the single detector signal. The disadvantages of using standard CCD sensors are connected with their time-integrating operation mode; therefore, no ac technique, mixing, or mode-lock detection are possible. To overcome these problems, synchronous illumination instead of the usual synchronous detection to obtain lock-in detection on every pixel of a CCD detector array or CMOS detector array can be used.[142] In the CMOS camera, each "smart pixel" consisting of photodetector and analog signal processing performs heterodyne detection in parallel, thus dramatically increasing the dynamic range compared to a CCD array. A corresponding two-dimensional "smart pixel detector array" that made it possible to record a data set of 58×58 pixels and 33 slices with an acquisition rate of 6 Hz is presented in Fig. 9.9.

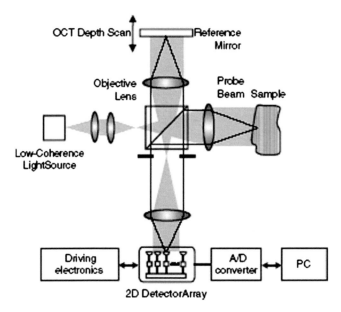

Figure 9.9 Parallel OCT setup with a two-dimensional 58×58 pixel CMOS detector array where each single detector performs heterodyne detection in parallel (Bourqin et al. 2001).[142]

To improve depth resolution, a thermal light source can be used.[142,1259,1260] A 100-W tungsten halogen thermal lamp in a modified Linnik microscope allowed a depth resolution of 1.2 μm to be obtained.[1259] Water immersion has been used to compensate dispersion; corresponding immersion-objective lenses with a NA of 0.3 provided a transverse resolution of about 1.3 μm. A three-dimensional OCT image of a *Xenopus laevis* tadpole eye has been synthesized from 300 tomographic images.

Contactless three-dimensional topology of the surface of human skin is necessary for the monitoring of wound and burn healing, observation of side effects of strong medicinal preparations, etc. Figure 9.10 shows a "coherent radar" based on a Michelson interferometer and a low-coherence light source [light-emitting diode (LED)].[919] Scanning the reference mirror in this device (with a scanning rate of 4 μm/s), one can obtain three-dimensional images of skin surface with a resolution within the limits of the coherence length of the light source. However, because the "coherent radar" has to scan the depth of the whole object with a limited velocity

(4 µm/s), the measuring time was long (about 150 s) and the field of illumination, in its turn limited by the power of the LED, was 7×10 mm. The influence of bulk scattering can be eliminated in this case if the skin is protected with a lightproof coating, e.g., graphite powder.[919]

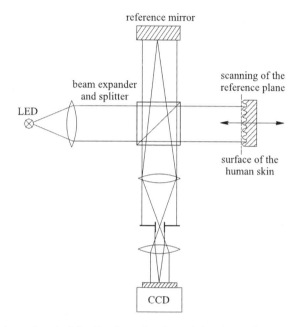

Figure 9.10 "Coherent radar" for the investigation of the three-dimensional topology of a skin surface.[919] LED, light-emitting diode; scanning of the reference plane is shown with the dashed line.

9.1.10 Optical coherence microscopy

Optical coherence microscopy (OCM) is a new biomedical modality for cross-sectional subsurface imaging of tissue combining the ultimate sectioning abilities of optical coherence tomography (OCT) and confocal microscopy (CM).[775,938] In OCM, spatial sectioning due to the tight focusing of the probing beam and pinhole rejection provided by CM is enhanced by additional longitudinal sectioning provided by OCT coherence gating.

Figure 9.11 illustrates the results of a comparative study of imaging potentialities of CM and full-field OCM.[1261] The significantly better quality of the reconstructed images was obtained in the latter case. The sample arm of the low-coherent interferometer consists of a high-speed scanning CM with a fast scanner (a resonant scanner or a rotating polygonal mirror) and a slow scanner (a galvanometric scanning mirror); the slow scanner was positioned at the image plane of the fast scanner. To exclude the mechanical scanning of the phase delay in the reference arm, a 40-MHz acoustooptic modulator (AOM) in combination with a fixed mirror was used for frequency shifting of the reference arm light due to double passage through AOM. In such a way, the reference arm length was fixed to match the optical path length of the sample arm. The setup was modified to CM by blocking the reference arm and detecting the dc centered signals from the sample arm. Both sets of images presented were recorded at eight frames per second with a polygonal mirror as a fast scanner. Each image is a single frame extracted from the video. Since the rotating polygonal mirror causes a path length change in the sample arm of the interferometer during the OCM image reconstruction, an additional frequency shift of the operating frequency of the AOM was introduced during the sample scanning. This shift was equal to 3 MHz from one side of the image to the other side. Thus, by moving the center frequency from 80 to 83 MHz, a full-field OCM image was captured.

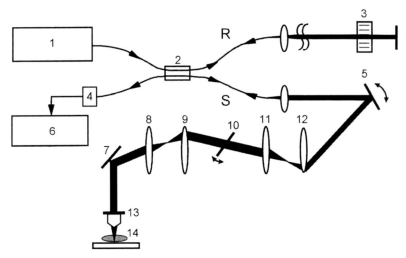

Figure 9.11 Optical scheme of a high-speed full-field optical coherence microscope:[1261] 1, low-coherence light source; 2, fiber-optical interferometer; 3, acoustooptic modulator; 4, detector; 5, resonant scanner; 6, computer-based data processing unit; 7, mirror; 8, 9, 11, and 12, lenses; 10, galvanometer; 13, microscope objective; 14, sample under study; R, reference arm; S, sample arm. By blocking the reference arm, this system can also be used as a confocal scanning microscope.

A compact OCM with a flexible sample arm and a remote optical probe for laboratory and clinical environments was developed.[775] To achieve an axial resolution of the cellular level, a light source with an effective bandwidth of 100 nm was used. The light source was comprised of two SLDs based on one-layer quantum-dimensional (GaAl)As heterostructures with shifted spectra. Radiations from both SLDs were coupled into a polarization-maintaining (PM) fiber by means of a multiplexer. The multiplexer was spectrally adjusted in order to achieve the minimum width of the autocorrelation function. The dynamic focusing was provided by scanning the output lens of the objective located at the very end of the sample arm. The lens movement was controlled by the electronic system, and aligning of the focal spot with the coherence gating during scanning up to depth of 0.5–0.8 mm into a tissue was provided. The spectral sidelobes, caused by nonuniformity of the light source spectrum, were suppressed.

9.1.11 Endoscopic OCT

Application of fiber-optical light-delivering and light-collecting cables allows one to build a flexible low-coherent imaging system providing the possibility of endoscopic analysis of human tissues and organs. In particular, an OCT system developed for endoscopic applications (high-speed *in vivo* intra-arterial imaging) is described in Ref. 907. A solid-state Cr^{+4}:Forsterite laser with Kerr lens mode locking was used as an illumination source with a median wavelength of 1280 nm and a bandwidth of 75 nm. Thus, the theoretical axial resolution of the system can be estimated as 10 μm [see Eq. (4.54)]; the actual depth resolution measured with a mirror as a standard technique for resolution evaluation gave an axial pixel size equal to 9.2 μm. The lateral resolution of this system, which depends on the spot size of the lens system used on the output tip of the light delivering fiber, was equal to 30 μm with the confocal parameter equal to 1.74 mm. Electronics allowed one to capture four frames per second for 512 transverse image pixels. The optical power incident on the imaged tissue was approximately 10 mW. The corresponding signal-to-noise ratio was 106 dB. The reference-arm phase-delay scanning device consisted of an oscillating galvanometer mirror, lens, and grating.

Various fiber-optical devices designed for endoscopic OCT imaging have been described.[717,1261] One of the examples is a fiber-optical scanning catheter used for intra-arterial imaging.[1261] Such a catheter consists of an optical coupling at its proximal end, a single-mode fiber as the light-delivering channel, and focusing and beam directing elements at the distal end. Beginning at the proximal end of the device, incident light from a fixed single-mode optical fiber is coupled through a narrow air gap into a second single-mode fiber that can rotate. The drive assembly of the catheter, located at the proximal end, uses an optical-fiber connector. A gear is attached to the connector and a shaft assembly, consisting of the connector, the fiber in the catheter, and the distal focusing elements. A dc motor is used to drive the shaft assembly through a gear mechanism. The beam is focused by a graded index lens and is directed by a microprism. The beam was scanned circumferentially by rotating the cable, fiber, and optical assembly inside the nonmovable housing. Power losses caused by suboptimal coupling and internal reflections within the catheter were 3–4 dB. For instance, an *in vivo* image of rabbit trachea, which was obtained with the described system, allows the differentiation between various tissue structures such as the pseudo-stratified epithelium, mucous, and surrounding hyaline cartilage.[1261]

A whole family of diagnostic endoscopic OCT devices suitable for studying different internal organs has been created.[1262] To probe the surface of an internal organ, a miniaturized electromechanical unit (optical probe) controlling and performing lateral scanning was developed. This probe is located at the distal end of the sample arm and its size provides fitting to the diameter and the curvature radius of standard biopsy channels of endoscopes. Figure 9.12(a) demonstrates the head of an endoscope for gastrointestinal investigations with the integrated OCT scanner.[717] A schematic diagram of the optical scanning probe and how it is positioned against a studied object is shown in Fig. 9.12(b). The probing beam is swung along the tissue surface with amplitude of 2 mm. The beam deviation system embodies the galvanometric principle, and the voltage with a maximum of 5 V is supplied to the distal end of the endoscope. The distance between the output lens and a sample varies from 5 to 7 mm; the focal spot diameter is 20 μm. The optical scanning probe and the part of the flexible sample arm that is inserted in the endoscope are both sealed; therefore, the conventional cleaning procedure and sterilization can be performed before applying the setup clinically. Implementation of an extended flexible arm of the OCT interferometer became feasible due to the use of polarization-maintaining fibers as a means for transportation of the low-coherence probing light. This allows the elimination of the polarization fading caused by polarization distortions at the sites of bending of the endoscope arm. The device features high-quality fiber polarizers and couplers. The "single-frame" dynamic range of the OCT scheme determined as the maximum variation of the reflected signal power within a single image frame attains 35–40 dB. With a scanning rate of 45 cm/s and the image depth of 3 mm (in free space units), an OCT image with 200×200 pixels is acquired in approximately 1 s. This acquisition rate is sufficient to eliminate the influence of moving of internal organs (moving artifact) on the image quality.

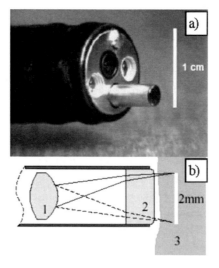

Figure 9.12 (a) Distal end of a gastroscope with OCT probe introduced through a biopsy channel.[717] (b) Schematic diagram of scanning unit: 1, output lens; 2, output glass window; 3, sample.

9.1.12 Speckle OCT

An original technique of coherent tomography that does not require transverse scanning and employs subject speckles is described in Ref. 906. Figure 9.13 shows a diagram of the experimental setup consisting of an SLD, mirror Mach-Zehnder interferometer, and a CCD camera. Radiation produced by the light source is focused onto the surface of an object. An incident light beam irradiates a surface at an angle of 45 deg. The light penetrates into the tissue and experiences scattering. Backscattered light emerges at the surface of the object. A fraction of this beam of light that propagates in the direction perpendicular to the surface of the sample is imaged by the CCD camera, where the beam is mixed with reference-wave radiation reflected from a mirror scanned with a constant rate. Observation of scattered light with an aperture of finite size gives subjective speckles in the image plane (the photosensitive surface of the CCD camera).

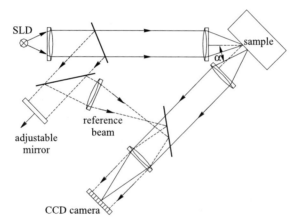

Figure 9.13 Coherent optical tomographic scheme based on subjective speckles.[906] SLD, superluminescent diode ($P = 4$ mW, $\lambda = 850$ nm, and $l_c = 30$ μm); α is the angle of radiation incidence on the surface of a sample.

In the case when partially coherent light is employed, one should consider different groups of photons passing through the image plane P_i. Each of these groups consists of photons that have traveled a definite path length $l_i \pm l_c$. Correspondingly, these photons produce their own coherent speckle pattern with intensity distribution $S_i(x, y)$. The resulting speckle pattern is produced by the incoherent superposition $\sum S_i$ of different speckle patterns. To locate regions inside an object from which photons with a definite path length $L = l_i$ come, one should superimpose a reference wave with the corresponding path length L. Then, only the photons from the chosen group P_i will ensure the required contrast V_I, which should be measured. Two sequential exposures are required to measure V_I. After the first exposition, the phase of the field in the reference beam is shifted by π, and exposure is repeated. The incoherent component remains unchanged in these two exposures and can be easily subtracted. Note that the sizes of subjective speckles should be adapted to the sizes of the pixels of the CCD camera. The maximum contrast is achieved when the coherence length l_c is as large as possible for a given resolution.

In an image processed in such a way, dark spots or speckle modulation of the surface image will be caused by the partial components of the scattered field that have run the same path length, but along different individual paths. A particular feature of the recorded image is a sharply curved edge of the impulse response, the "photon horizon." This curve defines the maximum penetration depth in the analyzed scattering system for each reference path length. Thus, different penetration depths can be visualized by a suitable adjustment of the reference path length. Figure 9.14 shows the character of these images obtained with different light sources.

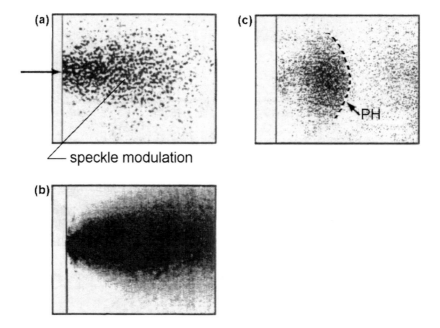

Figure 9.14 Images of the scattering process with different coherence lengths and different kinds of observation:[906] (a) Fully coherent illumination, direct observation. (b) Broadband illumination, direct observation. (c) Broadband illumination, extraction of the coherent part; dashed line shows the position of the "photon horizon" (PH).

The path-length resolution for this system is determined by the coherence length l_c of the light source used, and for a source with $l_c = 30$ μm, the resulting time resolution is about 100 fs.[906] The main advantage of this scheme is that the generation of a two-dimensional $(x - z)$ tomogram does not require transverse scanning of the beam. Being applied for the imaging of macroscopically inhomogeneous scattering phantoms and human skin *in vivo*, the "photon horizon" detection technique has demonstrated promising results.[1263] In the case of human skin, smaller exposure times are required than for a motionless scattering system because of the time integration of speckle-modulated images during the exposure. An exposure time of 40 ms is enough to provide adequate quality of "photon horizon" images of human tissues.

On the basis of the obtained results, it can be concluded that "photon horizon" detection makes it possible to get 2D surface images of strongly scattering media with a resolution better than 10 fs and in-depth range up to 350 μm. This gives the possibility to use such low-coherence imaging technique for the detection of pathological alterations of human skin (e.g., melanoma maligna).

The speckle OCT method was shown recently as an alternative to the Doppler OCT in 2D imaging of blood flow.[1186] Flow information can be extracted using speckle fluctuations in conventional amplitude OCT. Time-varying speckle is manifested as a change in OCT image spatial speckle frequencies. It was shown that over a range of velocities, the ratio of high to low OCT image spatial frequencies has a linear relation to flow velocity and that method is sensitive to blood flow of all directions without phase information needed. Using speckle OCT, 2D images of blood flow distributions for *in vivo* hamster skin were received.[1186]

9.2 Optical heterodyne imaging

Coherent heterodyne optical detection offers the following advantages over direct detection methods:[1,3,28,894,895,939,1212–1214,1264–1270]

- It has the highest sensitivity of any detection technique; the signal-to-noise ratio is several orders of magnitude better than that which can be achieved by incoherent methods.
- The extraordinary dynamic range, about 15 orders or more in magnitude of signal power for Hz-order bandwidths can be realized.
- Excellent selectivity or filtering capability in frequency domain and polarization.
- High spatial resolution, up to 60–80 lines per mm (with confocal optical arrangement).
- Substantial spatial selectivity or filtering capability due to highly directional antenna properties.
- It allows quantification of new tissue parameters by detecting the wavefront degradation within a given tissue.

The quantum limit of optical detection corresponds to the signal-limited shot noise when one employs a conventional photoelectronic detector. The minimal detectable signal power is given by[1266]

$$P_{\min} = h\nu B_d/\eta_q, \tag{9.4}$$

where $h\nu$ is the photon energy, B_d is the detection bandwidth, and η_q is the quantum efficiency of the detector. In the case of the direct detection technique, this limit is impossible to reach. Only the optical heterodyning detection and photon-counting methods allow one to realize this standard quantum detection limit.

The field of view of an optical heterodyne system that has an effective aperture A for signals at wavelength λ arriving within a single main antenna lobe expressed as a solid angle is defined by[1266]

$$\Omega \cong \lambda^2/A. \tag{9.5}$$

The antenna properties afford high spatial resolution for the detection and ranging of various bioobjects and their image formation, and excellent directionality to distinguish one specific direction from another.

Owing to these important features, which make it possible to detect and image very weak signals embedded in or hidden by appreciably large optical noise or background, the optical heterodyning technique has a good outlook for applications in tissue spectroscopy and imaging.[1,1264–1268] The coherent detection imaging (CDI) method based on the optical heterodyne detection technique was established in 1989 by Inaba (see Refs. 1 and 3). The CDI is a coherence gating method that discriminates the forward multiply scattering beam, preserving the direct geometrical correlation with the incident light beam, from the diffuse component of the transmitted light, which generally loses the properties of the incident beam, such as coherence, direction, and polarization due to multiple scattering.

The optical heterodyne detection method operates on the principle of mixing two optical waves at different frequencies [$(\omega_0 + \omega_1)$ and $(\omega_0 + \omega_2)$, where ω_0 is the optical frequency, ω_1 and ω_2 are radio modulation frequencies], on a square-law detector such as photodiode (see Fig. 9.15). The signal generated by the photodetector is the cross-product of the two optical fields as signal,

$$\sqrt{I_1(t, x, y)} \sin[(\omega_0 + \omega_1)t + \phi(t, x, y)], \tag{9.6}$$

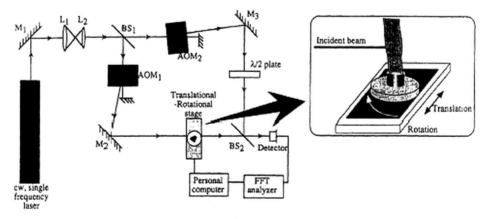

Figure 9.15 Schematic of a coherent detection imaging (CDI) system.[1267] L_1 and L_2, collimating lenses; M_1, M_2, and M_3, antireflection coated mirrors; BS_1 and BS_2, beam splitters, AOM_1 and AOM_2, acoustooptic modulators. The insert is a schematic of the finger mounted on the scanning apparatus.

and local oscillator,

$$\sqrt{I_2} \sin[(\omega_0 + \omega_2)t], \tag{9.7}$$

and can be expressed as

$$\sim \sqrt{I_1(t, x, y)}\sqrt{I_2} \cdot \sin[(\omega_1 - \omega_2)t + \phi(t, x, y)], \tag{9.8}$$

where $\sqrt{I_1(t, x, y)}$ and $\sqrt{I_2}$ are the amplitudes of the signal and local oscillator waves; $t, x,$ and y are the temporal and spatial coordinates; and $\phi(t, x, y)$ is the spatially and time-dependent phase shift caused by the spatial and temporal fluctuations of refraction of the tissue under investigation.

The signal amplitude is subject to temporal and spatial fluctuations caused by time-dependent and spatially dependent attenuation of light by the tissue. The attenuated intensity of the signal wave by a tissue of thickness $d(t, x, y)$ can be calculated as [see Eq. (1.1)]

$$I_1(t, x, y) \approx I_{10} \exp[-\mu_t(t, x, y)d(t, x, y)], \tag{9.9}$$

where I_{10} is the intensity of the signal wave incident at the object and $\mu_t(t, x, y)$ is the distribution of the attenuation coefficient.

The typical scheme of a CDI system is shown in Fig. 9.15. The well-collimated optical beam of a CW, single-frequency laser [Ar (514.5 nm), He:Ne (633 nm), Kr (647.1 nm), Ti:Al$_2$O$_3$ (tuned in the range 700–1000 nm) or a diode pumped Nd:YAG (1064 and 1319 nm)] with a power of about a few dozens of milliwatts is split into signal and local oscillator beams. The local oscillator beam and the signal beam are frequency shifted (modulated) by a pair of acoustooptic modulators to 80 and 80.05 MHz, respectively. The signal beam (about 0.8 mm in diameter), after passing through the object (a human finger) is mixed with the local oscillator beam at a silicon photodiode generating a signal at an intermediate frequency (IF) (the beat signal) [see Eq. (9.8)]. The IF signal is then fed to a fast Fourier transform analyzer that is interfaced to a personal computer. The dynamic range of the system, defined as P_{sat}/P_{min}, where P_{sat} is the optical power at which the detector is saturated and P_{min} is the minimal detected optical power, is about 140 dB. It should be noted that the signal power is proportional to the amplitude of the IF signal over the entire dynamic range of the heterodyne system. Very low optical power on the order of 10^{-17} W at a wavelength of 800 nm and a detection bandwidth limited to a few hertz could be detected.

Using the described CDI system, two-dimensional (projection) imaging was successfully performed for various *in vitro* and *in vivo* biological objects such as chicken legs and eggs, human tumor specimens, human teeth, animal bones, the head of an infant mouse, and a human finger.[1,1266–1268] The spatial resolution of the system is in the range 0.3 to 0.5 mm, depending on the optical and scanning arrangement. As an example, Figs. 9.15 and 9.16 illustrate the application of a CDI system for the tomographic study of a healthy human volunteer's finger. The index finger is mounted on a translational-rotational stepping motor stage (as shown in the insert of Fig. 9.15). The base of the finger was lightly bound with a silicon tube to reduce blood flow to the finger surface. This procedure was essential, especially when imaging at 715 nm, to minimize the influence of the Doppler effect on the IF signal caused by surface blood flow. The image displayed in Fig. 9.16(a) was reconstructed using 30 projections. Data for each projection consist of the averaged (32 times) amplitude of the IF signal for every 0.5-mm step of the translational scan across the finger joint. Averaging of the IF signal was necessary to minimize the speckle effect. After each translational scan, the finger was rotated by 6 deg and the translational scan was repeated. The finger was rotated by 180 deg for each data set. The data processing was carried out by the filtered back-projection method used in x-ray computed tomography (CT) image reconstruction.

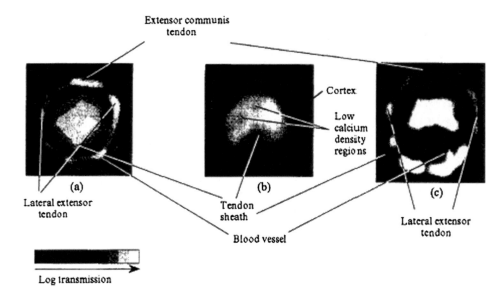

Figure 9.16 Computed tomographic images at the joint region corresponding to the head of the proximal phalanx of a healthy human index finger:[1267] (a) Laser CT image obtained with a Nd:YAG laser operating at 1064 nm. (b) Hard x-ray CT image. (c) MRI (T$_1$-weighted) image.

In Fig. 5.16, the CDI laser tomographic image of the finger (at the joint region corresponding to the head of the proximal phalanx) obtained at 1064 nm, as well as reference images obtained by conventional methods such as hard x-ray CT and magnetic resonance imaging (MRI), are shown. The diameter of the finger at the measurement plane was ∼14 mm and the incident power was ∼25 mW. Data for each image were collected during ∼25 min. The slice thickness of the x-ray and MRI images was 5 and 3 mm, respectively; whereas for the CDI images it was estimated as ∼0.5 mm. The use of a 2D heterodyne detector array would have helped through increased transverse resolution and reduced measurement time.[1267]

Because of the limited known data on the optical properties of human finger tissue components, it is difficult to identify all the structures in the CDI images. However, in comparison with x-ray and MRI images, some additional substructures are seen in the laser CDI images, which could correspond to ligaments, tendon sheath, etc. The usage of the ratio of images obtained at different wavelengths, as is widely employed in diffusion optical tomography, could help in identifying these substructures. A wavelength-dependent CDI should be able to provide both structural and functional information of the human finger and can be used for the early diagnosis of rheumatic arthritis, as was recently shown by utilization of another optical method.[1269] However, it should be noted that earlier optical tomographic studies of the human hand, which were done by the use of the CW[1269] and time-resolved (photon density waves)[502] imaging techniques, did not allow one to have as high a spatial resolution as that provided by the CDI method.

A method for real-time direct measuring of the mean square of the heterodyne beat amplitude (IF) and its proportionality to the overlap of the Wigner phase space distributions for a local oscillator and signal fields was demonstrated in Refs. 939 and 1270; such distributions give maximum information on the scattered field for imaging applications. A dynamic range of the experimental setup (see Fig. 9.17) of more than 13 orders of magnitude for a laser power of only 2 mW was realized. By increasing the local oscillator and input beams' power to 10 mW, a dynamic range of 15 orders is expected. The setup of the heterodyne experiments employs an He:Ne laser beam that is split into a 1-mW local oscillator (LO) and a 1-mW beam input to the sample. The beam transmitted through the sample is mixed with the LO at a 50 × 50 beamsplitter (BS2). Technical noise is suppressed by employing a standard balanced detection system.[939] The IF signal at 6 MHz is measured with an analog spectrum analyzer, the output signal of which was squared using a low-noise multiplier. A lock-in amplifier allows subtraction of the mean square signal

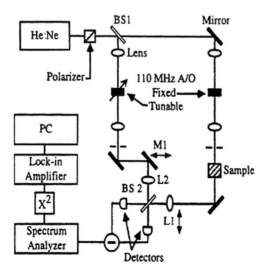

Figure 9.17 Heterodyne detection scheme.[1270]

and noise voltages with the input beam on and off. In this way, the real-time measurements of the mean square beat amplitude, $|V_B|^2$, were provided. Information about the Wigner phase space distribution of the transmitted light was obtained by measuring the beat intensity as a function of the shift in the LO center position (by translating mirror M1) and transverse momentum (angle) (by translating lens L1). The detector was in the Fourier plane of lenses L1 and L2, so that the LO position was fixed in the detection plane. The spatial resolution of the system was determined by the spatial width and diffraction angle of the LO.

The model experiments, which were performed by the authors of Ref. 939, provided a better understanding of the tomographic studies done by Inaba and coworkers for living tissues,[1266–1268] in particular, by linking the amplitude of the IF signal with tissue density and its coefficient of absorption. The beat intensity for propagation of a Gaussian beam through an Intralipid solution at fixed positions of the mirror M1 and the lens L1 corresponding to the maximal signal as a function of Intralipid solution concentration is shown in Fig. 9.18. The data clearly exhibit an initial rapid exponential decline with concentration, followed by a slower nonexponential decline. The spatial measurements showed that the fast decline corresponds to the transmission of the Gaussian beam, while the slower decline arises from a multiple scattering contribution. Using the concept of the Wigner phase space distributions, it was shown for a Gaussian input beam with a diffraction angle of about 1 mrad that at low Intralipid concentrations the amplitude of the beat signal decays exponentially due to absorption and scattering as

$$\sim \exp[-(\sigma_{\text{sca}} + \sigma_{\text{abs}})\rho_s d], \tag{9.10}$$

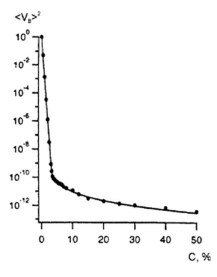

Figure 9.18 Beat intensity versus Intralipid concentration ($C \sim \rho_s$) in percent of 10% Intralipid solution in pure water, for 0.5-μm average particle size and a sample length of 1 cm.[1270]

where σ_{sca} and σ_{abs} are the scattering and absorption cross sections, respectively, ρ_s is the particle concentration, and d is the sample thickness. At an intermediate Intralipid concentration, when a transmitted beam broadens due to multiple scattering, the beat intensity must drop as

$$\sim \rho_s^{-2} \exp(-\sigma_{abs}\rho_s d). \tag{9.11}$$

At high concentrations, the scattering light distribution becomes isotropic and the beat intensity decay is defined mainly by the absorption as

$$\sim \exp(-\sigma_{abs}\rho_s d). \tag{9.12}$$

One expects that this simple picture will break down at a sufficiently high concentration, when the diffusion photon path is substantially larger than the thickness of the sample.

9.3 Summary

Thus, a brief review of the optical schemes of coherence tomographs and topographs demonstrates the possibilities of these instruments for the investigation of tissues. Despite the continuing development of OCT and the search for advanced tomographic schemes, the area of medical applications of OCT in its present form is continually growing. Currently, this area includes analysis of damaged skin zones of patients suffering from psoriasis, erythematosus lupus, scleroderma, and malignant melanoma; investigation of postburn keloid cicatrices and vascularization of subepidermal skin layers; monitoring of laser ablation of tissues; imaging of brain tissues and fundus of the eye; imaging of hard and soft tissue of the oral cavity; investigation of the mucous membrane of inner organs, etc.[1,3,8,13,17,18,28,45,76,77,84,102,108–111,116,126,127,129,135,136,138,139,141,142,343,717,775,865,901,902,909,931–939,1246,1247]

Conclusion

This tutorial deals only with some aspects of light-tissue interactions, being focused on noncoherent and coherent light scattering by random and quasi-ordered structures. The considered methods and presented results allow us to make certain conclusions and assumptions concerning the directions of further investigations and the development of laser diagnostic and other medical systems.

Results of numerous studies on light scattering emphasize the necessity of an in-depth evaluation of the optical properties of tissues with different structural organization. At present, light propagation in tissues is fairly well described in qualitative (and sometimes in quantitative) terms, which provides a sound basis for the implementation of different diagnostic, therapeutic, and surgical modalities. At the same time, the estimation of optimal irradiation doses or the choice of correct diagnostic clues sometimes presents great difficulty because of the lack of reliable criteria for the optical parameters of tissues.

Traditional spectrophotometry or angular and polarization measurements are useful to characterize tissues, but need to be improved if more sophisticated tissue models are to be obtained and applied to biomedical studies. Such models must take into consideration the spatial distribution of scatterers and absorbers, their polydisperse nature, and optical activity, along with the birefringence properties of the materials of which scatterers and base matter consist.

It is also necessary to further develop methods for the solution of inverse scattering problems with due regard for the real geometry of the object and the laser beam, which might be equally valid for the arbitrary ratio of the scattering and absorption coefficients. The inverse MC method is sometimes useful, but the fast computations required for practical medical diagnostics and dosimetry should be based on approximate solutions of the radiative transport equations.

Extensive studies are under way to better understand the role of photon-density waves and their use in phase modulation methods to obtain optical characteristics of tissues. They are expected to bring about novel algorithms for the reconstruction of three-dimensional tissue images and practical applications of diffuse optical tomography.

The polarization properties of tissues to which this tutorial is largely devoted are of primary importance for physiological polarization optics and early skin and epithelial cancer diagnostics. The scattering matrix technique has long been used in optics and is currently applied by many authors to investigate the properties of tissues and cell suspensions. The intensity matrix (Mueller matrix) is normally employed for this purpose, but the use of two-frequency lasers (e.g., a Zeeman laser)

591

or quadruple-channel OCT allows amplitude matrix elements to be measured, thus offering the possibility to simplify the inverse problem solution for many biological structures.

The designing of laser optoacoustic, acoustooptic, and optothermal imaging systems can be considered as a very prospective direction in biomedical optics that allows one to provide an impressive contrast and resolution in the imaging of both deep and superficial small tumors. It seems that these techniques should be optimal for the diagnosis of tissues at the middle depths (3 to 10 mm), where time-resolved diffusion and OCT methods are not effective.

Optical speckle techniques, in particular, the methods based on partially developed speckles emerging from the diffraction of focused beams, offer much promise for the investigation of the structure of tissues and the analysis of vibrations and motility (blood and lymph flow). The development of these techniques requires detailed research into the optics of speckles, speckle statistics, and the interference of light beams in dense scattering media.

Coherent optical methods are promising candidates for the development of new high-resolution and high-performance tomographic techniques allowing the imaging of subcellular structural and functional states of a tissue. Currently, investigation of superficial tissue layers with the use of optical coherent tomography (OCT) can provide important results and offer much promise for medical applications, especially for ophthalmology, early cancer detection of skin and the cervix, blood microcirculation analysis, and the endoscopic study of blood vessel wall and mucous of internal organs.

Methods based on the dynamic scattering of light are useful for the analysis of both weakly scattering and dense biological media. Many of these methods have already found their specific areas of biomedical applications. Considerable progress can be achieved with the use of the method of diffusion wave spectroscopy.

Heterodyne and homodyne interferometry, including speckle interferometry, has a wide range of biomedical applications. Laser Doppler blood flow and vibration measurements in tissues and organs are the most advanced field of applications for such techniques. The well-known advantages of confocal microscopy are very useful for the investigation of tissues and cells. Many coherent techniques and devices of medical diagnostics successfully employ the confocal principle of optical sectioning of an object under study. Two-photon and second-harmonic generation microscopies provide additional new possibilities in studying tissues and cells.

The optical immersion technique allows one to effectively control the optical properties of tissues and blood. Such control leads to an essential reduction of scattering and therefore causes much higher transmittance (optical clearing) and the appearance of a large amount of least-scattered (snake) and ballistic photons, allowing for the successful application of coherent-domain and polarization imaging techniques. It has great potentiality for noninvasive medical diagnostics using OCT due to the rather small thickness of tissue layers usually examined by OCT, which allows for fast impregnation of a target tissue at a topical application of an immersion liquid. It has been demonstrated that the body's interior tissues such as the blood vessel wall, esophagus, stomach, cervix, and colon can usually be imaged at a depth of about 1–2 mm. For more effective diagnosis using OCT, a higher penetration depth can be provided by the application of immersion substances. Evidently, reversible tissue optical clearing technology has valuable features to be applied not only to tissue spectroscopy and diagnostics, but also to a variety of laser and photothermal therapies and surgeries.

Glossary 1. Physics, Statistics, and Engineering

Abbe refractometer: an instrument for direct determination of the refractive index that uses the internal total reflection.

aberration: in optics this term is applied to certain defects in images formed in optical systems; aberration may be spherical, e.g., if the aperture of a concave spherical mirror (lens) is large, the rays at the periphery are brought to focus nearer the mirror than those rays which meet the mirror near the pole, giving a curve known as the caustic; the aberration in the case of lenses may also be **chromatic** when if the light contains more than one wavelength, e.g., white light, dispersion takes place, and since the lens is a series of small prisms the image will therefore be colored.

absolute (Kelvin) scale: a thermodynamic temperature scale in which the lower fixed point is absolute zero and the interval is identical with that on the Celsius scale.

absolute temperature: a temperature measured on an absolute scale.

absorbance: the ratio of the absorbed light intensity to the incident intensity; it is a dimensionless quantity.

absorbing medium: the medium that absorbs light at certain wavelengths or wavelength bands.

absorption: the transformation of light (radiant) energy to some other form of energy, usually heat, as the light transverses matter.

absorption band: a range of wavelengths for which a medium absorbs more strongly than at adjacent wavelengths.

absorption center: a particle or molecule that absorbs light.

absorption coefficient: in a nonscattering sample, the reciprocal of the distance d over which the light of intensity I is attenuated (due to absorption) to $I/e \approx 0.371$; the units are typically cm^{-1}.

absorption spectrum: the spectrum formed by light that has passed through a medium in which light of certain wavelengths was absorbed.

alternating current (ac): an electric current that reverses direction at regular intervals, having a magnitude that varies continuously in a sinusoidal manner.

acceptance angle: the maximum incident angle at which an optical fiber will transmit light by total internal reflection.

acoustic detector: a transducer that transforms an acoustic signal to an electrical one, such as a **microphone** or **piezoelectric transducer**; wideband piezoelectric transducers, being low-noise detectors, have proven to be most suitable for tissue spectroscopy and tomography.

acoustic (sound) waves: waves produced by a vibrating system that indexes the particles of the medium that vibrate in the same direction as the progress of the wave; successive points of high (compressions) and low (rarefactions) pressures are formed because of the regular impulses from the vibrating source; sound waves may be longitudinal progressive, longitudinal stationary, or transverse stationary; when sound is propagated through a medium (gas, liquid, or solid) it is transmitted by progressive longitudinal waves in that medium; sound waves interact with a complex and/or moving medium with their damping, refraction, diffraction, interference, and/or Doppler shift in frequency.

acoustooptical (AO) interaction (effect): the interaction of ultrasound and light beams within a homogeneous or inhomogeneous medium that occurs through a change in optical properties of the medium resulting from its compression by the ultrasound.

acoustooptical modulator: a device for intensity modulation of light at certain audio or radio frequencies using the acousto-optical effect.

acoustooptical tomography (AOT): optical tomography that is based on the acoustic [ultrasound (US)] modulation of coherent laser light traveling in tissue; an **acoustic wave** (AW) is focused into tissue and laser light is irradiating the same volume within the tissue; any light that is encoded by the ultrasound contributes to the imaging signal; axial resolution along the acoustic axis can be achieved with US-frequency sweeping and subsequent application of the Fourier transformation, whereas lateral resolution can be obtained by focusing the AW.

acquisition rate: the rate of acquiring experimental data.

acquisition time: the period of time of acquiring experimental data.

adaptive finite element method: a **finite element method**, where discretion of the underlying spatial domain (usually in two or three space dimensions) and establishment of discrete equations is done using an adaptive mesh refinement approach accounting for the complex structure of the object.

adding-doubling method: see **inverse adding-doubling (IAD) method**.

adiabatic expansion: a curve of pressure against volume, representing an adiabatic change (curves of equal **entropy**); these lines always slope more steeply than the isothermals they cross.

albedo: the ratio of scattering to the extinction cross section (or coefficient); ranges from zero for a completely absorbing medium to unity for a completely scattering medium.

algorithm: a procedure (a finite set of well-defined instructions) for accomplishing some task which, given an initial state, will terminate in a defined end-state; the computational complexity and efficient implementation of the algorithm are important in computing, and this depends on suitable data structures.

amplifier: an electronic (or optical) circuit that increases the voltage (or light power) of a signal fed into it by obtaining power from an external supply.

amplitude scattering matrix (S-matrix or Jones matrix): consists of four elements (the complex numbers) and provides a linear relationship between the incident and scattered field components; each element depends on scattering and azimuthal angles, and optical and geometrical parameters of the scatterer; both field amplitude and phase must be measured to quantify the amplitude scattering matrix.

analyzer: a **polarizer** that is placed in front of the detector.

anhydrous: having no water (for crystal, no water of crystallization).

angstrom (Å): a unit of length equal to 10^{-8} meter (m).

anisotropic crystal: a crystal for which some physical properties (mechanical, optical, magnetic, electrical, and etc.) are dependent on direction (not density and specific **heat capacity**); for example, for light propagation in transparent crystals (excluding crystals with cubic lattice) the light undertakes **birefringence** with mutually orthogonal polarization of rays propagating in different directions; for crystals with hexagonal, trigonal, and tetragonal structures (quartz, ruby, and calcite) birefringence is maximal along the direction that is perpendicular to the main axis of symmetry and is absent along this axis.

anisotropic scattering: a scattering process characterized by a clearly apparent direction of photons that may be due to the presence of large scatterers.

anode: a positive electrode that attracts anions (negative ions) during electrolysis.

antenna properties of the optical heterodyne system: the narrow field of view that provides a high spatial resolution for detection and ranging of the medium under study and its image formation.

anti-Stokes-Raman scattering: a photon interacts with a molecule in a higher vibrational level; the energy of the Raman scattered photons is higher than the energy of the incident photons.

aperture: the diameter of a circle through which light is allowed to pass; apertures may be varied and stated as fractions of the focal length, e.g., $f/6$ means that the diameter of the lens (aperture) is $1/6$ of the focal length f.

Ar (argon) laser: a laser with a lasing medium composed of ionized argon gas; the emission is mostly in the UV and visible ranges: (336.6–363.8) nm, 454.5 nm, 457.9 nm, 488 nm, 514 nm, and 528.7 nm.

arc lamp: a lamp that uses a luminous bridge formed in a gap between two conductors or terminals when they are separated.

artifact: (in microscopy) a structure that is seen in a tissue sample under a microscope but not in living tissue; may be caused by incorrect sample preparation or placement of the sample under the microscope.

A-scan: in-depth scanning perpendicular to the sample surface.

asymmetry parameter (skewness): a measure of asymmetry in a probability distribution function.

attenuated total reflectance Fourier transform infrared spectroscopy (ATR FTIR) method: based on a combination of the **total internal reflection** technique and Fourier transform infrared spectroscopy.

attenuation: a decrease in energy per unit area of a wave or beam of light: it occurs as the distance from the source increases and is caused by absorption or scattering.

attenuation (extinction) coefficient: the reciprocal of the distance over which light of intensity I is attenuated to $I/e \approx 0.37I$; the units are typically cm^{-1}.

autocorrelation: the correlation of an ordered series of observations with the same series in an altered order.

autocorrelation function: the characteristic of the second-order statistics of a random process that shows how fast the random value changes from point to point, e.g., the autocorrelation function of intensity fluctuations caused by scattering of a laser beam by a rough surface characterizes the size and the distribution of speckle sizes in the induced speckle pattern; the Fourier transform of the autocorrelation function represents the power spectrum of a random process.

autofluorescence: natural fluorescence of a tissue.

avalanche photodetector (APD): a type of silicon or germanium photodiode that uses a phenomenon of the electrical breakdown of a *p-n* junction, which induces collision ionization and the avalanche creation of electron-hole pairs; it increases the photocurrent up to 10^2 to 10^6.

back-projection algorithm: an approximate inverse solution to provide efficient tomographic imaging; in **x-ray** computed tomography the back-projection is reconstructed along the paths of x-ray propagation; in optical tomography the back-projection is reconstructed along the paths of **ballistic** and/or least scattering photon propagation; in **optoacoustical tomography** the back-projection is reconstructed along spherical shells that are centered at the detector and have a radius determined by the acoustic time of flight.

backscattering: the dispersion of a fraction of the incident radiation in a backward direction.

backscattering coefficient (volume-averaged): the sum of the particle cross sections weighted by their angular-scattering functions evaluated at 180 deg.

backscattering Mueller matrix: a **Mueller matrix** measured for light backscattered by a sample.

ballistic (coherent) photons: a group of unscattered and strictly straightforward scattered photons.

beam: a slender stream of light.

beat signal: the signal at the intermediate frequency produced by mixing the local and signal oscillator beams on a photodetector.

bend (bending): related to **infrared spectroscopy** that works because chemical bonds have specific frequencies at which they vibrate, corresponding to energy levels; simple diatomic molecules have only one bond, which may stretch; more complex molecules may have many bonds and vibrations can be conjugated, leading to infrared absorptions at characteristic frequencies that may be related to chemical groups; e.g., the nonlinear triatomic molecule in an H_2O atom can vibrate in three different ways: symmetrical and asymmetrical stretching and bending (scissoring); the atoms in a CH_2 group, commonly found in organic compounds, can vibrate in six different ways: symmetrical and asymmetrical stretching, bending (scissoring), rocking, wagging, and twisting.

biaxial crystal: an **anisotropic crystal** that is characterized by having two optic axes, i.e., two axes along which an incoming beam will remain collinear, and the light of both polarizations will propagate at the same speed; mica is an example.

bimodal distribution: a distribution that has two modes.

bioheat equation: describes the change in tissue temperature at a definite point in the tissue; it is defined by the thermal conductivity of tissue, the rate of metabolic heat generation, and the heat transfer caused by blood perfusion at this point.

birefringence: the phenomenon exhibited by certain crystals in which an incident ray of light is split into two rays, called an ordinary ray and an extraordinary ray, which are plane- (linear) polarized in mutually orthogonal planes.

BK7: bor-crown optical glass that is relatively hard and shows a good scratch resistance; very commonly used for manufacturing of high-quality optical components; it has the high linear optical transmission in the visible range down to 350 nm with a refractive index of $n_d = 1.51680$ (587.6 nm).

blocking filter: a filter that blocks the passage of a certain frequency (wavelength) band.

Boltzmann constant: the ratio of the universal gas constant to Avogadro's number, equal to 1.3803×10^{-16} erg per degree Celsius.

Born's approximation: the single-scattering approximation when the field affecting the particle does not essentially differ from that of the initial wave.

Bouguer-Beer-Lambert law: a exponential law describing attenuation of a collimated beam by a thin absorption layer with scattering.

boundary conditions: a stated restriction, usually in the form of an equation, that limits the possible solutions to a differential equation.

brightness: an attribute of visual perception in which a source appears to emit a given amount of light; in other words, brightness is the perception elicited by the luminance of a visual target; this is a subjective attribute/property of an object being observed.

Brownian motion: the irregular motion of particles suspended in a medium caused by the molecules of the medium bombarding these particles.

Brownian particles: small particles suspended in a liquid or a gas that undergo irregular motion caused by molecules of the medium bombarding the particles.

calcium fluoride (CaF_2): an insoluble ionic compound of calcium and fluorine; it occurs naturally as the mineral fluorite (also called fluorspar).

carbon nanoparticle: particles made of carbon with dimensions on the order of billionths of a meter; they have different physical, chemical, electrical, and optical properties than occur in bulk samples due in part to the increased surface-area-to-volume ratio at the nanoscale; carbon black is a form of amorphous carbon that has an extremely high surface-area-to-volume ratio and, as such, is one of the first nanomaterials to find common use; little is known about the interaction of carbon nanoparticles with human cells.

cathode: a negative electrode that attracts cations (positive ions) during electrolysis.

cavity-dumped mode-locked laser: a laser that produces high-energy ultrashort laser pulses by decreasing the pulse repetition rate (see mode-locked laser); the laser output mirror is replaced by an optical selector consisting of a couple of spherical mirrors and an acousto- or electrooptical deflector, which extracts a pulse from the cavity after it has passed over a few dozen cavity lengths; the pulse energy is accumulated between two sequential extractions: the pulse repetition rate can be tuned in the range from dozens of hertz to a few megahertz.

charge-coupled device (CCD): a solid-state electronic device that serves as an imaging chip and is used in video cameras and fast spectrometers.

Celsius temperature scale: a temperature scale for which the freezing point is at $0°$ and the steam point is at $100°$.

characteristic diffusion time: the reciprocal of an agent's diffusion coefficient diffused in a tissue; characterized by the time interval during which the applied agent concentration increases in the sample of ≈ 0.6 in the surrounding medium.

CH group: related to **infrared spectroscopy**, chemical bonds vibrate at the specific frequencies that correspond to energy levels; simple diatomic molecules have only one bond that may stretch; a CH group provides a stretching vibration mode.

chirality: the mirror-equal "right" or "left" modification of an object **optical activity** is one example of chirality, when the asymmetric structure of a molecule or crystal existing of two forms ("right" and "left") causes the substance (ensemble of these molecules or crystal) to rotate the plane of polarization of the incident linear polarized light: the pure "right" or "left" optically active substances have identical physical and chemical properties, but their biochemical and physiological properties can be quite different.

chromatic aberrations: the variation of either the focal length or the magnification of a lens system, with different wavelengths of light, characterized by prismatic coloring at the edges of the optical image and color distortion within it.

chromaticity coordinates: in the study of the perception of color, one of the first mathematically defined color spaces was the CIE XYZ color space (also known as CIE 1931 color space), created by the International Commission on Illumination (Commission Internationale de l'Éclairage—CIE) in 1931; the human eye has receptors for short (S), middle (M), and long (L) wavelengths, also known as blue, green, and red receptors; that means that one, in principle, needs three parameters to describe a color sensation; a specific method for associating three numbers (or tristimulus values) with each color is called a color space: the CIE XYZ color space is one of many such spaces; however, the CIE XYZ color space is special, because it is based on direct measurements of the human eye and serves as the basis from which many other color spaces are defined; the CIE1964 standard observer is based on the mean 10-deg color matching functions.

chromophore: a chemical that absorbs light with a characteristic spectral pattern.

clock frequency: in electronics and especially synchronous digital circuits, a clock signal is used to coordinate the actions of two or more circuits; for example, the clock rate (frequency) of a computer CPU is normally determined by the frequency of an oscillator crystal.

CMOS image sensors: an image sensor based on complementary metal-oxide-semiconductor (CMOS) camera that operates at lower voltages than a **CCD** camera, reducing power consumption for portable applications; analog and digital processing functions can be integrated readily onto the CMOS chip, reducing system package size and overall cost; each CMOS active pixel sensor cell has its own buffer amplifier and can be addressed and read individually—a commonly used cell has four transistors and a photoelement; all pixels on a column connect to a common amplifier.

carbon monoxide (CO) laser: a laser in which the lasing medium is CO gas with an IR emission from 5 to 6.5 µm.

carbon dioxide (CO$_2$) laser: a laser in which the lasing medium is CO$_2$ gas with an IR emission from 9.2 to 11.1 µm with the maximal efficiency at 10.6 µm.

coherence length: characterizes the degree of temporal coherence of a light source; $l_c = c\tau_c$, where c is the light speed and τ_c is the coherence time, which is approximately equal to the pulse duration of the pulsed light source or inversely proportional to the frequency bandwidth of a continuous-wave light source.

coherent detection imaging (CDI): a method based on the optical heterodyne detection technique: it is a coherence gating method that discriminates the forward multiply-scattering beam, preserving the direct geometrical correlation with the incident light beam against the diffuse component of transmitted light.

coherent light: light in which the electromagnetic waves maintain a fixed phase relationship over a period of time and in which the phase relationship remains constant for various points that are perpendicular to the direction of propagation.

collapse: catastrophic compression.

collimated beam: a beam of light in which all rays are parallel to each other.

collisional quenching rate: nonradiative relaxation of the excited molecular energy state (level) by collisions with surrounding molecules; in a condensed medium, the rate of collisional relaxation is significantly higher than the radiative one.

complex conjugation: the number corresponding to a given complex number that represents the given number's reflection with respect to the real axis.

confocal microscopy: microscopy that employs the confocal principle [two optically conjugate diaphragms (pinholes) or small-sized slits in the object and image planes] for the selection of scattered photons coming from a given volume; provides 3D imaging of living tissues.

contrast: is a distinction between two objects (or parts of an object) or colors; a large contrast is a big difference, and contrasting objects are boldly different; can refer to contrast (visibility) of regular **interference** or **speckle** patterns or the difference in color and light **intensity** between parts of an image (see **imaging contrast**).

contrast of the intensity fluctuations: the relative difference between light and dark areas of a speckle pattern.

contrasting agents: compounds used to increase the quality of x-ray, MRI, or optical tomography images.

constructive interference: the interference of two or more waves of equal frequency and phase, resulting in their mutual reinforcement and producing a single amplitude equal to the sum of the amplitudes of the individual waves.

cooled CCD: a highly sensitive CCD with the thermal noise suppressed by cooling the photosensitive chip.

correlation: the degree of co-relation between two or more attributes or measurements on the same group of elements; in probability theory and statistics, correlation indicates the strength and direction of a linear relationship between two random variables; in general statistical usage, correlation refers to the departure of two variables from independence, although correlation does not imply causation.

correlation coefficient: a number of different coefficients are used to characterize **correlation** between two random variables for different situations; the most well known is the Pearson product-moment correlation coefficient, which is obtained by dividing the covariance of the two variables by the product of their standard deviations.

correlation diffusion equation: describes the transport of temporal field correlation function in a system that multiply scatters laser radiation; may be valid for turbid samples with the dynamics of scattering particles governed by Brownian motion, and random and shear flow.

correlation length: the length within which the degree of correlation between two measurements of a spatially dependent quantity is high (close to unity); for example, L_c is the correlation length of the scattering surface of the spatial inhomogeneities (random relief).

coupler: an optical or acoustical device that interconnects optical or acoustical components with less loss of energy.

coupling gel: a gel that provides matching optical or acoustical properties from different elements of a device that minimizes the amplitude of light or acoustic reflecting signals (or both) from boundaries of the elements.

cross-correlation frequency (cross-correlation signal): a difference in the frequency of intensity modulated light at a certain wavelength and photodetector gain modulation; it carries the same phase and amplitude information as the original optical signal.

cross-correlation measurement device: a system that down converts a radio frequency prior to phase measurements.

crosstalk: the interrelations between measured signals induced by originally independent parameters (for example, by changes in blood volume and oxygenation); this factor is determined by calibration on a model (for example, a blood model).

continuous wave (CW): waves that are not intermittent or broken up into damped wave trains but, unless intentionally interrupted, follow one another without any interval of time between them.

CW laser: a laser producing CW waves.

CW RTT: the stationary radiation transfer theory that describes the intensity distribution of CW light in a scattering medium; it is based on the stationary integro-differential equation for the radiance- (or specific intensity) average power flux density at a point r in a given direction s (see **radiation transfer theory**).

dark-field illumination: used in dark-field microscopy for imaging of optically transparent nonabsorbing specimens, where illuminating light does not enter into ocular lenses and only light scattered by microparticles of the specimen creates the image; in the field of view of the microscope on the dark background, bright images of the specimen particles differing by their refractive index from the surrounding medium are seen.

decibel (dB): the engineering unit for the ratio of the input power, P_{in}, in a given device to the output power, P_{out}; it is convenient to measure the logarithm of the ratio $\log(P_{out}/P_{in})$, and the dB is a standard unit that is equal to 10 times that log: $10\log(P_{out}/P_{in})$ dB.

decorrelation of speckles: relates to statistics of the second order that characterize the size and distribution of speckle sizes and show how fast the intensity changes from point to point in the speckle pattern; decorrelation means that such changes of intensity tend to be faster.

deep RPS: a random-phase screen that induces phase fluctuations in a scattered field with a variance that is much more than unity.

deflectometry: a photorefractive technique based on the detection of refractive index gradients above and inside the sample using a laser probe beam.

deformation: in biomechanics, deformation is a change in shape due to an applied force; this can be a result of tensile (pulling) forces, compressive (pushing) forces, shear, bending, or torsion (twisting); deformation is often described in terms of strain.

0-degree hybrid (or splitter): a device that pertains to or denotes a current in one of two parallel circuits that have a single-phase current source and equal impedances and that produces currents of 0-degree phase shift.

90-degree hybrid (or splitter): a device that pertains to or denotes a current in one of two parallel circuits that have a single-phase current source but unequal impedances and that produces currents of 90-degree phase shift.

degree of polarization: the quantity that characterizes the ratio of the intensity of polarized light to the total intensity of light.

delta (δ)-Eddington approximation: a simple yet accurate method that was proposed for determining monochromatic radiative fluxes in an absorbing-scattering atmosphere; in this method, the governing phase function is approximated by a Dirac-delta function forward-scatter peak and a two-term expansion of the phase function; the fraction of scattering into the truncated forward peak is taken proportional to the square of the phase-function-asymmetry factor, which distinguishes the delta-Eddington approximation from others of a similar nature (http://adsabs.harvard.edu/abs/1976JAtS...33.2452J); one of the approximations of the actual phase function for tissue; in the diffusion approximation of RTT, it is the best function for simulating light transport in tissues characterized by anisotropic scattering.

demodulation: the separation and extraction of modulating low-frequency waves from a modulated carrier wave (high frequency or optical); the device or circuit used for demodulation is called a detector or demodulator.

depth of modulation: for amplitude modulation, this is the ratio of the amplitude of the alternating component of a signal to its mean value.

depolarization: deprivation (destruction) of light polarization.

depolarization length: the length of light beam transport in a scattering-depolarizing medium at which the polarization degree decays to the definite level compared with the totally polarized incident light.

destructive interference: the interference of two waves of equal frequency and opposite phase, resulting in their cancellation where the negative displacement of one always coincides with the positive displacement of the other.

developed speckles: the speckles that are characterized by Gaussian statistics of the complex amplitude, the unity **contrast of intensity fluctuations**, and a negative exponential function of the intensity probability distribution (the most probable intensity value in the corresponding speckle pattern is equal to zero, i.e., destructive interference occurs with the highest probability).

dichroism (diattenuation): a phenomenon related to **pleochroism** of a uniaxial crystal so that it exhibits two different colors when viewed from two different directions under transmitted light; pleochroism is the property possessed by certain crystals that exhibit different colors when viewed from different directions under transmitted light: this is one exhibition of the optical anisotropy caused by the anisotropy of absorption; the varieties of pleochroism are **circular dichroism**, different absorption for light with right and left circular polarization, and **linear dichroism**, different absorption for ordinary and extraordinary rays.

diffraction: a phenomenon associated with a wave motion when a wave train (optical, acoustical, thermal, **photon density**, etc.) passes the edge of an obstacle opaque to the wave motion; the phenomenon is a particular case of interference; the waves are bent at the edge of the obstacle, which acts as a source of secondary waves, all coherent; the interference between a primary wave and a secondary wave produces diffraction bands, which are interference bands.

diffraction of photon density wave (intensity wave): the bending of photon density waves around obstacles in their path; the phenomenon exhibited by wavefronts that, passing the edge of an opaque body, are modulated, thereby causing a redistribution of photon-density-wave amplitude within the front: it is detectable by the presence of minute bands with high and low amplitudes at the edge of a shadow; the phenomenon is a particular case of interference between primary and secondary photon density waves.

diffractometry: measuring techniques based on the phenomenon of wave diffraction.

diffuse photons: the photons that undertake multiple scatter with a broad variety of angles.

diffuse tomography: optical tomography based on reconstruction of the optical macro-inhomogeneity within a scattering medium using diffuse photon pathlength-gating techniques and **back-projection algorithms**.

diffusion: the process by which one gas mixes with another by the movement of the molecules of one gas into another and vice versa; diffusion also occurs when two miscible liquids or solids come in contact with a solvent; the term is also used to describe the passage of molecules through a porous membrane.

diffusion approximation (diffusion theory): the approximated diffusion-type solution of the **RTT**, which is accurate for describing photon migration in infinite, homogeneous, highly scattering media.

diffusion coefficient: the proportionality coefficient between mean-square displacement of a particle within time interval τ: $\langle \Delta r^2 \rangle \sim D\tau$; may be related to molecular or photon diffusion.

diffusion wave spectroscopy (DWS): spectroscopy based on the study of dynamic light scattering in dense media with multiple scattering and related to the investigation of particle dynamics within very short time intervals.

digital electronic autocorrelator: a device that reconstructs with a high accuracy the time-domain autocorrelation function of intensity fluctuations.

digital oscilloscope: a device that analyses with a high accuracy the waveform of **ac** signals.

diode laser: a semiconductor injection laser; **GaAs laser** (830 nm); $GaP_x As_{1-x}$ lasers emit light from 640 nm ($x = 0.4$) to 830 nm ($x = 0$); $Ga_x In_{1-x} As_y P_{1-y}$, lasers, at $y = 2.2x$ and for different values of x, emit in the range from 920 to 1500 nm; $Pb_x S_{1-x}$, $Sn_x Pb_{1-x} Te$, and $Sn_x Pb_{1-x} Se$ lasers, for different values of x, emit in the range from 2.5 to 49 μm.

diode-pumped Nd:YAG: an integrated solid-state laser with an Nd:YAG crystal as a lasing medium and optical pumping provided by a single laser diode or by a laser (light) diode array or matrix.

dipole moment of transition: the mutual displacement and charges of a two-charged particle system (model of molecule); defines the electrical field of the electrically neutral system on distances larger than its size and the action of external fields on the system; when the dipole moment changes, the system emits electromagnetic waves.

dispersion: the state of being dispersed, such as a photon trajectory (general); the variation of the index of refraction of a transparent substance, such as a glass, with the wavelength of light, the index of refraction increases as the wavelength decreases (optics); the separation of white or compound light into its respective colors, as in the formation of a spectrum by a prism (optics); the scattering of values of a variable around the mean or median of a distribution (statistics); a system of dispersed particles suspended in a solid, liquid, or gas (chemistry).

dissector: a transmitting television tube; it can be used as a coordinate-sensitive photodetector.

distribution size function: a function that describes the probability distribution of a particle size value over the size values in the system.

divergence: the "spreading" of a light beam in general, and of a laser beam as it moves away from the laser in particular.

Doppler effect: the apparent change in the frequency of a wave, such as a light wave or sound wave, resulting from a change in the distance between the source of the wave and the receiver.

Doppler interferometry: the dynamic dual-beam interferometry when the reference beam path length is scanned with a constant speed; the Doppler signal induced is the measuring signal for depth profiling of an object placed in the measuring beam; the method is used in partially coherent interferometry or tomography of tissues.

Doppler microscopy: **Doppler spectroscopy** of a medium at a microscopic scale.

Doppler spectroscopy: the spectroscopy based on the study of dynamic light scattering (**Doppler effect**) in media with single scattering and related to investigation of the dynamics (velocity) of particles from the measurements of the Doppler shifts in the frequency of the waves scattered by the moving particles.

double-balanced mixer: an electronic device that mixes two optically detected signals that have the same radio frequency but different amplitudes and phases.

double integrating sphere (DIS) technique: a technique for *in vitro* evaluation of the optical parameters of tissue samples (μ_a, μ_s, and g); it is often combined with collimated transmittance measurements; it implies either sequential or simultaneous measurement of three parameters: total transmittance T_t (using the **integrating sphere**), diffuse reflectance R_d (using the integrating sphere), and collimated transmittance T_c (using a distant detector behind the pinhole at the top of the integrating sphere).

dual-beam coherent interferometry: see **Doppler interferometry**.

dye laser: a laser in which the laser medium is a liquid dye; dye lasers emit in a broad spectral range (e.g., in the visible) and are tunable; wavelengths range from 340 to 960 nm, optical frequency doubling ranges from 217 to 380 nm, and parametric conversion ranges from 1060 to 3100 nm; emitted energy is from 1 mJ to 50 J in periodic pulse mode; mean power is from 0.06 to 20 W; and pulse duration is from 0.007 to 8 μs; pulse frequency from a single pulse to 1 kHz.

dynamic light scattering: light scattering by a moving object that causes a **Doppler shift** in the frequency of the scattered wave relative to the frequency of the incident light.

dynode chain of the PMT: a system of electrodes, each of which serves for the emission of secondary electrons in a vacuum tube.

elastic (static) light scattering: light scattering by static (motionless) objects that occurs elastically, without changes in photon energy or light frequency.

electromagnetic resonance: appears at interaction of the incident radiation with molecules attached to a rough metallic surface; is induced due to collective excitation of conduction electrons in the small metallic structures; also called surface plasmon resonance; **surface-enhanced Raman scattering (SERS)** is based on such electromagnetic effects.

electronic micrograph: micrographs of tissue and/or cell components received with the help of the **electronic microscope**.

electronic microscope: a parallel beam of electrons from an electron gun is passed through a very thin slice of tissue; differential scattering of the electron beam takes place, and an image of tissue microstructure is carried forward in the electron beam; an electron lens is used to focus the electron beam on a fluorescent screen, where a magnified image is formed; the image is registered using an optical camera; the resolving power of the electron microscope is very much greater than that of a light microscope.

electronic transition: if an electron in an atom is activated (given more energy) the electron moves to an energy level farther from the atom nucleus; if an electron moves back to a lower level, energy is given out as electromagnetic radiation.

electronic wave function: the magnitude of the wave function (ψ) represents the varying amplitude of the stationary wave system, in 3D, of an electron situated around a nucleus; associated with the stationary wave is a frequency, ν; ψ^2 is the density of electrons per unit volume; $\psi^2 d\mathrm{V}$ is the probability of finding the electron, when it is considered as a particle, in a volume $d\mathrm{V}$; the total volume of the orbital gives a probability of unity; the effective electrical charge associated with a volume dV is $-e\psi^2 dV$, where e is the charge of an electron; the four quantum numbers define possible states of the stationary waves.

electrophoresis: the movement of colloidal particles in an electric field; when two platinum electrodes connected to a dc supply are placed in a lyophobic sol (the disperse phase has no attraction for the continuous phase), the colloidal particles will move to either the cathode or anode depending on the charge on the particle; used for drug delivery in medicine.

emission spectrum: the emission obtained from a luminescent material at different wavelengths when it is excited by a narrow range of shorter wavelengths.

endoscopy: an optical technique and instrumentation for viewing internal organs.

energy: the product of power (watt, W) and time (sec, s); energy is measured in joule (J).

entropy: a measure of the amount of disorder in a system; the more disordered the system, the higher the entropy; an entropy change occurs when a system absorbs or emits heat; the change in entropy dS is measured as the heat change dQ divided by the temperature T at which the change takes place, $dS = dQ/T$.

erbium: yttrium aluminum garnet (Er:YAG) laser: a solid-state laser whose lasing medium is the crystal Er:YAG crystal with an emission in the mid-IR range of 2.79–2.94 μm.

evaporation: the process of changing a liquid into a vapor, usually by applying heat, or by the liquid taking heat from its surroundings; during this process the bulk of the liquid is reduced.

excitation spectrum: the emission spectrum at one wavelength is monitored, and the intensity at this wavelength is measured as a function of the exciting wavelength.

excimer laser: a laser whose lasing medium is an excited molecular complex, an excimer (molecule-dimer); the emission is in the UV; examples are: ArF laser, 193 nm; KrF laser, 248 nm; XeCl laser, 308 nm; and XeF laser, 351 nm.

excited state (energy level): electrons possess energy according to their position in relation to the nucleus of an atom; the closer the electron is to the nucleus the lower the energy; when the energy of an electron changes, it must do so in certain definite steps and not in a continuous way; the position in which electrons may be found according to their energy are called energy levels and sublevels; these levels are counted by their steps outward, and the numbers allotted to them are their quantum numbers.

extinction coefficient: see **attenuation coefficient**.

Fabry-Perot interferometer: the **interferometer** combined of two parallel mirrors (reflecting planes) displace each other by a distance of L (interferometer length); used as a precise optical filter in the super-resolution spectroscopy and as a cavity in lasers.

false color map: a map where each color is distributed through the map and the specific value of the measured parameter is prescribed, for example, the measured velocity of blood flow within the selected skin area; it is used for the fast qualitative estimation of parameter distribution and change.

far-field diffraction zone (far zone): the zone where Fraunhofer diffraction takes place; this is a type of diffraction in which the light source and the receiving screen are effectively at an infinite distance from the diffraction object, i.e., parallel beams of wave trains are used.

Faraday rotator: an optical device that rotates the polarization of light due to the Faraday effect, which in turn is based on a magneto-optic effect; it works because one polarization of the input light is in ferromagnetic resonance with the material that causes its phase velocity to be higher than the other.

fast Fourier transform (FFT) analysis: a fast algorithm for the expression of any periodic function as a sum of sine and cosine functions, as in an electromagnetic wave function.

F/D spectrometer: the spectrometer that uses the frequency-domain (photon-density wave) method for measuring the absorption and scattering spectra of an object (tissue).

femtosecond (fsec) (fs): -10^{-15} sec (s).

fiber: an optical waveguide that uses a phenomenon of total internal reflection for light transportation with low losses and is made from transparent glass, quartz, polymer, or crystal, usually with a circular cross-section; it consists of at least two parts, an inner part or **core** that has a higher refractive index and through which light propagates, and an outer part or **cladding** that has a lower refractive index and provides a totally reflecting interface between core and cladding.

fiber bundle: a flexible bundle of individual optical fibers arranged in an ordered or disordered manner and correspondingly named regular and irregular bundles.

fiber coupler: a fiber optical device that interconnects optical components.

fiber-optic catheter: a flexible single fiber or a fiber bundle used to move light into body cavities and back.

fiber-optic device: any type of device that uses fiber-optical components.

fiber-optic refractometer: a fiber-optic device used to measure the refractive index of a medium (tissue or biological liquid); such a device is usually used to explore the effect of disruption of the total internal reflection and is a robust instrument well suited for biomedical applications.

fiber-optic single-mode x-coupler: a fiber coupler that is made from a single-mode fiber and provides connections between four optical components; it is usually used as a key part of the integrated Michelson interferometer when it connects to a light source, a reference mirror, the reflecting surface under study, and a photodetector.

field of view: the extent of an object that can be imaged or seen through an optical system.

finite-difference method: in numerical analysis, finite differences play an important role they are one of the simplest ways to approximate a differential operator and are extensively used in solving differential equations.

finite-difference time-domain (FDTD): a numerical solution applied to a finite difference in space and time; numerical equivalent of the physical reality under investigation; for example, the solution of Maxwell's equations describing light scattering by a cell.

finite element method: a numerical method for finding an approximate solution to partial differential equations (PDEs) as well as integral equations, such as the **heat** or **radiation transfer equations**; the solution approach is based either on eliminating the differential equation completely (steady state problems), or rendering the PDE into an equivalent ordinary differential equation, which is then solved using standard techniques such as the **finite-difference method**, etc.

flow cytometry: a technique for counting, examining, and sorting microscopic particles (cells) suspended in a stream of fluid; it allows simultaneous multiparametric analysis of the physical and/or chemical characteristics of single cells flowing through an optical and/or electronic detection apparatus.

flowmeter: a device for measuring parameters of a flow, such as flow velocity; for instance blood flow velocity.

fluence rate (total radiant energy fluence rate): the sum of the **radiance** over all angles at a point $\bar{r}$; the quantity that is typically measured in irradiated tissues in units of watts per square centimeter.

fluorescence: the property of emitting light of a longer wavelength on absorption of light energy; essentially occurs simultaneously with the excitation of a sample.

fluorescence anisotropy: transition dipole moments have defined orientations within a molecule; upon excitation with linear polarized light, one preferentially excites those molecules, whose transition dipoles are parallel to the electric field vector of incident light; this selective excitation of an oriented population of molecules results in partially polarized fluorescence, which is described by fluorescence anisotropy.

fluorescence emission spectrum: a fluorescence spectrum measured at a certain excitation wavelength.

fluorescence excitation-emission map: a map presenting the combined data of **fluorescence emission** and **excitation spectra**, where excitation and emission wavelengths are presented on $x-y$ axes with corresponding fluorescence intensity values represented by the isometric lines on the map.

fluorescence excitation spectrum: the intensity of fluorescence measured at a certain emission wavelength as a function of the excitation wavelength.

fluorescence tomography: the **tomography** based on the detection of **fluorescence** signal fluorometer.

fluorophore: a **chromophore** that emits light with a characteristic spectral pattern at its excitation by a proper wavelength.

focal depth: every lens has a range of object positions that give an apparently focused image on a fixed screen; this range is called the depth of focus of the lens.

focal plane: a focusing plane that is perpendicular to the principle axis and also passes through the principle focus; rays parallel to each other, but at an angle to the principle axis, are brought to a focus in the focal plane.

focal spot: the spot obtained at the focus of a lens; the size of the spot depends on the lens and the wavelength, but its diameter is never smaller than the wavelength of light.

form birefringence: birefringence that is caused by the structure of a medium; for example, a system of long dielectric cylinders made from an isotropic substance and arranged in a parallel fashion shows birefringence of form.

forward scattering problem: the modeling of light propagation in a scattering medium by taking into account the experimental geometry, source, and detector characteristics and the known optical properties of a sample, and predicting the measurements and associated accuracies that result.

Fourier optical microscope: a microscope based on the principle that optical density spatial variations in the object plane of the microscope are converted by a **Fourier transform** into spatial frequency variations in the Fourier transform plane in the rear focal plane of the lens; if the optical density changes slowly across the object, the Fourier transform places most of the scattered light near zero angles (low spatial frequency) in the Fourier transform plane (a good model of a cell with clear cytoplasm); if the optical density changes rapidly across the object, the Fourier transform moves more of the energy to larger scattering angles (higher spatial frequency) in the Fourier transform plane (a good model of a cell with highly granular cytoplasm).

Fourier transform: an algorithm for the expression of any periodic function as a sum of sine and cosine functions.

Fourier transform infrared spectroscopy: spectroscopy based on light dispersion by using a Michelson interferometer with a tuned path length difference; in the IR it may provide a 10^2–10^3 higher signal-to-noise ratio, than a grating spectrometer.

fractal dimension: as the complexity of the object structure increases, its fractal dimension increases and is always higher than the topological dimension of the structure; for example, any structure described as a curve (tissue fiber) has a topological dimension of $D = 1$; however, if we make this curve more complex by bending it infinite times, its fractal dimension be equal to two when this curve will densely covers a finite area, or even to three when this curve will "pack" a cube.

fractal object: an object with a self-similar geometry, i.e., each arbitrarily selected part of it is similar to the whole object; **fractal dimension** of the object (structure) is always higher than its topological dimension.

frame grabber: an electronic device that provides video data acquisition and conversion to a digital form.

Franck-Condon principle: an **electronic transition** so fast that the vibrating molecule does not noticeably change its internuclear distance.

Fraunhofer diffraction: a type of diffraction in which the light source and the receiving screen are effectively at an infinite distance from the diffraction object, i.e., parallel beams of wave trains are used.

Fraunhofer diffraction approximation: a description of forward-direction scattering caused by large particles (on the order of 10 μm).

Fraunhofer zone: the zone where Fraunhofer diffraction takes place.

free diffusion: the **diffusion** process that occurs when molecules diffuse in the space free of any membranes and other barriers that hinder diffusion.

frequency-domain technique (method): a spectroscopic or imaging technique (method) that exploits an intensity-modulated light and narrow-band heterodyne detection.

Fresnel diffraction: a type of diffraction in which the light source and the receiving screen are both at a finite distance from the diffraction object, i.e., divergent and convergent beams of wave trains are used.

Fresnel reflection: the reflection of a beam of radiation, such as light, which takes place at the interface between two media of different refractive indexes; not all the radiation is reflected, some may be refracted.

Fresnel zone: the zone where Fresnel diffraction takes place.

GaAs laser: a laser based on the semiconductor material GaAs; the emission is in the NIR, at about 830 nm.

gas-microphone method: relates to opto-acoustic spectroscopy, when an object under study is surrounded by a gas (or combination of gases) that serves as an acoustic coupler between the object and an acoustic receiver such as a microphone; the spectroscopy of the surrounding gas, when an object's optical and acoustical properties are known or fixed, the spectroscopy of the surrounding gas can also be determined.

Gaussian correlation function: the correlation function described by a bell-shaped (Gaussian) curve.

Gaussian light beam: a light beam with a Gaussian shape for the transverse intensity profile; if the intensity at the center of the beam is I_o, then the formula for a Gaussian beam is $I = I_o \exp(-2r^2/w^2)$, where r is the radial distance from the axis and w is the beam "waist"; the intensity profile of such a beam is said to be bell shaped; a laser beam is a Gaussian one; a single-mode fiber also creates a Gaussian beam at its output.

Gaussian size distribution: **distribution size function** of a Gaussian shape.

Gaussian statistics (normal statistics): statistics when a bell-shaped (Gaussian) curve showing a distribution of probability associated with different values of a variate are valid.

Gegenbauer kernel phase function (GK): one of the approximations of the actual phase function for tissue; the **Henyey-Greenstein phase function** is a special case of the **GK**; **GK** is a good function for simulating light transport in a tissue characterized by a high scattering anisotropy, such as blood.

genetic inverse algorithm: genetic algorithms (GAs) are now widely applied in science and engineering as adaptive algorithms for solving practical problems; certain classes of problems are particularly suited for a GA based approach; the general acceptance is that GAs are particularly suited to multidimensional global search problems where the search space potentially contains multiple local minima; unlike other search methods, correlation between the search variables is not generally a problem; the basic GA does not require extensive knowledge of the search space, such as likely solution bounds or functional derivatives (http://gaul.sourceforge.net/intro.html).

Gladstone and Dale law: states that the mean value of the refractive index of a composition represents an average of the refractive indices of its components related to their volume fractions.

Glan-Taylor polarization prism: a type of prism which used as a polarizer or polarizing beamsplitter; the prism is made of two right-angled prisms of calcite (or other birefringent materials), which are separated on their long faces with an air gap; the optical axes of the calcite crystals are aligned parallel to the plane of reflection; total internal reflection of s-polarized light at the air gap ensures that only p-polarized light is transmitted by the device; because the angle of incidence at the gap can be reasonably close to Brewster's angle, unwanted reflection of p-polarized light is reduced.

gold nanoparticles: particles made of gold with dimensions typically of 10–50 nm; they have different chemical and optical properties than those that occur in bulk samples; due to the plasmon-resonant property, high surface reactivity, and their biocompatibility, gold nanoparticles can be used for *in vivo* molecular imaging and therapeutic applications, including optical detection of cancer and phototherapy.

goniophotometry (goniophotometric technique): the technique that measures of the angle-dependent light intensity distribution.

gradient index (GRIN) lens: focuses light through a precisely controlled radial variation of the lens material's index of refraction from the optical axis to the edge of the lens; this allows a GRIN lens with flat or angle polished surfaces to collimate light emitted from an optical fiber or to focus an incident beam into an optical fiber; end faces can be provided with an anti-reflection coating to avoid unwanted back reflection.

grating spectrograph: a spectrograph that uses diffraction grating to produce optical spectra; a diffraction grating is a band of equidistant, parallel lines, usually more than 5000 to the inch, ruled on a glass or polished metal surface for diffracting light to produce optical spectra with a high resolution.

group refractive index: the refractive index associated with the group velocity of a train of waves traveling in a dispersive medium; the group velocity, and correspondingly the group refractive index, depends on the mean wavelength of a train of waves and the rate of change in velocity with wavelength.

Grüneisen parameter: a dimensionless, temperature-dependent factor proportional to the fraction of thermal energy converted into mechanical stress.

halogen lamp: an iodine-cycle tungsten incandescent lamp that is the visible/near infrared (360 nm to >1 μm) light source for spectrophotometry.

hard sphere approximation: the model of mutually impenetrable (hard) spheres; the interparticle forces are zero, except for the fact that two neighboring particles cannot interpenetrate each other.

heat capacity: a measurable physical quantity that characterizes the ability of a body to store heat as it changes in temperature; defined as the rate of change of temperature as heat is added to a body at the given conditions and state of the body (foremost its temperature); expressed in units of joules per Kelvin.

heat (thermal) conduction: the process of heat transfer through a body without visible motion of any part of the body; the process takes place where there is a temperature gradient; heat energy diffuses through the body by the action of particles of high kinetic energy on particles of lower kinetic energy; for solids with covalent bonding, there will be molecules for which motion is restricted to vibrations about fixed positions, and the energy is transferred by high frequency waves.

heat transfer: a variety of processes that provide transfer of heat through a body and its surroundings, such as **heat (thermal) conduction**, heat convection, thermal radiation and absorption, latent heat of fusion, and vaporization.

helix: a connected series of concentric rings of the same radius, joined together to form a cylindrical shape.

hemoglobin spectrum: the main bands are the **Soret band**, the 400–440-nm segment, and the **Q band**, the 540–580-nm segment.

He-Ne (helium neon) laser: a gas laser whose medium is a mixture of He and Ne; lasers with the red emission (632.8 nm) are widely used; lasers with other wavelengths are also available: green (543 nm), yellow (594 nm), orange (604 and/or 612 nm), IR (1152, 1523, and/or 3391 nm).

Henyey-Greenstein phase function (HG): one of the practical semiempirical approximations of the scattering phase function.

hertz (Hz): a unit of frequency that is equal to 1 cycle per second; it is often used to indicate the pulse repetition rate of a laser (e.g., a 10-Hz laser emits 10 pulses per second); 1 kilohertz (kHz) is 10^3 Hz, 1 megahertz (MHz) is 10^6 Hz, 1 gigahertz (GHz) is 10^9 Hz, and 1 terahertz (THz) is 10^{12} Hz.

heterodyne microscopy: a microscopy technique that uses optical heterodyning to enhance the registered signal and **image contrast**.

heterodyne phase system: see **"cross-correlation" measurement device**.

heterodyne spectrum: the spectrum of the intensity fluctuations registered by a photodetector at the intermediate (beat) frequency as a central frequency of the measured spectrum; the intermediate frequency is chosen for technical reasons: it has the best signal-to-noise ratio.

heterodyne system with zero cross-phase detectors: the phase measuring system that uses amplitude modulation at two close radio frequencies, f_1 and f_2.

heterostructure: a semiconductor junction that is composed of layers of dissimilar semiconductor materials with nonequal band gaps; a quantum heterostructure's size restricts the movements of the charge carriers and forces them into a quantum confinement that leads to formation of a set of discrete energy levels with sharper density than for structures of more conventional sizes; important for the fabrication of short-wavelength light-emitting diodes and diode lasers.

histogram: a form that represents the distribution of experimental data as a bar diagram.

hologram: a negative produced by exposing a high resolution photographic plate to two interfering waves: the subject wave, which is formed by illumination of a subject by monochromatic, coherent radiation, as from a laser, and the reference wave, which goes directly from the same light source (laser); when a hologram is placed in a beam of coherent light, a true three-dimensional image of the subject is formed.

holographic microscopy: microscopy that uses holographic principles.

holography: the process or technique of making holograms.

homodyne phase system: a system that does not down convert the radio frequency prior to phase measurements.

homodyne spectrum: the self-beat spectrum of the intensity fluctuations registered by a photodetector; it is like the heterodyne spectrum, but with a central frequency equal to zero, and overlapping negative and positive spectrum wings.

homogeneous medium: a medium that has common physical properties, including optical properties, throughout.

humidity: a measure of the extent to which the atmosphere contains moisture (water vapor).

hydrated: being associated with water molecules.

hydrodynamic radius of a particle: the radius that is determined from the measurements of the translation diffusion coefficient for an ensemble of identical particles in the medium; it is larger than the initial one due to interactions with molecules of the medium.

hydrostatic pressure: the pressure at a point in a liquid is the force per unit area on a very small area round the point; if the point is at depth h in the liquid of density ρ, then pressure $p = \rho g h$; the pressure at a point in a liquid at rest acts equally in all directions; the force exerted on a surface in contact with a liquid at rest is perpendicular to the surface at all points; measured in N/m^2 or Pa (pascal).

hyperosmotic: a term that describes a liquid with a lower concentration of water and higher solute concentration than fluids in a tissue or cell; this term also means that if a cell is hyperosmotic, it absorbs water from the surroundings to dilute the higher solute concentration, thus making the cell isotonic to the environment.

hyperpolarizability: nonlinear **polarizability**, β, characterizes the nonlinear part of the induced dipole moment of the molecule, which is proportional to the squared external electric field with the coefficient of proportionality equal to $(1/2)\beta$.

Jabloski diagram: representation of molecular energy levels and transition rates using the potential curves that are plotted without regard to the variable nuclear distances.

Jones matrix: see **amplitude scattering matrix**.

image-carrying photons: a group of photons that produce an image of a certain macro-inhomogeneity within a scattering medium.

image reconstruction: see **TOAST** and **tomographic reconstruction**.

imaging (image) contrast: pointing differences between two or more objects or points on the object; a parameter that characterizes differentiation (visibility) of the visualized object(s) hidden in the scattering surroundings; contrasting parameters may be used: light intensity, reflectance, polarization degree, fluorescence intensity and life-time, refractive index, reconstructed absorption and scattering coefficients, etc.

imaging resolution: when two objects are close together they might not form two images on the retina of the eye or matrix detector that are distinguishable; the ability to detect two such images of two objects close together is measured by the resolving power of the instrument; for a microscope two objects are resolved if the angular separation of the objects is not less than λ/D, where λ is the wavelength of the light used and D is the diameter of the objective; the smallest separation of two objects, if they are to be resolved, is $0.61\lambda/\text{NA}$, where NA is the **numerical aperture**.

immersion medium (liquid): a liquid that provides optical matching between an objective and a biological object; it enhances the numerical aperture of the objective and the microscope resolution; in addition, optical matching reduces surface reflection and scattering, and consequently allows for the reception of higher **contrast** images.

immersion technique: the technique used to reduce light scattering in an inhomogeneous medium by matching the refractive index of the scatterers and ground substance; immersion liquids with an appropriate refractive index and rate of diffusion are usually used.

impedance: the measure of current flowing in an inductive or capacitative component of a circuit when an alternating potential difference is applied; the magnitude of the impedance varies with the frequency of the **ac**.

index of refraction: a number indicating the speed of light in a given medium as either the ratio of the speed of light in a vacuum to that in the given medium (**absolute index of refraction**) or the ratio of the speed of light in a specified medium to that in the given medium (**relative index of refraction**).

inelastic scattering: from quantum electrodynamics it follows that an individual light-scattering event is considered the absorption of the incident photon, which has the energy $h\nu$, momentum $(h/2\pi)\boldsymbol{k}$, and polarization p, by a particle of the scattering medium, and then the emission of the photon that has energy $h\nu'$, momentum $(h/2\pi)\boldsymbol{k}'$, and polarization p'; at $\nu \neq \nu'$ light scattering is accompanied by the redistribution of energy between the radiation and the medium, and is called inelastic, for example, **Raman scattering**; at $\nu = \nu'$, when no redistribution takes place light scattering is called elastic or **Rayleigh scattering**.

infrared spectroscopy: a spectroscopy of middle and far infrared wavelength range the that uses light-excited vibrational-energy states in molecules to get information about the molecular composition, molecular structures, and molecular interactions.

inhomogeneous medium: a medium with a regular or irregular spatial distribution of physical properties, including optical properties.

integrating sphere: a photometric sphere with a highly reflecting white or metallic coating and photodetector inside; used for the precise measurement of diffuse reflectance or total transmittance of scattering materials (tissues); integrating spheres are usually coated with materials that have smooth and high reflectance in the visible and NIR; barium-sulfate, **MgO**, **Spectralon**, and Zenith are most commonly used; for IR applications gold coatings are available.

integration time: the time interval over which measurements are taken; longer integration times allow more averaging in order to filter out background noise and boost the signal-to-noise ratio.

intensity: several measures of light are commonly known as intensity: radiant intensity is a radiometric quantity, measured in watts per steradian (W/sr); luminous intensity is a photometric quantity, measured in lumens per steradian (lm/sr), or candela (cd); **radiance** (**irradiance**) is commonly called "intensity," measured in watts per meter squared (W/m^2).

intensity probability density distribution function: a function that describes the distribution of probability over the values of the light intensity.

interference: the process in which two or more light, sound, or electromagnetic waves of the same frequency combine to reinforce or cancel each other, with the amplitude of the resulting wave being equal to the algebraic sum of the amplitude of the combining waves.

interference fringes: a series of alternating dark and bright bands produced as a result of light interference; with a monochromatic source of light, the bands (fringes) are alternately bright and dark; with white light, the interference bands are colored.

interference of photon density waves (intensity waves): the process in which two or more photon density waves of the same frequency combine to reinforce or cancel each other, with the amplitude of the resulting wave being equal to the algebraic sum of the amplitude of the combining waves.

interference of speckle fields (speckle-modulated fields): the interference of the fields in which amplitudes and phases are randomly modulated due to their interaction (scattering) with inhomogeneous (scattering) media.

interferometer: an instrument that splits a beam of light into a number of coherent beams and then superimposes the beams to obtain interference fringes; the instrument is used to accurately measure wavelengths of light, to examine the hyperfine structure of spectra, to test optical elements for refraction purposes; and to accurately measure distance, displacement, and vibrations.

internal conversion: as a transition between one set of atomic (or molecular) electronic excited levels to another set of the same spin multiplicity (for example, the second singlet state to the first singlet state); it is sometimes called "radiationless deexcitation," because no photons are emitted; it differs from **intersystem crossing** in that, while both are radiationless methods of deexcitation, the molecular spin state for internal conversion remains the same, whereas it changes for intersystem crossing.

intersystem crossing: a photophysical process; an isoenergetic nonradiative transition between two electronic states that have different multiplicities; it often results in a vibrationally excited molecular entity in the lower electronic state, which then usually deactivates to its lowest vibrational level.

invariant embedding method: a method applied to the propagation of various wave types (acoustic, gravity, and electromagnetic) in inhomogeneous media; this method is used to reduce the initial boundary value problems to problems with initial data, permitting the solution of both determinate and statistical problems; it is applicable to both stationary (linear and nonlinear) and nonstationary wave problems.

inverse adding-doubling (IAD) method (technique): a method that provides a tool for the rapid and accurate solution to the inverse scattering problem; it is based on the general method for the transport equation for plane-parallel layers; the term "doubling" means that the reflection and transmission estimates for a layer at certain ingoing and outgoing light angles may be used to calculate both the transmittance and reflectance for a layer twice as thick by means of superimposing one upon the other and summing the contributions of each layer to the total reflectance and transmittance; reflection and transmission in a layer that has an arbitrary thickness are calculated in consecutive order, first for the thin layer with the same optical characteristics (single scattering), then by consecutive doubling of the thickness for any selected layer; the term "adding" indicates that the doubling procedure may be extended to heterogeneous layers for modeling multilayer tissues or taking into account internal reflections related to abrupt changes in refractive index.

inverse MC (IMC) method: the iterative method based on the statistical simulation of photon transport in the scattering media; provides the most accurate solutions to inverse scattering problems; it takes into account the real geometry of the object, the measuring system, and light beams; the main disadvantage is the long computation time.

inverse scattering problem: an attempt to take a set of measurements and error estimates, and only a limited set of parameters describing the sample and experiment, and to deriving the remaining parameters; usually the geometry is known, intensities or their parameters are measured, and the optical properties or sizes of scatterers need to be derived; if these properties are considered to be spatially varying, then the resultant solutions can be presented as a 2D or 3D function of space, i.e., as an image.

ionizing radiation: either particle radiation or electromagnetic radiation in which an individual particle/photon carries enough energy to ionize an atom or molecule by completely removing an electron from its orbit; these ionizations, if enough occur, can be very destructive to living tissue and can cause **DNA** damage and mutations; examples of particle radiation that are ionizing may be energetic electrons, neutrons, atomic ions, or photons; electromagnetic radiation can cause ionization if the energy per photon, or frequency, is high enough, and thus the wavelength is short enough; the amount of energy required varies between molecules being ionized; **x rays** and gamma rays will ionize almost any molecule or atom; far ultraviolet, near ultraviolet, and visible light are ionizing to some molecules; microwaves and radio waves are nonionizing radiation.

IQ circuit: an in-phase quadrature demodulator, a device that allows one to measure the amplitude and phase of an ac signal using the $0°/90°$-phase mixing technique of the receiving and reference signals.

irradiance: a radiometric quantity, measured in watts per meter squared (W/m^2).

irreversible thermodynamics: if the change from initial state to final state of the system is so slow that the process can be assumed to be proceeding through a series of closely spaced quasi-equilibrium states, then such a process is called a reversible process, and the entire time evolution of each of the state variables can be obtained from the conventional theory of thermodynamics; but almost all the processes in which we are most interested are irreversible processes, and the system is not in an equilibrium state during the time the system is evolving; a general theory of nonequilibrium thermodynamics does not exist; the thermodynamics of steady-state processes is relatively well established at least when the system does not deviate from equilibrium substantially.

isobestic point: the point (wavelength) at the spectra having an identical absorption for different forms of molecules.

isotropic scattering: equality of scattering properties along all axes.

KDP: kalium dihydrophosphate; the material widely used in nonlinear optics, for example, for light modulation and frequency doubling.

Kirchhoff approximation: an approximate method for solving wave diffraction problems that is applicable for finding the diffracted field at wave diffraction on inhomogeneities with sizes much larger than the wavelength.

K-space spectral analysis: a near-field wave technique that relies on a series of 2D fast Fourier transforms and that is employed for fast image reconstruction.

Kubelka-Munk model: a two-flux model describing the transportation of radiation in a scattering medium; it employs simple relations for evaluating optical parameters using the diffuse transmittance and reflectance measurements.

laminar flow: occurs when a fluid flows in parallel layers with no disruption between the layers; in fluid dynamics, laminar flow is a flow regime characterized by high momentum diffusion, low momentum convection, and pressure and velocity independence from time; it is the opposite of **turbulent flow**.

Laplace transform: a technique for analyzing linear time-invariant systems such as electrical circuits, harmonic oscillators, optical devices, and mechanical systems; it gives a simple mathematical or functional description to the input or output of a system.

laser: acronym for light amplification by the stimulated emission of radiation; a device that generates a beam of light that is collimated, monochromatic, and coherent.

laser beam: a group of nearly parallel rays generated by a laser; a light beam with a Gaussian shape for the transverse intensity profile (see **Gaussian light beam**).

laser calorimetry: a measuring technique that detects a temperature rise in a sample induced by absorption of a laser beam.

laser Doppler anemometry: the technique of measuring the velocity of flows by the Doppler method using a laser (see **Doppler effect**, **Doppler microscopy**, and **Doppler spectroscopy**).

laser Doppler interferometry: the technique of measuring the velocity of the particles in a flow by the Doppler method using a laser interferometer when a particle's velocity is measured by its traversing of interference fringes.

laser flow cytometry: **flow cytometry** (see **cytometry** in Glossary 2) with laser excitation of **fluorescence**, **light scattering**, or **polarization** transform of cells under investigation.

laser heating: the heating of an object by laser radiation.

laser interferential retinometer: a device for determining retinal visual acuity in the human eye by projecting the interference fringes produced by a laser interferometer at the retina.

laser power: rate of radiation emission from a laser, normally expressed in watts (W), milliwatts (mW), or microwatts (μW).

laser radiation: the radiation emitted by a laser.

laser speckle contrast analysis (LASCA): the method that uses the spatial statistics of time-integrated speckles; the full-field technique for visualizing capillary blood flow.

latex: a suspension of micron-sized **polystyrene spheres**.

length of thermal diffusivity (thermal length): the length within a medium (tissue) characterizing the distance of heat diffusion at medium heating by a short localized laser or acoustic pulse.

lenslet: a set of spatially distributed lenslike (phase) irregular inhomogeneities.

Light: ultraviolet (UV), **UVC**: 100–280 nm; **UVB**: 280–315 nm; **UVA**: 315–400 nm: **visible**: 400–780 nm (**violet**: 400–450 nm; **blue**: 450–480 nm; **green**: 510–560 nm; **yellow**: 560–590 nm; **orange**: 590–620 nm; **red**: 620–780 nm); **infrared (IR) light, IRA**: 0.78–1.4 μm: **IRB**: 1.4–3.0 μm; **IRC**: 3–1000 μm; **near IR (NIR)**: 0.78–2.5 μm; **middle IR (MIR)**: 2.5–50 μm; and **far IR (FIR)**: 50–2000 μm.

light-emitting diode (LED) (light diode): a semiconductor device that emits light when the forward-directed current passes the p-n junction.

lifetime of the excited state: "lifetime" refers to the time the molecule (atom) stays in its excited state before emitting a photon; the lifetime is related to the rate of the excited state of decay, to the facility of the relaxation pathway, radiative and nonradiative; if the rate of spontaneous emission, or any other rate, is fast the lifetime is short; for commonly used fluorescent compounds the typical excited state decay times are within the range of 0.5 to 20 ns.

light guide: an assembly of optical fibers that are bundled but not ordered and that are used for illumination.

light scattering: a change in direction of the propagation of light in a turbid medium caused by reflection and refraction by microscopic internal structures.

light-scattering matrix [LSM (intensity or Mueller matrix)]: the 4×4 matrix that connects the **Stokes vector** of incident light with the **Stokes vector** of scattered light; it describes the polarization state of the scattered light in the far zone that is dependent on the polarization state of the incident light and structural and optical properties of the object.

linear regression: a regression method that allows for the linear relationship between the dependent variable Y and the p independent variables X and a random term ε.

Linnik microscope: see **Linnik-Tolansky interferometer**.

Linnik-Tolansky interferometer: a dual-beam interferometer with a beamsplitter, two reflecting surfaces, and two lenses for focusing beams on the surfaces.

liquid crystal: a substance that exhibits a phase of matter that has properties between those of a conventional liquid, and those of a solid crystal; for instance, a liquid crystal may flow like a liquid, but it has the molecules in the liquid arranged and/or oriented in a crystal-like way.

lithium niobate (LiNbO$_3$): a compound of niobium, lithium, and oxygen; it is a colorless solid material with a trigonal crystal structure; it is transparent for wavelengths between 350 and 5200 nanometers and is used for the manufacture of optical modulators and acoustic wave devices.

local oscillator: the radio- (or optical) frequency oscillator used in heterodyne detecting systems; a local oscillator is stable in frequency and amplitude and has a slightly different frequency than the receiving signal; it is used for converting a high-frequency receiving signal to an intermediate frequency by mixing the local oscillator signal and the receiving signal at an electronic (or photo) detector.

lock-in-amplifier: a low-frequency electronic device that provides synchronous detection of small signals that may have amplitudes a few orders lower than the noise level; the lock-in circuit contains the selective amplifier and a phase detector tuned to the modulation frequency of the detecting signal.

low-pass filter: a filter that rejects the high-frequency components.

low-step scattering: the scattering process in which, on average, each photon undertakes no more than a few scattering events (approximately less than five to ten).

LSM element: one of 16 elements of the light-scattering matrix; each element depends on the scattering angle and wavelength, and the geometrical and optical parameters of the scatterers and their arrangement.

luminescence: light not generated by high temperatures alone; it is different from incandescence, in that it usually occurs at low temperatures and is thus a form of cold body radiation; it can be caused by, for example, chemical reactions, electrical energy, subatomic motions, or stress on a crystal; the following kinds of luminescence are known: **fluorescence**, **phosphorescence**, bioluminescence, photoluminescence, **sonoluminescence**, chemoluminescence, electroluminescence, radioluminescence, mechanoluminescence, triboluminescence, piezoluminescence, thermoluminescence, et al.

Mach-Zehnder interferometer: a dual-beam, four-mirror (two serve as the beamsplitter and beam coupler, and two as reflectors) interferometer typically used as a refractometer, especially for objects occupying a large space.

magnetic resonance imaging (MRI): a noninvasive imaging technique that is based on magnetic resonance methods; it provides a wealth of information about inner structures of the body and, in particular, tumors.

matching substance: a substance used to reduce the boundary effects caused by the complex shape of a scattering object; the scattering properties of such a substance should be similar to the scattering properties of the object under study.

material dispersion: describes the separation of the different wavelengths in a given medium (material) that occurs because the waves are traveling at different velocities in that medium.

Matlab: a high-level language and interactive environment that enables one to perform computationally intensive tasks faster than with traditional programming languages such as C, C++, and Fortran.

mean free path length (MFP): the mean distance between two successive interactions with scattering or absorption experienced by a photon traveling in a scattering-absorption medium.

mechanical stress: the action on a body of any system of balanced forces that results in strain or deformation.

mercury arc lamp: a discharge arc lamp filled with mercury vapor at high pressure; it gives out very bright **UV** and **visible light** at some wavelengths, including 303, 312, 365, 405, 436, 546, and 578 nm.

meridional plane: planes that include the optical axis are meridional planes; it is common to simplify problems in radially symmetric optical systems by choosing object points in the vertical plane only; this plane is then sometimes referred to as the meridional plane.

MgO (magnesium oxide, or magnesia): a white solid mineral that occurs naturally as periclase and is a source of magnesium; it is formed by an ionic bond between one magnesium and one oxygen atom; it is used as a reference white color in photometry and colorimetry; the emissivity value is about 0.9; pressed MgO is used as an optical material; it is transparent from 300 nm to 7 μm; the refractive index is 1.72 at 1 μm.

Michelson interferometer: a dual-beam interferometer with a beamsplitter and two reflecting surfaces; it allows one to realize various types of interference and is widely used in metrology for measurements of lengths, displacements, vibrations, and surface roughness; recently an integrated fiber-optic prototype became very popular (see **fiber-optic single-mode X-coupler**); also widely used in tissue spectroscopy and imaging (see **dual-beam coherent interferometry** and **Doppler interferometry**).

micrometer (i.e., micron or μm): a unit of length that is 10^{-3} millimeter (mm) or 10^{-6} meter (m).

microphone: a device for transforming sound energy into electrical energy; the various types are the carbon microphone, crystal microphone, condenser microphone, and a moving coil or dynamic microphone.

microprofilometer: a device for measuring the roughness of a surface.

microscopy: any technique for producing visible images of structures or details too small to otherwise be seen by the human eye, using a microscope or other magnification tool; more specifically, it is a technique of using a microscope; there are three main branches of microscopy: optical, electron, and scanning probe microscopy; optical and **electronic microscopy** involves the **diffraction, reflection, or refraction** of radiation incident upon the subject of study, and the subsequent collection of this scattered radiation in order to build up an image; this process may be carried out by wide-field irradiation of the sample (for example, standard light microscopy and transmission electron microscopy) or by scanning a fine beam over the sample (for example, **confocal microscopy** and scanning electron microscopy); scanning probe microscopy involves the interaction of a scanning probe with the surface or object of interest.

microsecond (μsec) (μs): 10^{-6} sec (s).

micro-spectrophotometric technique: a technique that measures an object's transmittance or reflectance spectra with a high spatial resolution; usually a combination of a microscope and a grating spectrograph with an optical multichannel analyzer (cooled CCD or photodiode array) is used for such measurements.

Mie or **Lorenz-Mie scattering theory**: an exact solution of Maxwell's electromagnetic field equations for a homogeneous sphere.

millisecond (msec) (ms): 10^{-3} sec (s).

minimal erythema dose (MED): the minimal single dose of UV radiation, expressed as energy per unit area J/cm^2, producing a clearly marginated erythema at the irradiated skin site after 24 hours for UVB and 48 hours for UVA.

M-mode OCT image: an image obtained from repeated **A-scans** without moving the incident beam, so it can easily visualize living-object movement and time-dependent changes of tissue structure.

mode-locked laser: a multimode laser with synchronously irradiating modes; the regime is obtained by applying an intracavity high-frequency modulator, with a typical pulse duration of up to a subpicosecond range and a repetition frequency of dozens of megahertz.

modulation: the process of varying the characteristics of an optical wave motion by superimposing on it the characteristics of a second (audio- or radiofrequency) wave motion; there are three main types of modulation: amplitude modulation, frequency modulation, and phase modulation.

modulation frequency: the frequency of the modulating wave.

molecular hyperpolarizability: **see hyperpolarizability**.

monochromatic light: light of one color (wavelength) only or a very limited range of wavelengths; produced by a CW single-frequency (single longitudinal mode) laser.

monodisperse model: a model presenting a disperse medium as monodisperse, such as an ensemble of scatterers with an equal size and refractive index for each scatterer: a healthy eye cornea is a good example of a monodisperse model.

monodisperse system: a disperse system (medium) with a single characteristic parameter, such as an ensemble of scatterers with an equal size and refractive index for each scatterer; a healthy eye cornea is a good example of a monodisperse system, because it consists of dielectric rods with the same refractive index and radius dispersed in a homogeneous ground substance.

Monte Carlo method: a numerical method of statistical modeling; in tissue studies it provides the most accurate simulation of photon transport in samples with a complex geometry, accounting for the specificity of the measuring system and light beam configurations.

Mueller matrix: a 4×4 matrix that transforms an incident **Stokes vector** into the corresponding output Stokes vector of the sample; it fully characterizes the optical polarization properties of the sample; it can be experimentally obtained from measurements with different combinations of source **polarizers** and detection analyzers; at least 16 independent measurements must be acquired to determine a full Mueller matrix.

multichannel optoelectronic near-infrared system for time-resolved image reconstruction (MONSTIR): the noninvasive imaging technique developed at University College London for studying infant brain function that is based on the detection of transmitted pulsed NIR radiation.

multichannel plate: an integrated optical system used for optical amplification (image intensification).

multichannel plate-photomultiplier tube (MCP-PMT): a photomultiplier tube that is integrated with a multichannel plate.

multiflux model: the simplest multiflux model describing transportation of radiation in a scattering medium that employs only two fluxes is the **Kubelka-Munk model**; a more general approach is the discrete ordinates method (or many-flux theory) when the transport equation (**RTT**) can be converted into a matrix differential equation by considering the radiance at many discrete angles; by increasing the number of angles, the matrix solution should approach the exact solution; for laser beam applications, the four-flux model makes use of two diffuse fluxes and forward and backward coherent fluxes; a three-dimensional six-flux model is also available.

multifrequency multiplex (time division multiplex): a process in which measurements on many modulation frequencies are provided concurrently.

multilayered tissue: a tissue that consists of many layers with different structural and optical properties, such as skin, the bladder wall, and wall of bladder.

multimode fiber: a single fiber that allows the excitation (direction) of many modes (rays); e.g., for a fiber with a core diameter of 50 μm, numerical aperture, NA = 0.2, and an excitation wavelength of 633 nm, the number of excited modes is equal to 1250.

multiphoton absorption process: a process that needs a very high density of photons (0.1–10 MW/cm^2) from a ps-to-fs-pulsed light source; this is because the virtual absorption of a photon of nonresonant energy lasts only for a very short period (10^{-15}–10^{-18} s); during this time a second photon must be absorbed to reach an excited state (http://www.fz-juelich.de/ibi/ibi-1/Two-Photon_Microscopy/).

multiphoton fluorescence: relies on the quasi-simultaneous absorption of two or more photons (of either the same or different energy) by a molecule; during the absorption process, an electron of the molecule is transferred to an excited-state molecular orbit; the molecule (i.e., the fluorophore) in the excited state has a high probability ($>10\%$) to emit a photon during relaxation to the ground state; due to radiationless relaxation in vibrational levels, the energy of the emitted photon is lower compared to the sum of the energy of the absorbed photons (http://www.fz-juelich.de/ibi/ibi-1/Two-Photon_Microscopy/).

multiphoton fluorescence scanning microscopy: the microscopy that employs detection of multiphoton fluorescence at the scanning of a laser beam, inducing the multiphoton signal (see also **two-photon fluorescence microscopy**).

multiple scattering: a scattering process in which, on average, each photon undertakes many scattering events (approximately more than five to ten).

multiwavelength multiplex (wavelength division multiplex): a process in which measurements on many wavelengths are provided concurrently.

nanometer (nm): a unit of length equal to 10^{-9} meter (m).

nanoparticle: a microscopic particle whose size is measured in **nanometres** (nm); it is defined as a particle with at least one dimension <100 nm.

nanosecond (nsec) (ns): 10^{-9} sec (s).

narrow-band filter: an electronic device that selectively damps oscillations of frequencies out of the narrow band while not affecting oscillations of frequencies within this band.

Nd:YAG (neodymium:yttrium aluminium garnet) laser: a solid-state laser whose lasing medium is the crystal Nd:YAG with emission in the NIR at 1064 nm; other less intensive lines at 946, 1319, 1335, 1338, 1356, and 1833 nm are also available.

network analyzer: a two-channel electronic system for measuring amplitude frequency characteristics of a four-terminal network producing modulation swept in the wide-frequency range (for example, 0.3–1000 MHz) and analyzing the detected signal synchronously in the same frequency range.

Nomarski polarizing-interference microscope: an optical microscope with differential interference contrast that incorporates a common path interferometer based on a polarizing prism.

non-Gaussian statistics: a statistically nonuniform process in which the statistical characteristics of the scattered light essentially depend on the observation angle and the degree of nonuniformity of an object.

nonlinear polarization of a material: if the dielectric polarization density (dipole moment per unit volume) P is not linearly proportional to the electric field E, the medium is termed nonlinear and is described by the field of nonlinear optics; to a good approximation (for sufficiently weak fields, assuming no permanent dipole moments are present), P is usually given by a Taylor series in E whose coefficients are the **nonlinear susceptibilities**.

nonlinear regression technique: related to a model $y = f(x, \theta) + \varepsilon$, based on multidimensional x, y data, where f is some nonlinear function with respect to unknown parameters θ; at a minimum, one may like to obtain the parameter values associated with the best fitting curve.

nonlinear susceptibility: the electric susceptibility of a dielectric material is a measure of how easily it polarizes in response to an electric field; this, in turn, determines the electric permittivity of the material and thus influences many phenomena in that medium (for instance, the speed of light); linear susceptibility is defined as the constant of proportionality (which may be a tensor) relating an electric field E, and nonlinear, relating to E^2, E^3, etc., to the induced dielectric polarization density P.

non-Newtonian flow: the flow of a fluid in which the viscosity changes with the applied strain rate; as a result, non-Newtonian fluids may not have a well-defined viscosity.

nonradiative relaxation (nonradiative energy transfer): the relaxation of an excited molecule (losing energy) without emission of light, when the molecule's energy is transformed into the heat, which raises the temperature of the body absorbing the energy by increasing the kinetic energy of the particles composing the body.

nonuniform medium: see **inhomogeneous medium**.

numerical aperture (NA): the light-gathering power of an objective or optical fiber; it is proportional to the sine of the acceptance angle.

objective speckles: the speckles formed in a free space and usually observed on a screen placed at a certain distance from an object.

open-circuit: in electronics, the absence of a load through which electric current would otherwise flow; this can be represented by an infinitely large resistance or **impedance**.

optical activity: the ability of a substance to rotate the plane of polarization of plane- (linear) polarized light (see **chirality**).

optical anisotropy: the difference of optical properties of materials caused by the dependence of light velocity (**refractive index**) on the direction of light propagation and **polarization of light** (see **anisotropic crystal**); it manifests as **birefringence**, **dichroism**, and **optical activity**, as well as **depolarization** at light scattering in a medium, polarized fluorescence, etc.; optical anisotropy may be induced in an optically isotropic medium at external action (mechanical, electrical, magnetic, etc.) that changes its local symmetry; related effects are **photoelasticity**, Kerr effect, Faraday effect, Cotton-Mouton effect, and nonlinear optical activity.

optical attenuator: a device for decreasing the intensity of light; optically neutral or color filters are usually used as attenuators with a fixed or stepwise-variable attenuation; for polarized light an attenuator with a continuously variable attenuation of the rotating polarizer (analyzer) is used.

optical autocorrelator: a device for measuring the autocorrelation function of intensity fluctuations of a scattered optical field.

optical birefringence: see **birefringence**.

optical breakdown: a breakdown in air (and in other transparent media) that is initiated by intense light; the required intensity for optical breakdown depends on the pulse duration; for example, for 1-ps pulses an optical intensity of $\approx 2 \times 10^{13}$ W/cm^2 is required; the high optical intensities can be reached in pulses as generated, e.g., in a **Q-switched laser** (with nanosecond durations) or in a **mode-locked laser** and amplified in a regenerative amplifier (for pulse durations of **picoseconds** or **femtoseconds**).

optical calorimetry: a measuring technique that detects of a temperature rise in a sample induced by light-beam absorption.

optical clearing: controlling optical properties of a scattering medium resulting in the increase of its optical transmittance.

optical coherence interferometry (OCI): see **Doppler interferometry** and **dual-beam coherent interferometry**.

optical coherence microscopy: an **optical microscope** based on a short-focused **OCT**.

optical coherence tomography (OCT): a technique that is based on **Doppler interferometry** in which a partially coherent light source is used and, in addition to the reference beam path length scanning (z-scan) that provides depth profiling of an object, transverse (x–y) scanning for 3D images is used; the integrated **single-mode fiber-optic Michelson interferometer** is usually used in OCT; the method is widely used for subsurface tomography of tissue.

optical coherent reflectometry: see **optical coherence tomography (OCT)**.

optical conjugate: two optical points, lines, etc. that are so related as to be interchangeable in certain optical properties; an optical system that provides two points so that a source at one point is brought to focus at the other, and vice versa.

optical darkening effect: controlling of optical properties of a scattering medium resulting in the decrease of its optical transmittance.

optical depth: a measure of transparency, and is defined as the fraction of radiation that is scattered and/or absorbed on a path; the optical depth τ expresses the quantity of light removed from a beam by scattering and/or absorption during its path through a medium.

optical detector: a device that converts optical energy to an electric signal.

optical diffusion tomography: an optical **tomography** that is based on the measurements of CW, pulsed, or modulated light beam transmittance or spatially-resolved reflectance of scattering media with an object (i.e., a tumor) hidden in it; to provide 3D images, synchronous light beam-detector scanning devices or systems with multiple fixed-position light sources and detectors are used; the **back-projection algorithm** is used to provide image reconstruction along the paths of ballistic and/or least scattering photon propagation; the method is used for tomography of thick tissues (breast, brain, arm).

optical fiber cladding: see **fiber**.

optical fiber core: see **fiber**.

optical fiber coupler: see **fiber coupler**.

optical fiber dispersion: in optics, **dispersion** is a phenomenon that causes the separation of a wave into spectral components with different wavelengths, due to the dependence of the wave's speed on its wavelength; dispersion is sometimes called chromatic dispersion to emphasize its wavelength-dependent nature; there are generally two sources of dispersion: material dispersion, which comes from the frequency-dependent response of a material to waves, and waveguide dispersion, which occurs when the speed of a wave in a waveguide (optical **fiber**) depends on its frequency; the transverse modes for waves confined laterally within a finite waveguide generally have different speeds (and field patterns), depending upon the frequency (that is, on the relative size of the wave, the wavelength, compared with the size of the waveguide); dispersion in an optical fiber results in signal degradation, because the varying delay in arrival time between different components of a signal "smears out" the signal in time; a similar phenomenon is modal dispersion, caused by a waveguide that has multiple modes at a given frequency, each with a different speed; a special case of this is polarization mode dispersion, which comes from a superposition of two modes that travel at different speeds due to random imperfections that break the symmetry of the waveguide.

optical Fourier transform: the transform when spatial variations in optical density in the object plane are converted into **spatial frequency** variations in the Fourier transform plane in the rear focal plane of a lens.

optical image: a reconstructed image of an object expressed in terms of local optical parameters, such as absorption and/or scattering coefficients.

optical Kerr effect: the double refraction of light in certain substances that is produced by an external electric field, including high-frequency fields up to the frequencies of IR light.

optical Kerr gate: a transparent cell filled with a substance that shows a Kerr effect and contains two electrodes placed between two polarizers: the cell serves as a high-speed optical shutter.

optical length (optical path length): in a medium with a constant refractive index, the product of the geometric distance and the refractive index; in a medium with a varying refractive index, the integral of the product of an element of length along the path and the local refractive index; optical length is proportional to the phase shift that a light wave undergoes along a path.

optical mean free path (MFP): see **mean free path length**.

optical medical tomography: see **optical diffusion tomography** and **optical coherence tomography (OCT)**.

optical microscopy: see **microscopy**.

optical multichannel analyzer (OMA): a spectrometric instrument that senses incident radiation in several channels at the same time, sorts the radiation from deep ultraviolet to the infrared, and digitizes and stores the information so that it can be processed and analyzed individually by channel.

optical parameters: the physical parameters that characterize the optical properties of an object.

optical parametric oscillator (OPO): a parametric oscillator that oscillates at optical frequencies; it converts an input laser wave (called a "pump") into two output waves of lower frequency (ω_s, ω_i) by means of nonlinear optical interaction; the sum of the output wave frequencies is equal to the input wave frequency: $\omega_s + \omega_i = \omega_p$; the OPO essentially consists of an optical resonator and a nonlinear optical crystal; the optical resonator serves to resonate at least one of the output waves.

optical path: see **optical length (optical path length)**.

optical phantom: a medium that models the transport of visible and infrared light in tissue and is needed to evaluate techniques, to calibrate equipment, to optimize procedures, and for quality assurance.

optical retarder: a device that provides an optical retardation: phase shift or optical path difference; such retarders as the half- or the quarter-wavelength plates provide, respectively, the half-wave or the quarter-wave phase difference.

optical sectioning (slicing): the process of extracting the optical image of a thin layer of tissue; the image is used for **tomographic reconstruction** of a whole body organ.

optical transition: typically an **electronic transition**, where energy is given out as electromagnetic radiation in the optical range.

optical transition lifetime: the radiative lifetime, which is determined by the emission cross section for transition to a lower-lying energy level.

optically thick sample: optical thickness is the depth of a material or medium in which the intensity of light of a given wavelength is reduced by a factor of $1/e$

because of absorption and/or scattering; a sample with high thickness and/or high turbidity that correspond to a few optical thickness depths is optically thick.

optically thin (transparent) sample: a sample with low thickness and/or low turbidity that corresponds to one or less than one optical thickness depth is optically thin.

optoacoustic (OA) interaction: the generation of acoustic waves by the interaction of pulsed or intensity-modulated optical radiation with a sample; actually, several effects can be responsible for such interaction, e.g., the optical inverse piezoelectric effect, optical electrostriction, or optothermal effect.

OA method: the detection of acoustic waves generated via OA interaction with a sample (the term OA primarily refers to the time-resolved technique utilizing pulsed lasers and measuring profiles of pressure in tissue).

OA microscopy: **microscopy** based on detection of an OA signal induced by a sharply focused laser beam.

OA spectroscopy: spectroscopy based on the detection of an OA signal induced by a monochromatic light source (laser) with tuned wavelength.

OA tomography: the **tomography** that is based on the **OA method**.

optode: a transducer that is attached to the distal tip of a fiber-optic sensor; the interaction between the optode and the body is monitored by the fiber-optic sensor.

optogeometric technique: the detection of surface deformation in solids and volume changes in fluids induced by an **optothermal interaction**.

optothermal interaction: the generation of heat waves by the interaction of pulsed or intensity-modulated optical radiation with a sample.

optothermal method: the detection of heat waves generated via interaction of pulsed or intensity-modulated optical radiation with a sample.

optothermal radiometry (OTR): the detection of time-dependent infrared thermal emissions induced by the **optothermal interaction** of light with a sample.

Ornstein-Zernice equation: an equation for the **radial distribution function** $g(r)$ of classical many-particle systems; thermodynamic properties of such systems are determined by the interaction between the particles from which the system is built up; if one knows the radial distribution function, one can calculate all thermodynamic properties of the considered system; light scattering properties of such systems also can be calculated.

osmotic phenomenon: the tendency of a fluid to pass through a semipermeable membrane into a solution where its concentration is lower, thus equalizing the conditions on either side of the membrane.

osmotic pressure: the hydrostatic pressure produced by a solution in a space divided by a differentially permeable membrane due to a differential in the concentrations of a solute.

osmotic stress: the force that a dissolved substance exerts on a semipermeable membrane through which it cannot penetrate, when it is separated from a pure solvent by the membrane.

overtone: a sinusoidal component of a waveform of greater frequency than its fundamental frequency; the term is usually used in acoustics.

oxymetry: the measurement of tissue or blood oxygenation.

packing dimension: one of the most important notions of **fractal dimension**.

packing factor: the fraction of volume in a medium structure that is occupied by particles; it is dimensionless and always less than unity; for practical purposes, a medium structure is often determined by assuming that particles are rigid spheres.

packing function: an analytical expression for molecular (particle) overlap as a function of position; it can be calculated by means of Fourier transforms; overlap functions between pairs of symmetry elements can be combined to give a crystallographic packing function.

paraxial approximation: an approximation used in ray tracing of light through an optical system.

paraxial region: the region where paraxial rays, lying close to the axis of an optical system, propagate.

partial-coherence interferometry: see **Doppler interferometry**, **dual-beam coherent interferometry**, **optical coherence interferometry (OCI)**.

partially-coherence tomography: see **optical coherence tomography (OCT)**.

percolation: concerns the movement and filtering of fluids through porous materials; recent percolation theory, an extensive mathematical model of percolation, has brought new understanding and techniques to a broad range of topics in physics and materials science.

perfusion pump: a fluid propulsion system that provides, for instance, long-term controlled-rate delivery of drugs such as chemotherapeutic agents or analgesics.

permeability coefficient: permeability (P) of molecules across a biological (cell) membrane can be expressed as $P = K D / \Delta x$, where K is the partition coefficient, D is the **diffusion coefficient**, and Δx is the thickness of the membrane; the diffusion coefficient (D) is a measure of the rate of entry into the cell cytoplasm depending on the molecular weight or size of a molecule; K is a measure of the solubility of the substance in lipids; a low value of K describes a molecule like water that is not soluble in lipid.

perturbation method: a method used to find an approximate solution to a problem that cannot be solved exactly, by starting from the exact solution of a related problem; is applicable if the problem at hand can be formulated by adding a "small" term to the mathematical description of the exactly solvable problem; leads to an expression for the desired solution in terms of a power series in some "small" parameter that quantifies the deviation from the exactly solvable problem.

perovskite laser: a neodymium: yttrium aluminum perovskite laser (Nd:YAP); a laser using an yttrium-aluminum-perovskite crystal doped with neodymium as a lasing medium emitting on the wavelength $\lambda = 1341$ nm.

phantom: a standard experimental tissue model (see **optical phantom**).

phase-contrast microscopy (phase microscopy): a microscopy that translates the difference in the phase of light transmitted through or reflected by an object into the difference of intensity in the image.

phase-delay measurement device: see **"cross-correlation" measurement device** and **heterodyne phase system**.

phase function: see **scattering phase function**.

phase fluctuations of the scattered field: the fluctuations that are induced by different optical paths for different parts or time periods of a wavefront interacting with an inhomogeneous, generally dynamic medium.

phase lag: a phase shift relative to the incident light modulation phase.

phase object: an object that introduces the difference in phase of the light transmitted through or reflected by an object.

λ/4-phase plate: see **optical retarder**; a device that provides an optical phase shift of 90° ($\pi/2$ radians) or an optical path difference equal to a quarter of the wavelength; a thin plate of birefringent substance, such as calcite or quartz, is cut parallel to the optical axis of the crystal with a specific thickness that is calculated to give a phase difference of 90° ($\pi/2$ radians) between the emergent ordinary ray and the emergent extraordinary ray for light of a specified wavelength; quarter-wave plates are usually constructed for the wavelengths of sodium light (589 nm); if the angle between the plane of polarization of light incident upon the plate and the optic axis of the plate is 45°, then circularly polarized light is produced and emerges from the plate; if the angle is other than 45°, elliptically polarized light is produced.

phase shift (phase difference): the difference in phase between two wave forms; the phase difference is measured by the phase angle between the waves: when two waves have a phase shift (difference) of 90° (or $\pi/2$ radians), one wave is at maximum amplitude when the other wave is at zero amplitude; with a phase difference of 180° (π radians), both waves have zero amplitude at the same time, but one wave is at a crest when the other wave is at a trough.

phase or amplitude cancellation (phased array) method: the basis for this method is the interference of photon-density waves [see **interference of photon density waves (intensity waves)** and **photon-density wave**]; it uses either duplicate sources and a single detector or duplicate detectors and a single source so that the amplitude or phase characteristics can be nulled and the system becomes a differential.

phase plate: see **λ/4-phase plate** and **optical retarder**.

phased-array technique: a spectroscopic or imaging technique that utilizes the **interference of photon density waves (intensity waves)**.

phonon: a quantized mode of vibration occurring in a rigid crystal lattice, such as the atomic lattice of a solid; the study of phonons is an important part of solid state physics, because phonons play a major role in many of the physical properties of solids, including a material's thermal and electrical conductivities; in particular, the properties of long-wavelength phonons give rise to sound in solids—hence the name "phonon," i.e., "voice" in Greek; in insulating solids, phonons are also the primary mechanism by which heat conduction takes place.

phosphorescence: **luminescence** that is delayed with respect to the excitation of a sample.

photoacoustic (PA) method: see **optoacoustic (OA) method**; the term PA primarily describes spectroscopic experiments with CW-modulated light and a photoacoustic cell.

photoacoustic microscopy (PAM): a microscopy utilizing the photoacoustic method and a photoacoustic cell for signal detection.

photobiochemical reaction: a chemical reaction in living matter that is induced by light.

photodetector: see **optical detector**.

photocathode: a cathode that has the property of emitting electrons when activated by light or other radiation.

photoconductive detector: a **photodetector** in which an electric potential is applied across the absorbing region and causes a current to flow in proportion to the irradiance if the photon energy exceeds the energy gap between the valence and the conduction band; for the visible wavelength range—cadmium sulfide, for IR—lead sulfide, silicon doped with arsenide (Si:As), and mercury-cadmium-telluride (HgCdTe) are used as photoconductive materials.

photoelasticity: stress-induced **birefringence** and **dichroism** of a medium.

photomechanical waves: see **laser-generated stress waves**.

photomultiplier [photomultiplying tube (PMT)]: an extremely sensitive detector of light and other radiation consisting of a tube in which the electrons released by radiation striking a photocathode are accelerated to successive dynodes that release several electrons for each incident electron, greatly amplifying the signal obtainable from small quantities of radiation.

photon: a quantum of electromagnetic radiation, usually considered as an elementary particle that has its own antiparticle and that has zero rest mass and charge and a spin of 1.

photon absorption cross section: the ability of a molecule to absorb a photon of a particular wavelength and polarization; although the units are given as an area, it does not refer to an actual size area, at least partially because the density or state of the target molecule will affect the probability of absorption; quantitatively, the number dN of photons absorbed, between the points x and $x + dx$ along the path of a light beam is the product of the number N of photons penetrating to depth x times the number ρ of absorbing molecules per unit volume times the absorption cross section σ_{abs}: $dN/dx = -\rho\sigma_{abs}N$.

photon-correlation spectroscopy: a noninvasive method for studying the dynamics of particles on a comparatively large time scale; the implementation of the single-scattering regime and the use of coherent light sources are of fundamental importance in this case; the spatial scale of testing a colloid structure (an ensemble of biological particles) is determined by the inverse of the wave vector; **quasi-elastic light scattering** spectroscopy, **spectroscopy of intensity fluctuations**, and **Doppler spectroscopy** are synonymous terms related to **dynamic light scattering**.

photon-counting system: a system that makes use of a specific method of photoelectron signal processing and provides sequential detection of single photons; **photomultipliers (PMT)** or **avalanche photodetectors (APD)** are usually used for photon counting; the technique is applicable for detecting very weak signals.

photoelasticity: an experimental method to determine stress distribution in a material; unlike the analytical methods of stress determination, photoelasticity gives a fairly accurate picture of stress distribution even around abrupt discontinuities in a material; the method serves as an important tool for determining the critical stress points in a material and is often used for determining stress concentration factors in irregular geometries.

photon-density wave: a wave of progressively decaying intensity; microscopically, individual photons migrate randomly in a scattering medium, but collectively they form a photon-density wave at a modulation frequency that moves away from a radiation source.

photon diffusion coefficient: see **diffusion coefficient**.

photon (intensity) diffusion wave: see **photon-density wave**.

photon scattering cross section: the ability of a particle to scatter a photon of a particular wavelength and polarization; although the units are given as an area, it does not refer to an actual size area; quantitatively, the number dN of photons scattered, between the points x and $x + dx$ along the path of a light beam, is the product of the number N of photons penetrating to depth x times the number ρ of scattering particles per unit volume times the scattering cross section σ_{sca}: $dN/dx = -\rho \sigma_{sca} N$.

photon shot noise: the noise caused by the irregularity of photoelectron emission; it induces random errors in a photoelectron measuring system; the mean square of photoelectron current fluctuation is defined by the average photocurrent i and the photodetector's bandwidth B_D : $i^2 = 2ei B_D$, where e is the charge of the electron; it is difficult to achieve the shot noise limit in practice.

photon transport: a process of photon travel in a homogeneous or inhomogeneous medium with possible macroinhomogeneities; a photon changes its direction due to reflection, refraction, diffraction, or scattering and can be absorbed by an appropriate molecule on its way.

photonic crystal: a periodic optical (nano)structure that affects the propagation of electromagnetic waves (EM) in the same way as the periodic potential in a semiconductor crystal affects the electron motion by defining allowed and forbidden electronic energy bands; the absence of allowed propagating EM modes inside the structures, in a range of wavelengths called a photonic band gap, gives rise to distinct optical phenomena, such as inhibition of spontaneous emission, high-reflecting omni-directional mirrors, and low-loss-waveguiding amongst others; since the basic physical phenomenon is based on **diffraction**, the periodicity of the photonic crystal structure has to be in the same length-scale as half the wavelength of the EM waves, i.e., ~ 300 nm for photonic crystals operating in the visible part of the spectrum; photonic crystals occur in nature, including biological tissues.

photorefractive technique: the detection of refractive index gradients above and inside a sample using thermal blooming, thermal lensing, probe beam refraction, or interferometry and deflectometry.

photosensitizer: a substance that increases the absorption of another substance at a particular wavelength band.

photothermal flow cytometry: **flow cytometry** that uses photothermal detection abilities.

photothermal microscopy: **microscopy** based on the detection of the photothermal signal induced by a sharply focused laser beams.

photothermal radiometry (PTR): see **optothermal radiometry (OTR)**.

picosecond: (psec) (ps)–10^{-12} sec (s).

piezoceramics: a piezoelectric material that is used to make electromechanical sensors and actuators; lead zirconate titanate (PZT) ceramics are an example; there are several different formulations of the PZT compound, each with different electromechanical properties.

piezodeflector: a device for light beam deflection at certain audio- or radiofrequencies using an acoustooptical effect.

piezo-driver: a device that uses the inverse piezoelectric effect in certain asymmetric crystals, which is obtained by applying a potential difference to a crystal; an alteration in the size of the crystal takes place.

piezoelectric transducer: a device that uses the piezoelectric effect in certain asymmetric crystals; the effect is obtained by applying external pressure to a crystal; positive and negative charges are produced on opposite faces of the crystal, giving rise to a potential difference between the faces; the potential difference operates in the opposite direction if tension is applied instead of pressure; this potential difference is the signal detected in a crystal microphone.

piezooptical coefficient: characterizes the efficiency of stress-induced **birefringence** and **dichroism** of a medium (see **photoelasticity**); indicates whether the material is good or not for stress sensors (**piezoelectric transducer**) and **acousto-optical modulators**.

PIN photodetector: a photodetector based on *p-i-n* semiconductor structure that has a fast response.

pixel: the smallest element of an image that can be individually displayed.

Planck curve (function): gives the intensity radiated by a blackbody as a function of frequency (or wavelength) for a definite body temperature; a blackbody is an object that absorbs all the electromagnetic energy that falls on the object, no matter what the wavelength of the radiation; the area under the curve increases as the temperature is increased (the Stefan-Boltzmann law); the peak in the emitted energy moves to the shorter wavelengths as the temperature is increased (Wien's law).

Planck's constant: (denoted as h) a physical constant that is used to describe the sizes of quanta; it plays a central role in the theory of quantum mechanics and is named after Max Planck, one of the founders of quantum theory; a closely-related quantity is the reduced Planck constant [also known as Dirac's constant $(\hbar = h/2\pi)$]; Planck's constant is also used in measuring energy emitted by light photons, such as in the equation $E = h\nu$, where E is energy, h is Planck's constant, and ν is frequency.

Pockel's cell: a piezoelectric crystal with two plane electrodes for applying an external electric field, placed between two crossed polarizers; the basis of its function is the linear electro-optical effect, which relates to a change in the refractive index of a crystal caused by an external electric field: the phase shift between ordinary and extraordinary rays linearly depends on the electrical field strength; the cell is widely used as an external laser or other light source intensity modulator, as well as an internal laser modulator for giant pulse **Q-switching**.

point spread function (PSF): describes the response of an imaging system to a point source or point object; another commonly used term for the PSF is a system's impulse response; the degree of spreading (blurring) of the point object is a measure for the quality of an imaging system; in functional terms it is the spatial domain version of the modulation transfer function; it is a useful concept in Fourier optics, electron microscopy and other imaging techniques such as 3D microscopy (i.e., **confocal microscopy** and **fluorescence** microscopy).

polarimetry: measurement of the polarization properties of light.

polarizability: the relative tendency of a charge distribution, like the electron cloud of an atom or molecule, to be distorted from its normal shape by an external electric field, which may be caused by the presence of a nearby ion or dipole; the electronic polarizability α is defined as the ratio of the induced dipole moment p of an atom to the electric field E that produces this dipole moment: $p = \alpha E$.

polarization of light (polarized light): a state, or the production of a state, in which rays of light exhibit different properties in different directions; **linear (plane)**: when the electric field vector oscillates in a single, fixed plane all along the beam, the light is said to be linearly (plane) polarized; **elliptical**: when the plane of the electric field rotates, the light is said to be elliptically polarized because the electric field vector traces out an ellipse at a fixed point in space as a function of time; **circular**: when the ellipse happens to be a circle, the light is said to be circularly polarized.

polarization anisotropy: an inequality of polarization properties along different axes.

polarization-gating techniques: techniques for selecting diffuse photon groups with different path lengths, in particular **ballistic** or least-scattering photons, based on their polarization properties; used in polarization-sensitive diffuse or **coherence optical tomography** and spectroscopy.

polarization optical spectroscopy: optical **spectroscopy** using polarizied light as a probe beam and/or detection of transmitted, scattered, or re-emitted polarized light.

polarization optical tomography: optical **tomography** using polarizied light as a probe beam and/or detection of transmitted, scattered, or re-emitted polarized light.

polarizer: a device, often a crystal or prism, that produces polarized light from unpolarized light.

polydisperse system: a **disperse** system (medium) with multiple values of characteristic parameters, such as an ensemble of scatterers with different sizes and refractive indices; a cataract eye lens is a good example of a polydisperse system, because it consists of dielectric balls (aggregated α-crystallins) with various refractive indices and radii dispersed in a homogeneous ground substance.

polydispersion: differently sized (and/or with different refractive indices) dispersed particles suspended in a solid, liquid, or gas.

polymer fiber: a fiber made from transparent polymer materials (see **fiber**).

polystyrene microspheres (beads): used for quality control, calibration, and sizing; widely used for cleanroom certification, filter testing, light-scattering experiments, tissue phantoms design, cell labeling, etc.; nanobeads ranging from 40 to 950 nm, microbeads ranging from 1.00 to 9.00 μm, and megabeads ranging from 10.0 to 175.0 μm are available on the market.

polyvinydene fluoride (PVDF): belongs to piezoelectric materials that are used to make electromechanical sensors and actuators.

porosity: a measure of the void spaces in a material, measured as a fraction, between 0–1.

potassium chromate (K_2CrO_4): the nonscattering, homogeneously absorbing liquid used for constructing phantoms.

power: the rate of energy delivery; it is normally measured in watts, that is, joules per second power-size distribution.

preamplifier: an electronic amplifier that precedes another amplifier to prepare an electronic signal for further amplification or processing.

pressure: the force per unit area applied on a surface in a direction perpendicular to that surface; pressure is scalar and has units of pascals, $1\ Pa = 1\ N/m^2$; pressure is transmitted to solid boundaries or across arbitrary sections of fluid normal to these boundaries or sections at every point.

pressure transient: the analysis of **pressure** changes over time.

probe beam: a light or **laser beam** used for an object or material probing.

probability: the relative frequency with which an event occurs or is likely to occur.

probability density function (probability density distribution): a function that describes the distribution of probability over the values of a variable.

propagation constant: the logarithmic rate of change, with respect to distance in a given direction, of the complex amplitude of any electromagnetic field component.

pulse laser: a laser that generates a single pulse or a set of pulses; a laser with **Q-switching** produces the so-called giant pulses, the **mode-locked laser** produces ultrashort pulses with a high repetition rate.

pump-beam (pulse): laser (light) beam (or pulse) used for the nonlinear material pumping or for interactions of the optical fields with matter in order to provide lasing or spectroscopy.

Q-switching: sometimes known as giant pulse formation, is a technique by which a laser can be made to produce a pulsed output beam; the technique allows the production of light pulses with extremely high (gigawatt) peak power, much higher than would be produced by the same laser if it is operating in a continuous wave mode; compared to **mode-locking**, Q-switching leads to much lower pulse repetition rates, much higher pulse energies, and much longer pulse durations; both techniques are sometimes applied at once.

quadrature mixer: an electronic device that mixes signals with different frequencies by the act of squaring.

quantum detection limit: the limit of detection that is defined by the quantum fluctuations of any light source, including a laser, associated with spontaneous emission and defined by the temperature of the medium that emits the light being detected; such fluctuations, as in the case of **photon shot noise**, cause the irregularity of photoelectron emissions hat induce random errors in a photoelectron measuring system; the mean square of photoelectron current fluctuations also is defined by the average photocurrent i and the photodetector's bandwidth B_D: $i^2 = 2ei\,B_\text{D}$, where e is the charge of electrons, but the average photocurrent I is proportional to the mean power of quantum fluctuations: it is also difficult to achieve the quantum detection limit in practice.

quantum dot: a semiconductor nanostructure that confines the motion of conduction band electrons, valence band holes, or excitons (bound pairs of conduction band electrons and valence band holes) in all three spatial directions; the confinement can be due to electrostatic potentials (generated by external electrodes, doping, strain and impurities), the presence of an interface between different semiconductor materials (e.g., in core-shell nanocrystal systems), the presence of the semiconductor surface (e.g., semiconductor nanocrystal), or a combination of these; a quantum dot has a discrete quantized energy spectrum; a quantum dot contains a small finite number (of the order of 1–100) of conduction band electrons, valence band holes, or excitons, i.e., a finite number of elementary electric charges.

quantum efficiency of the detector: the ratio of the number of electrons emitted by a photodetector to the number incident at the detector's surface photons.

quantum flux: see **intensity**.

quantum yield: for a radiation-induced process quantum yield is the number of times that a defined event (usually a chemical reaction step) occurs per photon absorbed by the system; a measure of the efficiency with which absorbed light produces some effect; since not all photons are absorbed productively, the typical quantum yield is less than one; quantum yields greater than one are possible for photo-induced or radiation-induced chain reactions, in which a single photon may trigger a long chain of transformations; in optical spectroscopy, the quantum yield is the probability that a given quantum state is formed from the system initially prepared in some other quantum state; for example, a singlet to triplet transition quantum yield is the fraction of molecules that, after being photoexcited into a singlet state, cross over to the triplet state; the fluorescence quantum yield is defined as the ratio of the number of photons emitted to the number of photons absorbed.

quantum-well laser: a diode laser with a quantum-dimension heterostructure as a lasing medium; owing to a high gain, it has a high slope of the watt/ampere characteristic.

quasi-ballistic photons: photons that migrate within a scattering medium along trajectories that are close but not the same as for **ballistic photons**.

quasi-crystalline approximation: first introduced by Lax to break the infinite heirarchy of equations that results in studies of the coherent field in discrete random media; it simply states that the conditional average of a field with the position of one scatterer held fixed is equal to the conditional average with two scatterers held fixed; successful for a range of concentrations from parse to dense and for long and intermediate wavelengths.

quasi-elastic light scattering: see **dynamic light scattering**.

quasi-monochromatic wave: a wave that has a very narrow but nonzero frequency (or wavelength) bandwidth; it can be presented as a group of monochromatic waves with a slightly different wavelength.

quasi-Newton inverse algorithm: the algorithm for finding an extreme point; it builds up an approximation of the inverse Hessian of the function; it is often regarded as the most sophisticated for solving unconstrained problems.

quasi-ordered medium: a medium that has a structure very close to the ordered one but nevertheless is not completely ordered, which is caused by specific interactions between molecules and molecular structures; many of the natural media, including water and some living tissues, are examples of quasi-ordered media.

quasi-periodic (process, signal, function, fluctuations): almost periodic (process, signal, function, fluctuations); almost periodic, it is a property of dynamical systems that appear to retrace their paths through phase space, but not exactly.

radar graph: similar to line graphs, except that they use a radial grid to display data items; a radial grid displays scale value grid lines circling around a central point, which represents zero; higher data values are farther from the center point; the radar graph type gets its name because it resembles a radar screen; the radial grid is not circular but an equilateral polygon.

radial distribution function $g(r)$: the pair distribution function that is a statistical characteristic of the spatial arrangement of the scatterers; used to describe light scattering in a correlated disperse system.

radiance: see **intensity**, **irradiance**.

radiation dosimetry: the measurement or calculation of a radiation dose; the quantity of radiation absorbed by a given mass of material, especially tissue, is dependent upon the strength and distance of the light source and the duration of exposure.

radiation transfer equation (RTE): the integro-differential equation (the Boltzmann or linear transport equation), which is a balance equation describing the flow of particles (e.g., photons) in a given volume element that takes into account their velocity c, location r, and changes due to collisions (i.e., scattering and absorption).

radiation transfer theory (RTT): the theory based on the **radiation transfer equation (RTE)** allowing one to calculate light distributions in the scattering media with absorption.

radio frequency (RF): the part of the electromagnetic spectrum between about 10^6 and 10^9 Hz.

Raman amplifier: based on the **stimulated Raman scattering (SRS)** phenomenon, this process, as with other stimulated emission processes, allows all-optical amplification; **optical fiber** is almost exclusively used as the nonlinear medium for SRS, which is therefore characterized by a resonant frequency downshift of ∼13 THz; the SRS amplification process can be readily cascaded, thus accessing essentially any wavelength in the fiber low-loss guiding window.

Raman scattering: the change in wavelength of light scattered while passing through a transparent medium; the collection of new wavelengths is characteristic for the molecular structure of the scattering medium and differs from the fluorescence spectrum in being much less intense and unrelated to an absorption band of the medium; the frequencies of new lines are combinations of the frequency of the incident light and the frequencies of the molecular vibrational and rotational transitions.

Raman shifter: a device based on **stimulated Raman scattering** phenomenon; typically 1st, 2nd, and 3rd Stokes components are induced by a nonlinear medium pumped by a laser whose wavelength should be shifted; for example, the optimum conversion in the $Ba(NO_3)_2$ crystal at pump with a Ti:Sapphire laser (815 to 900 nm) provides a 1047 cm^{-1} shift and extends the laser tuning range to 1300 nm; gaseous and liquid Raman cells are also available; however, among the most efficient Raman crystals suitable for a wide range of pumping pulse durations from picoseconds to nanoseconds; $Ba(NO_3)_2$, $KGd(WO_4)_2$, and $BaWO_4$ are known.

Raman spectroscopy: a spectroscopic technique used in condensed matter (physics, chemistry, and biology) to study **vibrational**, **rotational**, and other low-frequency modes in a system; it relies on **inelastic scattering**, or **Raman scattering** of monochromatic light, usually from a **laser** in the **visible**, **near infrared**, or near **ultraviolet** range; **phonons** or other excitations in the system are absorbed or emitted by the laser light, resulting in the energy of the laser **photons** being shifted up or down; the shift in energy gives information about the phonon modes in the system; **infrared spectroscopy** yields similar, but complementary, information.

random medium: a specific state of a nonuniform (inhomogeneous) medium characterized by the irregular spatial distribution of its physical properties, including its optical properties.

random phase screen (RPS): a specific state of a random medium characterized by random spatial variations of the refractive index, which induces corresponding variations in the phase shift of the optical wave transmitted through or reflected by the RPS.

raw experimental data: experimental data before processing.

Rayleigh-Debye theory (approximation): the theory that addresses the problem of calculating the scattering by a special class of arbitrarily shaped particles; it requires that the electric field inside the particle be close to that of the incident field and that the particle can be viewed as a collection of independent dipoles that are all exposed to the same incident field.

Rayleigh distribution: the probability distribution of a random variable x described by the probability density function $p(x) = (x/a^2)\exp(-x^2/2a^2)$, $x \geq 0$; $p(x) = 0$, $x < 0$; the distribution has a positive asymmetry; its mode is at the point $x = a$; the mean value and variance are, respectively, equal to $\langle x \rangle = (\pi/2a)$ and $\sigma^2 = (4 - \pi)a^2/2$.

Rayleigh-Gans theory (approximation): see **Rayleigh-Debye theory (approximation)**.

Rayleigh (resolution) limit: the resolution of an optical device with a circular aperture is limited by the diffraction of light through that aperture; as the aperture increases in diameter, the diffraction spot gets smaller, which increases the resolution of the instrument; in the case of a circular aperture, the diffraction pattern has the shape of a disk surrounded by rings, which is called the Airy disk; if the images of two point sources of light overlap such that the centers of the images are closer than the radius of the Airy disk, the images are considered to be unresolvable; this definition can be written as: $\Delta\theta = 1.22\lambda/D$, where $\Delta\theta$ is the minimum resolvable angular separation of the two objects, λ is the wavelength of the light, and D is the diameter of the aperture.

Rayleigh (scattering) theory: the theory that addresses the problem of calculating scattering by small particles (with respect to the wavelength of the incident light) when individual particle scattering can be described as if it is a single dipole, the scattered irradiance is inversely proportional to λ^4 and increases as a^6, and the angular distribution of the scattered light is isotropic.

reduced scattering coefficient: a lumped property incorporating the **scattering coefficient** μ_s and the **scattering anisotropy factor** g: $\mu'_s = \mu_s(1 - g)$ [cm^{-1}]; μ'_s describes the diffusion of photons in a random walk of step size of $1/\mu'_s$ [cm] where each step involves isotropic scattering; this is equivalent to the description of photon movement using many small steps, $1/\mu_s$, which each involve only a partial (anisotropic) deflection angle if there are many scattering events before an absorption event, i.e., $\mu_a \ll \mu'_s$ (diffusion regime, see **diffusion approximation**); μ'_s is useful in the diffusion regime, which is commonly encountered when treating how visible and near-infrared light propagates through tissues (http://omlc.ogi.edu/classroom/).

reflectance (reflection coefficient): the ratio of the intensity reflected from a surface to the incident intensity; it is a dimensional quantity.

reflecting spectroscopy: the spectroscopy that uses the light back-reflected (scattered) by an object for spectral analysis.

refraction: the change in direction of a ray of light, sound, heat, or the like, in passing obliquely from one medium into another in which its speed is different; the ability of the eye to refract entering light, forming on the retina; the determination of the refractive condition of the eye.

refractive index: see **index of refraction**.

refractive index mismatch: a difference in the index of refraction of two media in contact; a scattering medium that contains scattering particles whose index of refraction is mismatched relative to the index of refraction of the ground substance [see **immersion medium (liquid)** and **immersion technique**].

repetition rate: the number of pulses per second; the repetition rate is measured in hertz.

reproducibility: one of the main principles of the scientific method, and refers to the ability of a test or experiment to be accurately reproduced, or replicated, by someone else working independently.

resistor: a two-terminal electrical or electronic component that resists an electric current by producing a voltage drop between its terminals in accordance with Ohm's law.

retarder: see **optical retarder**.

reverberation: the persistence of sound in a particular space after the original sound is removed; when sound is produced in a space, a large number of echoes build up and then slowly decay as the sound is absorbed by the walls and air, creating reverberation, or reverb.

Riccati-Bessel function: only slightly different from a spherical Bessel function, this function arises in the problem of scattering of electromagnetic waves by a sphere, known as **Mie** scattering.

root mean square (rms): a measure of dispersion in a frequency distribution, it is equal to the square root of the mean of the squares of the deviations from the arithmetic mean of the distribution.

rotational diffusion: the molecular rotational motion is usually only the rotational rocking near the equilibrium orientation; they depend on the interactions with their neighbors, and by jumping in time they are changing orientation; the energy of activation is required for changing the angle of orientation; the Brownian rotational motion can be valid only for comparatively big molecules with the slow changing of orientation angles; in this case the differential character of rotational motion is valid and the rotational diffusion equation can be written; the interaction of molecules between each other can be considered as the friction foresees with the moment P proportional to the angle velocity Ω, $P = \xi\Omega$, where ξ is the rotational coefficient of friction that can be connected with the rotational diffusion coefficient, $D_R = kT/\xi$; in the case of a small macroscopic sphere with radius a, $\xi = 8\pi a^3 \eta$, where η is the coefficient of viscosity (http://aph.huji.ac.il/feldman/diel/Diel_Lecture9.ppt).

rotational state (level): the particular pattern of energy levels (and hence of transitions in the rotational spectrum) for a molecule is determined by its symmetry: linear molecules (or linear rotors), symmetric tops (or symmetric rotors), spherical tops (or spherical rotors), and asymmetric tops; rotational spectroscopy (using microwave and/or Raman spectroscopic techniques) studies the absorption and emission electromagnetic radiation by molecules associated with a corresponding change in the rotational quantum number of the molecule.

scatterer: an inhomogeneity or a particle of a medium that refracts or diffracts light or other electromagnetic radiation; light is diffused or deflected as a result of collisions between the wave and particles of the medium; sometimes it is a rough surface or a random-phase screen, also called a scatterer.

scattering: the process in which a wave or beam of particles is diffused or deflected by collisions with particles of the medium it transverses.

scattering angle: related to a photon scattered by a particle so that its trajectory is deflected by a deflection (scattering) angle θ in the **scattering plane** and/or by the azimuthal angle of scattering φ in the plane perpendicular to the scattering plane.

scattering anisotropy factor: the amount of forward direction retained after a single scattering event; if a photon is scattered by a particle so that its trajectory is deflected by an angle θ, then the component of the new trajectory aligned in the forward direction is presented as $\cos\theta$; there is an average deflection angle, and the mean value of $\langle\cos\theta\rangle$ is defined as the anisotropy (http://omlc.ogi.edu/classroom/).

scattering coefficient: a particle with a particular geometrical size redirects incident photons into new directions and so prevents the forward on-axis transmission of photons, this process constitutes **scattering**; the scattering coefficient $\mu_s[\text{cm}^{-1}]$ describes a medium containing many scattering particles at a concentration described as a volume density $\rho[\text{cm}^3]$; the scattering coefficient is essentially the cross-sectional area $\sigma_{sca}[\text{cm}^{-1}]$ per unit volume of medium: $\mu_s = \rho\sigma_{sca}$ (http://omlc.ogi.edu/classroom/).

scattering indicatrix: an angular dependence of the scattered light intensity; for thin samples, the normalized scattering indicatrix is equal to the **scattering phase function**.

scattering medium: a medium in which a wave or beam of its particles is diffused or deflected by collisions with particles.

scattering phase function: the function that describes the scattering properties of the medium and is, in fact, the probability density function for scattering in the direction $\bar{s}'$ of a photon traveling in the direction $\bar{s}$; it characterizes an elementary scattering act: if scattering is symmetric relative to the direction of the incident wave, then the phase function depends only on the scattering the angle θ (angle between directions $\bar{s}$ and $\bar{s}'$).

scattering plane: a plane defined by positions of a light source, a scattering particle, and a detector.

scattering spectrum: the spectrum of scattered light; it can be differential, measured, or calculated for a certain scattering angle, or integrated within an angle (field) of view of the measuring spectrometer.

second-harmonic generation (SHG): (also called frequency doubling) a nonlinear optical process in which photons interacting with a nonlinear material are effectively "combined" to form new photons with twice the energy, and therefore twice the frequency and half the wavelength of the initial photons; in the past several years, SHG has been extended to biological applications: to the imaging of molecules that are intrinsically second-harmonic-active in live cells, such as collagen, and for studying biological molecules by labeling them with second-harmonic-active tags, in particular as a means to detect conformational change at any site and in real time.

self-beating: the signal produced by photomixing the electric components of a scattered field.

semilogarithmic scale: for example, functions of the kind $y = Be^{-\varphi x}$ are used to describe the attenuation of light intensity with distance x and may be plotted on a semilogarithmic scale; taking $\log_{10}$ of both sides gives $\log_{10} y = \log_{10} B - \varphi x$ $\log_{10} e = \log_{10} B - 0.434 \varphi x$; plotting $\log y$ vs x will therefore give a straight line of slope $0.434 \times \varphi$.

shear rate: the rate of shear **deformation**; for the ease of it is just a gradient of velocity.

shear stress: a **stress** state where the stress is parallel or tangential to a face of the material, as opposed to normal stress when the stress is perpendicular to the face; for a Newtonian fluid wall shear stress is proportional to shear rate, where the coefficient of proportionality is the viscosity of the fluid.

short-circuit: in electronics, a circuit with a load of an infinitely low resistance or **impedance** through which electric current flows.

shot noise: a type of electronic noise that occurs when the finite number of particles that carry energy, such as electrons in an electronic circuit or photons in an optical device, is small enough to give rise to detectable statistical fluctuations in a measurement; it is important in electronics and photoelectronics; the strength of this noise increases with the average magnitude of the current or intensity of the light; often, however, as the signal increases more rapidly as the average signal becomes stronger, shot noise often is only a problem with small currents and light intensities.

Siegert formula: the formula that, for Gaussian statistics, relates the intensity **autocorrelation function** to the first-order autocorrelation function.

signal oscillator: used in heterodyne detecting systems for notation of the receiving signal, which can be presented as the radio- (or optical) frequency oscillator and has a slightly different frequency than that for the **local oscillator** [used for converting a high-frequency receiving signal to an intermediate frequency by mixing the local oscillator signal and receiving signal at an electronic (or photo-) detector].

signal-to-noise ratio: the ratio of a received (the detector) signal (an electric impulse) to noise (an electric disturbance in a measuring system that interferes with or prevents reception of a signal).

single-frequency laser: a laser that generates a single frequency (one longitudinal mode).

single-integrating sphere "comparison" technique: this technique uses a single integrating sphere containing no baffles but three ports that can be opened for light transmission, or closed or covered by a sample or reference standards for the calibration (comparison) of reflectance or transmittance measurements; an additional two ports are used to illuminate samples by a collimated light beam and collect the scattered light by a fiber bundle placed at the "north pole" of the sphere; this technique has an advantage over the conventional **double integrating sphere technique** in that no corrections are required for sphere properties.

single-mode fiber: a fiber in which only a single mode can be excited; for a fiber with a numerical aperture 0.1 and a wavelength of 633 nm the single mode can be excited if the core diameter is less than 4.8 μm.

single-mode fiber-optic Michelson interferometer: a **Michelson interferometer** integrated with a **fiber-optic single-mode X-coupler**, which optically connects a light source, reference mirror, object, and photodetector.

single-mode laser: a laser that produces a light beam with a Gaussian shape of the transverse intensity profile without any spatial oscillations (see **Gaussian light beam**); in general, such lasers generate many optical frequencies (so-called longitudinal modes), which have the same transverse Gaussian shape.

single-photon counting mode: see **photon-counting system**.

single scattering: the scattering process that occurs when a wave undertakes no more than one collision with particles of the medium in which it propagates.

single scattering approximation: the approximation that assumes tissue is sufficiently thin that single scattering accurately estimates the reflection and transmission for the slab.

singlet state: one of the two ways in which the **spin** of two electrons in an atom or molecule can be combined in atomic physics, the other being a **triplet state**; a single electron has spin $1/2$, and a pair of electron spins can be combined to form a state of total spin 1 (triplet state) and a state of spin 0 (singlet state); singlet state is an excited state of a molecule that, upon absorbing light, can release energy as heat or light (**fluorescence**) and thus return to the initial (ground) state; it may alternatively assume a slightly more stable, but still excited state (triplet state), with an electron still dislocated as before but with reversed spin.

singular eigenfunction method: the method for rigorous solving of the transport equation which is solved using the Green's functions in terms of the singular eigenfunctions and their orthogonality relations together with the appropriate boundary conditions; the convergence of the numerical results is fast and the analytical expressions are simple for solving numerically.

small angular (angle) approximation: a useful simplification of the laws of trigonometry, which is only approximately true for finite angles, but correct in the limit as the angle approaches zero; it involves linearization of the trigonometric functions (truncation of their Taylor series); this approximation is useful in many areas of physical science, including optics, where it forms the basis of the **paraxial approximation**.

S-matrix: see **LSM [light-scattering matrix (intensity or Mueller matrix)]**.

snake photons: photons that travel in near-forward paths, having undergone few scattering events, all of which are in the forward or near-forward direction; consequently, they retain the image bearing characteristics to some extent.

soft scattering particles: the refractive index of these particles, n_s, is close to the refractive index of the ground (interstitial) substance, n_0 ($n_s \geq n_0$).

sonoluminescence: the emission of short bursts of light (**luminescence**) from imploding bubbles in a liquid when excited by sound.

sonophoresis: a process that exponentially increases the absorption of topical compounds (transdermal delivery) into the epidermis, dermis, and skin appendages; it occurs because ultrasound waves stimulate microvibrations within the skin epidermis and increase the overall kinetic energy of molecules making up topical agents; it is widely used in hospitals to deliver drugs through the skin; pharmacists compound the drugs by mixing them with a coupling agent (gel, cream, ointment) that transfers ultrasonic energy from the ultrasound transducer to the skin; the ultrasound probably enhances drug transport by cavitation, microstreaming, and heating; it is also used in physical therapy.

Soret band: a very strong absorption band in the blue region of the optical absorption spectrum of a haem protein.

spatial correlation: see **correlation**; valid for the spatial variables.

spatial frequency: a spatial harmonic in the Fourier transform of a periodic or aperiodic (random) spatial distribution.

spatial resolution: a measure of the ability of an optical imaging system to reveal the details of an image, i.e., to resolve adjacent elements.

spatially modulated laser beam: a laser beam with regular interference fringes or irregular speckle modulation.

spatially resolved reflectance technique (SRR): a technique that uses two or more fibers to illuminate an object and collect the back-reflected light; the positions of the illuminating and light-collecting fibers can be fixed or scanned along the object's surface perpendicular or have some angle to the object's surface.

specific heat capacity: also known simply as specific heat; an intensive quantity, meaning it is a property of the material itself, and not the size or shape of the sample; its value is affected by the microscopic structure of the material; commonly, the amount is specified by mass; for example, water has a mass-specific heat capacity of about 4186 joules per Kelvin per kilogram; volume-specific and molar-specific heat capacities are also used; the specific heat of virtually any substance can be measured, including pure elements, compounds, alloys, solutions, and composites.

speckle: a single element of a speckle structure (pattern) that is produced as a result of the interference of a large number of elementary waves with random phases that arise when coherent light is reflected from a rough surface or when coherent light passes through a scattering medium.

speckle contrast: see **contrast of the intensity fluctuations**.

speckle correlometry: a technique based on the measurement of the intensity **autocorrelation function**, characterizing the size and distribution of speckle sizes in a speckle pattern, caused, for example, by the scattering of a coherent light beam from a rough surface: the statistical properties of the scattering object's structure can be deduced from such measurements.

speckle interferometry: the technique that uses the **interference of speckle fields**.

speckle photography: the measuring technique that uses a set of sequential photos of the speckle pattern taken at different moments or with different exposures: this is a full-field technique and can be used to study the dynamic properties of a scattering object (see **LASCA**); the updated instruments make use of computer-controlled CCD cameras for averaging and storing the speckle patterns.

speckle statistics of the first order: the statistics that define the properties of speckle fields at each point.

speckle statistics of the second order: the statistics that show how fast the intensity changes from point to point in a speckle pattern, i.e., they characterize the size and the distribution of speckle sizes in the pattern.

speckle structure: see **speckle**.

Spectrolon: a very white reflective plastic used as the "white reference" in spectral measurements and in **integrating sphere** spectrometers.

spectrophotometry: the spectroscopic method and instrument for making photometric comparisons between parts of spectra.

spectroscopy: the science that deals with the use of the spectroscope and with spectrum analysis.

spectroscopy of intensity fluctuations: see photon-correlation spectroscopy.

spectrum: the range of frequencies or wavelengths.

spectrum analysis: to ascertain the number and character of the constituents combining to produce a signal spectrogram.

spectrum analyzer: an instrument for making the spectrum analysis of a signal.

specular: pertaining to or having the properties of a mirror.

spin: the angular momentum intrinsic to a body; in classical mechanics, the spin angular momentum of a body is associated with the rotation of the body around its own center of mass; in quantum mechanics, spin is particularly important for systems at atomic length scales, such as individual atoms, protons, or electrons; such particles and the spin of quantum mechanical systems ("particle spin") possesses several nonclassical features, and for such systems, spin angular momentum cannot be associated with rotation but instead refers only to the presence of angular momentum.

standard deviation: see **rms (root mean square)**.

statistics: a mathematical science pertaining to the collection, analysis, interpretation or explanation, and presentation of data; it is applicable to a wide variety of academic disciplines.

statistical approach: an approach based on **statistics** as a mathematical science.

statistically significant: a result is called significant if it is unlikely to have occurred by chance; "a statistically significant difference" simply means there is statistical evidence that there is a difference; the significance of a result is also called its p-value; the smaller the p-value, the more significant the result is said to be; popular levels of significance are 5%, 1%, and 0.1%.

stepper motor: a machine that converts electrical energy into mechanical energy by steps; used as the computer-controlled mechanical drivers of optical stages.

stimulated Raman scattering (SRS): a phenomenon that occurs when a lower frequency "signal" photon induces the inelastic scattering of a higher-frequency "pump" photon in a nonlinear optical medium; as a result, another "signal" photon is produced, with the surplus energy resonantly passed to the **vibrational states** of the nonlinear medium; this process is the basis for the **Raman amplifier**.

Stokes parameters: the four numbers I, Q, U, and V representing an arbitrary polarization of light; I refers to the irradiance or intensity of the light; the parameters Q, U, and V represent the extent of horizontal linear, 45° linear, and circular polarization, respectively.

Stokes-Raman scattering: the energy of the Raman scattered photons is lower than the energy of the incident photons.

Stokes shift: the difference (in wavelength or frequency units) between positions of the band maxima of the absorption and **luminescence** spectra (or **fluorescence**) of the same **electronic transition**; when a molecule or atom absorbs light, it enters an excited electronic state; the Stokes shift occurs because the molecule loses a small amount of the absorbed energy before re-releasing the rest of the energy as fluorescence, depending on the time between the absorption and the reemission; this energy is often lost as thermal energy.

Stokes vector: the vector formed by the four **Stokes parameters**.

Stokes wave: the induced (scattered) wave that has a frequency less than the frequency of the incident radiation.

streak camera (synchroscan streak camera): an instrument for recording the temporal profile of light intensity with a high time resolution (of about 10 ps), displaying it as a spatial profile; synchronous scanning controlled by a reference (**trigger beam**) light pulse is provided.

stress: internal distribution of force per unit area that balances and reacts to external loads applied to a body; stress is a second-order tensor with nine components, but can be fully described with six components due to symmetry in the absence of body moments; stress is often broken down into its shear and normal components as these have unique physical significance; stress can be applied to solids, liquids, and gases; static fluids support normal stress (hydrostatic pressure) but will flow under **shear stress**; moving viscous fluids can support shear stress (dynamic pressure); solids can support both shear and normal stress, with ductile materials failing under shear and brittle materials failing under normal stress; all materials have temperature dependent variations in stress related properties, and **non-Newtonian** materials have rate-dependent variations.

stress amplitude: the value of a **stress**.

stress distribution: **stress** is a second-order tensor with nine components.

structure function: the function that describes the second-order statistics of a random process and is proportional to the difference between values of the **autocorre-**

lation function for zero and arbitrary values of the argument; the structure function is more sensitive to small-scale oscillations.

subject arm of an interferometer: the arm on which an object under study is placed.

subjective speckles: the speckles produced in the image space of an optical system (including an eye).

superluminescent diode: a very bright diode light source with a broad linewidth; it is usually manufactured using a laser diode technology (heterostructure, waveguide, etc.), but without reflecting mirrors (there is an antireflection coating at the diode faces or their out-of-parallelism is provided).

surface-enhanced Raman scattering (SERS): a strong increase in **Raman** signals from molecules if those molecules are attached to submicron metallic structures; for a rough surface due to excitation of **electromagnetic resonances** by the incident radiation, such enhancement may be of a few orders; both the excitation and Raman scattered fields contribute to this enhancement; thus, the SERS signal is proportional to the fourth power of the field enhancement factor.

surface plasmon resonance: also known as surface plasmon polaritons, surface plasmon resonances are surface electromagnetic waves that propagate parallel to a metal/dielectric interface; for electronic surface plasmons to exist, the real part of the dielectric constant of the metal must be negative, and its magnitude must be greater than that of the dielectric; this condition is met in the visible-IR wavelength region for air/metal and water/metal interfaces (where the real dielectric constant of a metal is negative and that of air or water is positive); the excitation of surface plasmons by light is denoted for planar surfaces as for nanometer-sized metallic structures which is called localized surface plasmon resonance; typical metals that support surface plasmons are silver and gold, but metals such as copper, titanium, or chromium can also support surface plasmon generation; surface plasmons have been used to enhance the surface sensitivity of several spectroscopic methods, including **fluorescence**, **Raman scattering** (see **surface-enhanced Raman scattering**), and **second harmonic generation**.

symmetric molecule: refers to molecular geometry; molecules have fixed equilibrium geometries—bond lengths and angles—about which they continuously oscillate through vibrational and rotational motions; a symmetric molecule contains identical bonds; for example, trigonal planar, tetrahedral and linear bonding arrangements often lead to symmetrical, nonpolar molecules that contain polar bonds.

symmetric vibrational mode: for example, a moving linear triatomic molecule being in a movement, when each atom oscillates or vibrates along a line connecting them, may be in symmetric or antisymmetric vibrational mode relative to a central atom.

systematic errors: the errors caused by finite tissue volume, curved surfaces, tissue inhomogeneity when scanning, finite source and detection size, uncertainty in their relative positions, etc.; they can be much larger than random errors induced by a **shot noise**.

swept-laser source: a rapidly tunable laser over a broad optical bandwidth.

tensor: a tensor has slightly different meanings in mathematics and physics; in the mathematical fields of multilinear algebra and differential geometry, a tensor is a multilinear function; in physics and engineering, the same term usually means what a mathematician would call a tensor field: an association of a different (mathematical) tensor with each point of a geometric space, varying continuously with position; in the field of diffusion tensor imaging, for instance, a tensor quantity that expresses the differential permeability of organs to water in varying directions is used to produce scans of the brain; perhaps the most important engineering examples are the **stress** tensor and strain tensor, which are both second-rank tensors, and are related in a general linear material by a fourth-rank-elasticity tensor; the rank of a particular tensor is the number of array indices required to describe such a quantity.

therapeutic (or diagnostic) window: the spectral range from 600 to 1600 nm within which the penetration depth of light beams for most living tissues and blood is the highest; certain phototherapeutic and diagnostic modalities take advantage of this range for visible and NIR light.

thermal blooming: a major effect in high-power laser beams transmitting through gaseous mediums as well as the atmosphere; due to this nonlinear heating effect, the beam pattern is deformed through the propagation path.

thermal diffusivity: the ratio of **heat (thermal) conductivity** to volumetric **heat capacity** in **heat transfer** analysis; expressed in units of $m^2\,s^{-1}$.

thermal expansion coefficient: the energy that is stored in the intermolecular bonds between atoms changes during **heat transfer**; when the stored energy increases, so does the length of the molecular bond; as a result, solids typically expand in response to heating and contract on cooling; this response to temperature change is expressed as its coefficient of thermal expansion; the coefficient of thermal expansion is used in two ways: as a volumetric thermal expansion coefficient (liquids and solid state) and as a linear thermal expansion coefficient (solid state).

thermal image: pictures created by heat, received by a thermal imager, rather than light; it measures radiated IR energy and converts the data to corresponding maps of temperatures; instruments provide temperature data at each image pixel; images may be digitized, stored, manipulated, processed, and printed out.

thermal length: the length of thermal diffusivity that chacterizes the distance in a medium where heat is diffused during the heating laser pulse.

thermal lensing: the virtual lens that is induced in a transparent material by its local heating, particularly by laser beam absorption; the local changes in the refractive index of a sample induce such a lens for some period; such an effect can be used to estimate tissue optical and thermal properties if a probing laser beam is applied.

thermal relaxation time: the time to dissipate the heat absorbed during a laser pulse.

thermoelastic effect: the generation of mechanical stress (acoustic) waves via the time-dependent thermal expansion of a sample.

third harmonic generation: if a narrow-band optical wave pulse at a frequency ω propagates through a nonlinear medium with a nonzero Kerr **nonlinear susceptibility** $\chi^{(3)}$ due to nonlinearity, one will get a signal at a frequency 3ω (see **second harmonic generation**).

three-photon fluorescence microscopy: the microscopy that employs both **ballistic** and scattered photons at the wavelength of the **third harmonic** of incident radiation; it possesses the same advantages as **two-photon fluorescence microscopy** but ensures a somewhat higher spatial resolution and provides an opportunity to excite chromophores with shorter wavelengths.

time-correlated single-photon counting technique: the time-resolved single-photon counting method and instrument (see **photon-counting system**) used for receiving low-intensity light pulses.

time-dependent radiation transfer theory (RTT): the theory that is based on the time-dependent integro-differential equation (the Boltzmann or linear transport equation), which is a balance equation describing the time-dependent flow of particles (e.g., photons) in a given volume element that takes into account their velocity c, location $\bar{r}$, and changes due to collisions (i.e., scattering and absorption).

time-domain technique: a spectroscopic or imaging technique that uses ultrashort laser pulses.

time-gating: a method for selecting photon groups with different arriving times to a detector within a selected and moveable time window; used in diffuse optical tomography and spectroscopy; may be purely electrical or optical, or a combination of both.

time-of-flight: the mean time of photon travel between two points that account for refractive index and scattering properties of the medium.

time-share control: the regime that ensures that, at one time, an optical signal of only one wavelength passes through the whole system.

tissuelike phantom: see **phantom**.

tissue optical parameters (properties) control: any kind of physical or chemical action, such as mechanical stress or changes in osmolarity, which induces reversible or irreversible changes in the optical properties of a tissue [see **immersion medium (liquid)**, **immersion technique**, **matching substance**, and **mechanical stress**].

TOAST (time-resolved optical absorption and scattering tomography): an image reconstruction package developed at University College London that employs a finite-element-method-forward model and an iterative reconstruction algorithm.

tomographic reconstruction: the mathematical procedure of obtaining 3D images by which the size, shape, and position of a hidden object can be determined.

tomography: imaging by sections or sectioning (the Greek word tomos, meaning "a section" or "a cutting"); a device used in tomography is called a tomograph, while the image produced is a tomogram; the method is used in medicine, biology, and other sciences; in most cases it is based on the mathematical procedure called **tomographic reconstruction**; there are many different types of tomography, including functional magnetic resonance imaging (fMRI), **magnetic resonance imaging (MRI)**, **optical coherence tomography (OCT)**, optical projection tomography (OPT), positron emission tomography (PET), single photon emission computed tomography (SPECT), x-ray tomography.

total internal reflection: the reflection of light at the interface between media of different refractive indexes, when the angle of incidence is larger than a critical angle (determined by the media).

transfer matrix method: this method assumes that light consists of various plane waves traveling with oblique angles through the sample; the latent bulk image is obtained by first calculating the vertical amplitude dependence of the field, resulting from the excitation with one plane wave of definite amplitude; it is is applicable for the analysis of a stratified medium; using the vector version of the transfer matrix algorithm, an arbitrarily polarized light can be simulated.

transillumination digital microscopy (TDM): the light transillumination **microscopy** based on the usage of fast and high resolution CCD cameras and corresponding software; it is applicable for *in vivo* **flow cytometry**.

transition matrix (T-matrix) approach: this approach is similar to the Mie theory used for nonspherical objects such as spheroids; the T-matrix for the spherical particles is diagonal.

transmittance: the ratio of the intensity transmitted through a sample to the incident intensity; it is a dimensionless quantity.

trigger beam: the part of a laser beam used to synchronize the measuring system (for example, the streak camera).

triplet state: see **singlet state**.

t-**test (Student's *t*-test)**: a test for determining whether an observed sample mean differs significantly from a hypothetical normal population mean.

tunable laser: most lasers emit at a particular wavelength; in tunable lasers, one can vary the wavelength over some limited spectral range.

turbidity: a cloudiness or haziness of material (biological fluid or tissue), caused by individual particles (suspended scatterers) that are generally invisible to the naked eye, thus being much like milk.

turbulent flow: a flow regime characterized by chaotic, stochastic property changes; this includes low-momentum diffusion, high-momentum convection, and rapid variation of pressure and velocity in space and time; flow that is not turbulent is called **laminar flow**; the dimensionless Reynolds number characterizes whether flow conditions lead to laminar or turbulent flow; e.g., for pipe flow, a Reynolds number above about 2300 will be turbulent.

two-frequency Zeeman laser: a laser with the active medium placed in the axial magnetic field; the laser produces two laser lines with a small frequency separation (about 250 kHz) and mutually orthogonal linear polarizations.

two-photon fluorescence microscopy: the microscopy that employs both **ballistic** and scattered photons at the wavelength of the second harmonic of incident radiation coming to a wide-aperture photodetector exactly from the focal area of the excitation beam.

ultrashort laser pulse: the pulses usually produced by mode-locked lasers (picosecond and subpicosecond range) or their modifications, such as synchronously optically pumped or colliding-pulse mode-locked dye (CPM laser) lasers (femtosecond range), or the titanium-sapphire laser with passive mode locking via a Kerr lens (KLM laser) (10–100 fs).

ultrasonic transducer: a device that converts energy into **ultrasound**; refers to a **piezoelectric transducer** that converts electrical energy into sound; alternative methods for creating and detecting ultrasound include magnetostriction and capacitive actuation; it is used in many applications including medical ultrasonography, and nondestructive testing.

ultrasound: mechanical vibrations with frequencies in the range of 2×10^4 to 10^7 Hz.

uniaxial crystal: an **anisotropic crystal** that exhibits two refractive indices: an "ordinary" index (n_o) for light polarized in the x or y directions, and an "extraordinary" index (n_e) for polarization in the z direction; a uniaxial crystal is "positive" if $n_e > n_o$ and "negative" if $n_e < n_o$; light polarized at some angle to the axes will experience a different phase velocity for different polarization components and cannot be described by a single index of refraction; this is often depicted as an index ellipsoid.

variance: the square of the standard deviation.

vector RTT: **radiation transfer theory (RTT)** accounting for the polarization properties of light and its interaction with a scattering medium.

vibrational spectrum: see **vibrational transition**.

vibrational transition: denotes an energetic transition of a molecule with the change of vibrational quantum number (energetic state, or level); at vibrational transitions, only the absorption or emission of infrared light (**vibrational spectrum**) is possible.

vibronic spectrum: see **vibronic transition**.

vibronic transition: denotes the simultaneous change of a vibrational and electronic quantum number (energetic state or level) in a molecule; according to the separability of electronic and nuclear motion in the Born-Oppenheimer approximation, the vibrational transition and electronic transition may be described separately; the selection rule for vibrational transitions is described by the **Franck-Condon principle**; most processes lead to the absorption and emission of relatively broad bands of visible light (**vibronic spectra**), resulting in the colorful world around us.

vibrometer: an instrument that measures amplitudes and frequencies of the mechanical vibrations of an object.

viscosity of the medium: viscosity arises from the friction between one layer of a fluid in motion relative to another layer of the fluid; it is caused by the cohesive forces between molecules; the viscosity of glycerol is high, but the viscosity of water or ethanol is low.

volume fraction: a fraction dealing with mixtures in which there is a large disparity between the sizes and refractive indices of the various kinds of molecules or particles; it provides an appropriate way to express the relative amounts of the various components; in any ideal mixture, the total volume is the sum of the individual volumes prior to mixing; in nonideal cases the additivity of volume is no longer guaranteed; volumes can contract or expand upon mixing and molar volume becomes a function of both concentration and temperature; this is why mole fractions are a safer unit to use.

waist of a laser beam: the narrowest part of a Gaussian beam.

water-binding mode: denotes biological molecule interaction with water molecules and corresponding changes in molecular-water complex spectra.

watt: a unit of power; one watt is equal to one joule per second.

wavelength: distance between two adjacent peaks in a wave.

wavelet transformation: the representation of a signal in terms of scaled and translated copies (known as "daughter wavelets") of a finite length or fast decaying oscillating waveform (known as the "mother wavelet"); in formal terms, this representation is a wavelet series that is a representation of a square-integrable (real or complex valued) function by a certain orthonormal series generated by a wavelet.

weakly scattering RPS: the RPS whose variance in induced-phase fluctuations in the scattered field is much less than unity.

Wigner phase space distribution function: the complex function that defines the coherence property of an optical field for a given position depending on the wave vector.

white noise: a random signal (or process) with a flat power spectral density; the signal's power spectral density has equal power in any band, at any center frequency, having a given bandwidth; white noise is considered analogous to white light, which contains all frequencies.

Wollaston prism: an optical device invented by William Hyde Wollaston that manipulates polarized light; it separates randomly polarized or unpolarized light into two orthogonal, linearly polarized outgoing beams; a prism consists of two orthogonal calcite prisms cemented together on their base (typically with Canada balsam) to form two right triangle prisms with perpendicular optic axes; outgoing light beams diverge from the prism, giving two polarized rays, with the angle of divergence determined by the prisms' wedge angle and the wavelength of the light; commercial prisms are available with divergence angles from $15°$ to about $45°$.

xenon arc lamp: a discharge arc lamp filled with xenon; it gives out very bright UV and visible light in the range from 200 nm to >1.0 μm.

x ray (or Röntgen rays): a form of electromagnetic radiation with a wavelength in the range of 10 to 0.01 nm; primarily used for diagnostic radiography and crystallography; it is a form of ionizing radiation and can be dangerous.

zigzag or snake photons: low-angle scattered photons having zigzag (or snake) trajectories.

ZnSe crystal: a crystal for **ATR** (attenuated total reflectance) spectroscopy; insoluble with a refractive index of 2.4, a long-wavelength cut-off frequency of 525 cm^{-1}, a depth of penetration at 1000 cm^{-1} of 1.66 μm, and a pH range of samples under study of 5–9.

z-scan: see **A-scan**.

Sources

This glossary was compiled using mostly Refs. 1–7, 25, 40, 75, 87, 129, 130, 132, 135, 136, and the following sources:

1. *Webster's New Universal Unabridged Dictionary*, Barnes & Noble Books, New York, 1994.
2. A. Godman and E. M. F. Payne, *Longman Dictionary of Scientific Usage*, reprint edition, Longman Group, Harlow, UK, 1979.

3. A. M. Prokhorov (ed.), *Physical Encyclopedic Dictionary*, Soviet Encyclopedia, Moscow, 1983.
4. A. M. Prokhorov (ed.), *Physical Encyclopedia*, vol. 1, Soviet Encyclopedia, Moscow, 1988.
5. A. M. Prokhorov (ed.), *Physical Encyclopedia*, vol. 2, Soviet Encyclopedia, Moscow, 1990.
6. A. M. Prokhorov (ed.), *Physical Encyclopedia*, vol. 3, Big Russian Encyclopedia, Moscow, 1992.
7. A. M. Prokhorov (ed.), *Physical Encyclopedia*, vol. 4, Big Russian Encyclopedia, Moscow, 1994.
8. http://en.wikipedia.org

Glossary 2. Medicine, Biology, and Chemistry

abdominal fat: the adipose tissue that contains fat cells and that is found around the abdominal organs such as the **intestines, kidneys**, and **liver**.

abdominal organs: the organs contained in the abdominal region of the body; the **diaphragm** separates the abdomen from the **thorax**; the abdomen is posterior to the thorax; viscera other than the **heart** or lungs (e.g., **intestines, kidneys, liver**) are abdominal organs.

ablation: the removal of tissue.

abrasive cream: a cream, containing abrasive (hard mineral) particles, that allows one to provide **skin peeling** and make its relief more smooth and penetrative for **liposomes** and **nanospheres**.

acanthocyte: shrunken **erythrocyte**, also known as a spur cell, the term is derived from the Greek word "acanthi" meaning "thorn"; the acanthocyte cell has five to ten irregular, blunt, fingerlike projections that vary in width, length, and surface distribution; acanthocytes form when erythrocyte **membranes** contein excess **cholesterol** compared to **phospholipid** content, which is caused by the increase in blood cholesterol content or the presence of abnormal plasma lipoprotein composition.

acetic acid: a colorless, pungent, water-miscible liquid, CH_3COOH, used in the production of numerous esters that are solvents and flavoring agents.

acetowhitening effect: the effect caused by acetic acid when used during **colposcopy** to enhance differences in the diffuse reflectance (whitening) of normal and diseased regions of the cervical **epithelium**; transient whitening of tissue after the application of acetic acid serves as a simple and inexpensive method for identifying areas that may eventually develop into **cervical cancer**.

actinic keratosis: a scaly or crusty bump that forms on the skin surface; also called solar keratosis, sun spots, or precancerous spots.

acyl group: a functional group derived by the removal of one or more hydroxyl group from an oxoacid; in organic chemistry, the acyl group is usually derived from a carboxylic acid in the form of RC O OH; it therefore has the formula $RC(=O)-$, with a double bond between the carbon and oxygen atoms (i.e., a carbonyl group), and a single bond between R and the carbon.

adenocarcinoma: a **malignant tumor** originating in **glandular epithelium**.

adenoma: a **benign tumor** originating in **glandular epithelium**.

adenomatous: related to **adenoma** and to some types of **glandular hyperplasia**.

adenosine triphosphate (ATP): a **coenzyme** of fundamental importance found in the cells of all organisms; it provides a means of storing energy for many cellular activities.

adipose tissue: a modification of **areolar tissue** in which globules of **oil** are deposited in some of the cells (**fat cells**); the cells tend to be grouped together and, in mammals, occur in the tissues under the skin and around the **abdominal organs** (**kidneys, liver**, etc.).

administration: in medicine the route of administration means the path by which a medicinal substance is brought into contact with the body.

adventitia: the external covering of an organ or other structure, derived from **connective tissue**; especially the external covering of a **blood vessel**.

African frog (*Xenopus laevis*): a frog that occurs naturally in southern Africa; there is a substantial population has been introduced in California.

agarose: made from agar, agarose is a gelatinlike product of certain seaweeds; it is used for solidifying certain culture media, as a substitute for **gelatin**, as an emulsifier, etc.

agglomeration: whereby moist sticky particles collide due to turbulence in a medium and adhere to each other.

aggregation: a heterogeneous mass of independent but similar units (molecules, cells, etc.); the term implies the formation of a whole without an intimate mixing of constituents.

albumin: a group of water-soluble proteins coagulated by heat; they occur in egg-white, **blood serum**, milk, and in other animal and plant tissues.

albumin blue: is a dye used as a quantitative assay to measure albumin levels in biological samples including serum and urine; the intensity of the fluorescent signal is directly proportional to the albumin concentration of the sample; albumin-bound dye has a greatly increased excitation, thus the background caused by the emission of any free dye is minimal; albumin-bound dye absorbs light at 590 nm and emits fluorescence at 620 nm; long wavelength albumin blue dyes—absorption at 633 nm (AB633) and 670 nm (AB670), are also available and used for selective detection of human serum albumin in plasma and blood.

alcohol (spirit): an organic compound that contains one or more hydroxyl groups ($-OH$): the alcohols are hydroxy derivatives of alkanes; they can be classified according to the number of hydroxyl groups: monohydric, C_2H_5OH, **ethanol**; dihydric, $C_2H_4(OH)_2$, **ethylene glycol**; trihydric, $C_3H_5(OH)_3$, **glycerol**.

Alzheimer's disease: a neurodegenerative disease characterized by progressive cognitive deterioration together with declining activities of daily living and neuropsychiatric symptoms or behavioral changes.

amide: the organic functional group characterized by a carbonyl group (C=O) linked to a nitrogen atom (N); in the midinfrared spectral domain, bands due to the amide I, II, III, and A vibrations have been shown to be sensitive to the secondary structure content of **proteins**.

amino acid: any molecule that contains both amine and carboxyl functional groups; in biochemistry, this term refers to alpha amino acids; these are molecules where the amino and carboxylate groups are attached to the same carbon, which is called the α-carbon; the various alpha amino acids differ in which the side chain is attached to their α-carbon; this can vary in size from just a hydrogen atom in glycine, through a methyl group in alanine, to a large heterocyclic group in **tryptophan**; these amino acids are components of **proteins**; there are twenty standard amino acids used by cells in protein biosynthesis and these are specified by the general genetic code; these amino acids can be biosynthesized from simpler molecules; only obtained from food, essential amino acids are: histidine, isoleucine, leucine, lysine, methionine, **phenylalanine**, threonine, **tryptophan**, and valine.

δ-aminolevulenic acid (ALA): a "pro drug" that leads to the endogenous synthesis of **protoporphyrin IX** in the **cells** and **tissues**: it is applying ALA either systematically or topically.

amphiphilic: a chemical compound possessing both **hydrophilic** and **hydrophobic** properties.

anabolism: the synthesis of complex organic compounds from simpler organic compounds, e.g., the synthesis of proteins from amino acids; the process requires energy, mainly supplied in the form of **ATP** (see **metabolism**).

aneurysm: a permanent cardiac or arterial dilation usually caused by a weakened **vessel wall**.

angioplasty: the "reshaping" of blood vessels to improve blood flow.

anhydrous: having no water.

animal model: refers to a nonhuman animal with a disease or **injury** that is similar to a human condition; these test conditions are often termed as animal models of disease; the use of animal models allows researchers to investigate disease states in ways that would be inaccessible in a human patient; performing procedures on the nonhuman animal imply a level of harm that would not be considered ethical to inflict on a human.

anionic: relating to anions, negatively charged ions.

antibiotic: a drug that kills or prevents the growth of **bacteria**; antibiotics are one in a larger class of antimicrobials that include antiviral, antifungal, and antiparasitic drugs; they are relatively harmless to the host and therefore can be used to treat infections; the term "antibiotic" is also applied to synthetic antimicrobials, such as the sulfa drugs; antibiotics are generally small molecules with a molecular weight less than 2000 Da; they are not **enzymes**.

antigen: any foreign protein, or certain other large molecules, that, when present in a host's tissues, stimulates the production of a specific antibody by the host, a response leading to rejection of the antigen by the host; an antigen invades or is injected into an individual.

antritis: an antral (**antrum**) disorder of which examples include acute antritis; it is also known as or related to acute maxillary sinusitis and nodular antritis that is defined as antral gastritis with endoscopic findings characterized by a miliary pattern and prominent lymphoid follicles in biopsy specimens.

antrum: a general term for cavity or chamber, which may have a specific meaning in reference to certain organs or sites in the body; the antrum of the stomach (gastric antrum) is a portion before the outlet that is lined by mucosa and does not produce acid; the paranasal sinuses can be referred to as the frontal antrum, ethmoid antrum, and maxillary antrum.

anxiety: a generalized anxiety disorder is characterized by excessive, exaggerated anxiety and worry about everyday life events.

aorta: the large **artery** that leaves the left ventricle; it conducts the whole of the arterial blood supply to all parts of the body other than the **lungs**; in humans, it carries blood at the rate of 4 dm^3 per minute.

apatite [hydroxyapatite (HAP)]: the natural HAP crystal, $Ca_5OH(PO_4)_3$; dental **enamel** consists of 87–95% HAP crystals; **bone** consists of 50–60% HAP crystals.

aphakis subject: a subject missing **crystalline lens** of the eye.

apoptosis: the natural, programmed death of a cell (type I cell-death, compare **autophagy**) in response to an external signal a chain of biochemical reactions leading are induced to the death of cells no longer needed (as in embryonic development).

aqueous humor: a watery fluid, similar in composition to **cerebrospinal fluid**; it fills the anterior chamber of the **eyeball** behind the **cornea**; the **iris** and **crystalline lens** lie in it; it is continually secreted by the **ciliary body** and absorbed; it helps maintain the shape of the eyeball and assists in the refraction of light.

areolar tissue: a soft, sometimes spongelike **connective tissue** that consists of an amorphous polysaccharide-containing and jellylike ground matrix in which a loose network of **white fibers**, **yellow fibers**, and **reticulin fibers** are embedded; **fibroblasts** form in and maintain the matrix; areolar tissue is found all over the vertebrate body, binding together organs (by **mesenteries**) and **muscles** (by **sheaths**), and occurring as **subcutaneous tissue**; its function is to support or fill in the space between organs or between other tissues; the **fibrous** nature of the matrix is modified by variation in the concentration of white, yellow, or reticulin fibers; this alters the characteristics of toughness, elasticity, and inextensibility to suit the function of the tissue; many modifications of areolar tissue occur.

arm: in anatomy an arm is one of the upper limbs of a two-legged animal; the term arm can also be used for analogous structures, such as one of the paired upper limbs of a four-legged animal; anatomically, the term arm refers specifically to the segment between the shoulder and the elbow; the segment between the elbow and **wrist** is the **forearm**.

arteriole: a branch of an artery with a diameter less than 1/3 mm; arteriole walls are formed from smooth muscle under the control of the autonomic nervous system; their function is to control blood supply to the **capillaries**.

arteriosclerosis: an arterial disease that occurs especially in the elderly; it is characterized by inelasticity and thickening of the **vessel walls**, with lessened **blood flow**.

artery: a blood vessel conducting blood from the **heart** to tissues and organs; it is lined with endothelium (smooth flat cells) and surrounded by thick, muscular, elastic walls containing white and yellow **fibrous tissue**.

artifact: see Glossary 1.

astrocyte: a starlike cell of the macroglia of nerve tissue.

astrocytoma: a rather well-differentiated **glioma**, which consists of cells that look like **astrocytes**.

atheroma: fatty degeneration of the inner walls (**intima**) of the arteries in **arteriosclerosis**.

atherosclerotic plaque: a fibrous tissue that also contains **fat** and sometimes calcium; it accumulates in **arteries** and leads to the occlusion of the vessel.

atrium: (plural: atria) refers to a chamber or space; the **blood** collection chamber of a **heart**; in humans there are two atria, one on either side of the heart; on the right side is the atrium that holds blood that needs **oxygen**; it sends blood to the right ventricle, which sends it to the **lungs** for oxygen; after it comes back, it is sent to the left atrium; the blood is pumped from the left atrium and sent to the ventricle where it is sent out of the heart to all the rest of the body.

auscultation: hearing sounds of different body structures with diagnostic purposes.

autophagy (autophagocytosis): a process where cytoplasmic materials are degraded through the lysosomal machinery; the process is commonly viewed as **organelles** and long-lived **proteins** sequestered in a double-**membrane vesicle** inside the **cell**, where the contents are subsequently delivered to the **lysosome** for degradation; autophagy is part of everyday normal cell growth and development; for example, a liver-cell **mitochondrion** lasts around ten days before it is degraded and its contents are reused; autophagy also plays a major role in the destruction of bacteria, viruses, and unnecessary proteins that have begun to aggregate within a cell and may potentially cause problems; when autophagy involves the total destruction of the cell, it is called autophagic cell death (also known as cytoplasmic cell death or type II cell death); this is one of the main types of programmed cell death (compare **apoptosis**); it is a regulated process of cell death in a multicellular organism, or in a colony of individual cells such as yeast.

axillary: pertaining to the cavity beneath the junction of the arm and the body, better known as the armpit.

axon (or nerve fiber): a long, slender projection of a nerve cell, or neuron, that conducts electrical impulses away from the neuron's cell body or soma.

bacteria: (singular: bacterium) unicellular microorganisms; they are typically a few micrometers long and have many shapes including spheres, rods, and spirals; bacteria are ubiquitous in every habitat on Earth, growing in soil, acidic hot springs, radioactive waste, seawater, and deep in the earth's crust; some bacteria can even survive in the extreme cold and vacuum of outer space; there are typically 40 million bacterial cells in a gram of soil and a million bacterial cells in a milliliter of fresh water.

baroreflex: in cardiovascular physiology, the baroreflex or baroreceptor reflex is one of the body's homeostatic mechanisms for maintaining blood pressure; it provides a negative feedback loop in which an elevated blood pressure reflexively causes blood pressure to decrease; similarly, decreased blood pressure depresses the baroreflex, causing blood pressure to rise.

Barrett's esophagus: refers to an abnormal change (metaplasia) in the cells of the lower end of the **esophagus** thought to be caused by damage from chronic acid exposure, or reflux esophagitis; it is considered to be a premalignant condition and is associated with an increased risk of esophageal **cancer**.

basal cell carcinoma: the most common **skin cancer**; risk is increased for individuals with a high cumulative exposure to UV light via sunlight; treatment is with surgery, **topical** chemotherapy, x-ray, **cryosurgery**, **photodynamic therapy**; it is rarely life-threatening but if left untreated can be disfiguring, cause bleeding and produce local destruction (e.g., **eye**, **ear**, nose, **lip**).

baseline: information gathered at the beginning of a clinical study from which variations found in the study are measured; a person's health status before he or she begins a clinical trial; baseline measurements are used as a reference point to determine a participant's response to the experimental treatment.

basement membrane: a very thin sheet of **connective tissue** below the **epithelia**; it usually contains polysaccharide and very fine fibers of **reticulin** and **collagen**.

benign tumor: a nonmalignant **tumor**.

beta-carotene: the provitamin for **retinol** (**vitamin A**); it is converted to retinol in the **liver**; an antioxidant.

bile: a secretion of the **liver**; it is a bitter, slightly alkaline liquid, yellowish-green to golden-brown in color, consisting of bile salts, bile pigments, and other substances dissolved in water; its function is to assist in the digestion of fat and to act as a vehicle for the rejection of toxic or poisonous substances.

bilirubin: a yellow **bile** pigment formed in the breakdown of **heme**.

biocompatible: the quality of not having toxic or injurious effects on biological systems.

biofilm: a complex aggregation of microorganisms marked by the excretion of a protective and adhesive matrix; biofilms are also often characterized by surface attachment, structural heterogeneity, genetic diversity, complex community inter-actions, and an **extracellular** matrix of polymeric substances.

biological cell: an individual unit of protoplasm surrounded by a plasma **membrane** and usually containing a nucleus; a cell may exhibit all the characteristics of a living organism, or it may be highly specialized for a particular function; cells vary considerably in size and shape, but all have the common features of metabolism; every living organism is composed of cells, and every cell is formed from existing cells, usually by division, but also by fusion of sex cells; a cell may contain more than one nucleus; in **prokaryotic** cells, the genetic material is not contained in a **nucleus**.

biopsy: a medical test involving the removal of cells or tissues for examination; when only a sample of tissue is removed, the procedure is called an incisional biopsy or core biopsy; when an entire lump or suspicious area is removed, the pro-cedure is called an excisional biopsy; when a sample of tissue or fluid is removed with a needle, the procedure is called a needle aspiration biopsy; biopsy specimens are often taken from part of a lesion when the cause of a disease is uncertain or its extent or exact character is in doubt; pathologic examination of a biopsy can determine whether a lesion is benign or malignant, and can help differentiate be-tween different types of cancer; the margins of a biopsy specimen are also carefully examined to see if the disease may have spread beyond the area biopsied; "clear margins," or "negative margins," means that no disease was found at the edges of the biopsy specimen; "positive margins" means that disease was found and addi-tional treatment may be needed.

biospeckle: **speckle** (see Glossary 1) formed by coherent light scattering from a cell or tissue.

bladder: a membranous sac that contains or stores fluid.

blister: when the outer (epidermis) layer of the skin separates from the fiber layer (dermis), a pool of lymph and other bodily fluids are collected between these layers while the skin regrows from underneath; this is a defense mechanism of the human body; blisters can be caused by chemical or physical **injury**; an example of chemical injury would be an allergic reaction; physical injury can be caused by heat, frostbite, or friction.

blood: a fluid tissue contained in a network of vessels or sinuses in humans and animals; the vessels or sinuses are lined with **endothelium**; blood is circulated through the network by muscular action of the vessels or the **heart**; it transports oxygen, **metabolites**, and hormones; it contains soluble colloidal proteins (**blood plasma**) and **blood corpuscles**; it assists in temperature control in mammals.

blood cell: see **blood corpuscle**.

blood corpuscle: one of the various types of cells that circulate in **blood plasma**; also called **blood cell**: **red blood cell (RBC) (erythrocyte)**, **white blood cell (WBC) (leukocyte)**, **platelet**, **thrombocyte**.

blood flow: blood movement along a **blood vessel**.

blood microcirculation: the peripheral **blood** circulation, which is provided by the **capillary** network.

blood perfusion: blood pumping (supplying) through an organ or a tissue.

blood plasma: the clear, waterlike, colorless liquid of **blood**; blood plasma is formed by removing all blood corpuscles from blood; plasma can be clotted.

blood vessel: a tube through which **blood** flows either to or from the **heart**; a general term for a conducting vessel for blood: **artery**, **vein**, **arteriole**, **venule**, and **capillary**.

blood volume: the total blood content within the region of a tissue; includes volumes of both oxygenated and deoxygenated blood.

bone: a connective tissue forming the skeleton; it consists of cells embedded in a matrix of bone salts and **collagen fibers**; the bone salts (mostly calcium carbonate and phosphate) form about 60% of the mass of the bone and give it its tensile strength; the bone cells are interconnected by fine protoplasmic processes situated in narrow channels in the bone, and are nourished by the blood stream; this vascular nature of bone differentiates it from **cartilage**.

brain: the coordinating center of the nervous system.

breathing: the transport of oxygen into the body and carbon dioxide out of the body; aerobic organisms require oxygen to create energy, via respiration, in the form of energy-rich molecules such as glucose.

burn: a type of **injury** to the **skin** caused by heat, cold, electricity, chemicals, or radiation (e.g., a sunburn).

burn scars: there are three major types of burn related scars: keloid, hypertrophic, and contractures; keloid scars are an overgrowth of scar tissue, the scar will grow beyond the site of the **injury**, these scars are generally red or pink and will become a dark tan over time; hypertrophic scars are red, thick, and raised, however they differ from keloid scars in that they do not develop beyond the site of injury or **incision**; a contracture scar is a permanent tightening of skin that may affect the underlying muscles and tendons, limiting mobility and possibly damaging or causing degeneration of the nerves (http://www.burnsurvivor.com/scar_types.html).

butanediol: 1,4-butanediol ($C_4H_{10}O_2$) is an **alcohol** derivative of the alkane butane, carrying two hydroxyl groups; it is a colorless viscous liquid; its molecular mass is 90.12 g/mol; its melting point is 20°C and its boiling point is 230°C.

butilene glycol: 1,3 butilene glycol used in **cosmetics**; prevents loss of moisture/gain of moisture; very safe and nonirritating; FDA and Food Chemical Codex III approved; gives rigidity and gloss to lipsticks; retains fragrance on the skin; gives better smoothness, elasticity, and gloss to hair; it provides more inhibition of microorganisms than other glycols; it has lower oral toxicity than other glycols/**glycerin**; its boiling point: 207.5°C, neutral pH.

calcification: the deposition of lime or insoluble salts of calcium and magnesium in a **tissue**.

calf: the fleshy part at the back of the lower part of a human leg.

cancer: a general term applied to a **carcinoma** or a **sarcoma**; the typical symptoms are a **tumor** or swelling, a discharge, pain, an upset in the function of an organ, general weakness and loss of weight.

canine: a dog or any animal of the Canidae, or dog family, including the wolves, jackals, hyenas, coyotes, and foxes.

capillary: a minute hairlike tube (diameter about 5–20 μm) with a wall consisting of a single layer of flattened cells (**endothelium**); the wall is permeable to substances such as water, oxygen, **glucose**, amino acids, carbon dioxide, and to inorganic ions; the capillaries form a network in all tissues; they are supplied with oxygenated **blood** by **arterioles** and pass deoxygenated blood to **venules**; their function is the exchange of dissolved substances between blood and tissue fluid.

capsular: pertaining to a capsule a membranous, again sac or integument.

carbohydrates: simple molecules that are straight-chain aldehydes or ketones with many hydroxyl groups added, usually one on each carbon atom that is not part of the aldehyde or ketone functional group; carbohydrates are the most abundant biological molecules, and fill numerous roles in living things, such as the storage and transport of energy (starch, **glycogen**) and acting as structural components (cellulose in plants, chitin in animals); additionally, carbohydrates and their derivatives play major roles in immune system function, fertilization, pathogenesis, blood clotting, and development; the basic carbohydrate units are called monosaccharides, such as **glucose**, galactose, and fructose; the general chemical formula of an unmodified monosaccharide is $(C \cdot H_2O)n$, where n is any number of three or greater.

carbonyl: a carbon atom double-bonded to an oxygen atom: $C=O$.

carcinoma [carcinoma *in situ* (CIS)]: a malignant growth of abnormal epithelial cells.

cardiovibrations (heartbeats): the rhythmic vibrations and sound of the heart pumping blood; it has a double beat caused by the sound of ventricles contracting, followed by a shorter, sharper sound of the semilunar valves closing; the atria do not contribute to the sound of the beat.

carious (caries): caries is a multifunctional dental disease; the following factors influence its progression: dental plaque **biofilm** (contains bacteria that are both acid-producing and survive at low pH, *Mutans streptococci* are believed to be the most important bacteria in the initiation and progress of dental caries); the availability of **glucose** that drives bacterial **metabolism** to produce lactic acid; generally caries is initiated in the **enamel** but it may also begin in **dentine** or cementum; when acid challenges occur repeatedly the eventual collapse of enough enamel crystals will result in cavitation.

caries (carious): in human anatomy, the common carotid **artery** supplies blood to the head and neck; it divides in the neck to form the external and internal carotid arteries.

cartilage: a strong, resilient, skeletal tissue; its simplest and most common form consists of a matrix of a polysaccharide-containing protein in which cartilage cells are embedded (chondroblasts); the matrix is hyaline cartilage, which is without structure and **blood vessels**; it is translucent and clear, and occurs in the cartilaginous rings of the trachea and bronchi; elastic cartilage (yellow fibrocartilage) contains **yellow fibers** in the matrix; it occurs in the external ear and in the epiglottis; white fibrocartilage contains **white fibers** in the matrix; it occurs in the disks of cartilage between the vertebrae; all types of cartilage contain chondroblasts, which deposit the matrix and become enclosed in the matrix as chondrocytes.

catabolism: the decomposition of chemical substances within an organism; the substances they are usually complex organic substances, and their products are simpler organic substances; the process is typically accompanied by a release of energy (see **metabolism**).

cataract: an abnormality of the eye characterized by opacity of the **crystalline lens**.

cataractogenesis: the process of **cataract** formation.

cationic: relating to cations, positively charged ions.

cavitation: a general term used to describe the behavior of voids or bubbles in a liquid.

cell: see **biological cell**.

cell fixation: killing, making rigid, and preserving a **cell** for microscopic study.

cell membrane: the thin, limiting covering of a **cell** or cell part; regulates the ingress of substances into a cell or its parts and may have some other functions that depend on the cell's specialization.

cellular organelle: a part of a **cell** that is a structural and functional unit, e.g., a flagellum is a locomotive organelle, a **mitochondrion** is a respiratory organelle; organelles in a cell correspond to organs in an organism.

centriole: a barrel shaped microtubule structure found in most animal **cells**; the walls of each centriole are usually composed of nine triplets of microtubules; an associated pair of centrioles, spatially arranged at right-angles, constitutes the compound structure known to cell biologists as the centrosome; centrioles are very important in the cell division process.

cerebellum: a region of the **brain** that plays an important role in the integration of sensory perception and motor output; many neural pathways link the cerebellum with the motor **cortex**, which sends information to the **muscles** causing them to move, and the spinocerebellar tract, which provides feedback on the position of the body in space; the cerebellum integrates these pathways, using the constant feedback on body position to fine-tune motor movements.

cerebrospinal fluid (CSF): the clear liquid that fills the cavities of the brain and spinal cord and the spaces between the arachnoid and pia matter; the fluid moves in a slow current down the central canal and up the spinal meninges; it is a solution of blood solutes of low molar mass, such as **glucose** and sodium chloride, but not of the same concentration as in the **blood**; it contains little or no protein and very few cells; its function is to nourish the **nervous tissue** and to act as a buffer against shock the total quantity of CSF in humans is about 100 cm^3.

cervical cancer: a malignant tumor of the **cervix uteri**; e.g., a **cervical intraepithelial neoplasia (CIN)**, CIN I and CIN II are precancerous; CIN III stage corresponds to **carcinoma *in situ* (CIS)**; the next stage is an invasive cancer.

cervical intraepithelial neoplasia (CIN): CIN I and CIN II are precancerous, and correspond to slight and moderate **displasia**; CIN III stage corresponds to marked displasia or **carcinoma *in situ* (CIS)**.

cervical smear: a thin specimen of the cytologic material taken from the female cervical channel; it is usually received as a smear on a glass plate, and is fixed, and stained before being examined.

cervical tissue: the multilayered tissue consisting of the upper epithelial, basal (basal **membrane**), and stromal layers; depending on the area of the **cervix**, the epithelium may be in one of two forms: squamous or columnar.

cervix uteri: the narrow opening to the uterus; a short tube leading from the vagina to the uterus.

chemical potential: in thermodynamics, the amount by which the energy of the system would change if an additional particle was introduced, with the **entropy** (see Glossary 1) and volume held fixed; if a system contains more than one species of particle, there is a separate chemical potential associated with each species, defined as the change in energy when the number of particles of that species is increased by one; the chemical potential is a fundamental parameter in thermodynamics and it is conjugate to the particle number.

chemiluminescence: the emission of light (luminescence) without emission of heat as the result of a chemical reaction.

chest: the part of the trunk between the neck and the abdomen, containing the cavity and enclosed by the **ribs**, sternum, and certain vertebrae, in which the **heart**, **lungs**, etc., are situated.

cholesterol: a sterol (a combination steroid and **alcohol**) and a **lipid** found in the **cell membranes** of all body **tissues**, and transported in the **blood** plasma of all animals.

choroid: the membranous, pigmented middle layer of the **eyeball** between the **sclera** and the **retina**; it contains numerous **blood vessels**; its function is to absorb light to prevent internal reflection in the eyeball and to provide nourishment for the retina; the choroid is continuous with the **iris** in the front of the eye.

chromatin: the complex of **DNA** and **protein** found inside the nuclei of eukaryotic **cells**; the **nucleic acids** are in the form of double-stranded DNA (a double helix); the major proteins involved in chromatin are histone proteins, although many other chromosomal proteins have prominent roles too; the functions of chromatin are to package DNA into a smaller volume to fit in the cell, to strengthen the DNA, to allow **mitosis** and meiosis, and to serve as a mechanism to control expression.

chromatin filaments: chromatin fibers condensed to 30 nm and consisting of nucleosome arrays in their most compact form.

cicatrix: the new tissue that forms over a **wound** and later contracts into a scar.

ciliary body: a thickened circular structure at the edge of the **choroid** and at the border of the **cornea**; the **iris** and the suspensory ligaments are attached to it; it contains the **ciliary muscle** used in accommodation; it secretes **aqueous humor**.

ciliary muscle: a smooth **muscle** that affects zonular fibers in the **eye** (fibers that suspend the **lens** in position during accommodation), enabling changes in lens shape for light focusing.

ciliary pigmented epithelium: the darkly colored **melanin**-pigmented epithelial layer of **ciliary body**.

clinical trials: the application of the scientific method to human health; researchers use clinical trials to test hypotheses about the effect of a particular intervention upon a **pathological** disease condition; well-run clinical trials use defined techniques and rigorous definitions to answer the researchers' questions as accurately as possible; the most commonly performed clinical trials evaluate new drugs, medical devices, biologics, or other interventions on patients in strict scientifically controlled settings, and are required for regulatory authority approval of new therapies; trials may be designed to assess the safety and efficacy of an experimental therapy, to assess whether the new intervention is better than standard therapy, or to compare the efficacy of two standard or marketed interventions; the trial objectives and design are usually documented in a clinical trial protocol; synonyms are clinical studies, research protocols, and medical research.

coagulation: the process of coagulation or of causing something to coagulate, to cause particles (components) to collect together in a compact mass, e.g., the coagulation of egg white is brought about by heat.

coenzyme: the small organic nonprotein molecules that carry chemical groups between **enzymes**; many coenzymes are phosphorylated water-soluble vitamins; however, nonvitamins may also be coenzymes such as **ATP**; coenzymes are consumed in the reactions in which they are substrates, for example: the coenzyme **NADH** is converted to NAD+ by oxidoreductases; coenzymes are, however, regenerated and their concentration maintained at a steady level in the **cell**.

cold cataract: a temperature-induced reversible **cataract**.

collagen: a tough, inelastic, **fibrous** protein; when bolied, it forms a **gelatin**; on adding **acetic acid** it swells up and dissolves; collagen is formed and maintained in tissues by **fibroblasts**; it forms **white fibers** in **connective tissue**; the tropocollagen or "collagen molecule" subunit is a rod about 300-nm long and a 1.5-nm diameter, made up of three polypeptide strands and subunits with regularly staggered ends that spontaneously self-assemble into even larger arrays in the **extracellular** spaces of tissues; there is some covalent cross linking within the triple helices, and a variable amount of covalent cross linking between tropocollagen helices, to form the different types of collagen found in different mature tissues—similar to the situation found with the α-**keratins** in **hair**; a distinctive feature of collagen is the regular arrangement of **amino acids** in each of the three chains of collagen subunits; in **bone**, entire collagen triple helices lie in a parallel, staggered array; 40-nm gaps between the ends of the tropocollagen subunits probably serve as nucleation sites for the deposition of long, hard, fine crystals of the mineral component, which is (approximately) **hydroxyapatite** with some phosphate, which turns certain kinds of **cartilage** into **bone**; collagen gives bone its elasticity and contributes to fracture resistance.

collagen fibers: bundles of **collagen fibrils** (see **white fibers, white fibrous tissue**).

collagen fibrils: **collagen** molecules packed into an organized overlapping bundle.

collagen secondary structure: a **collagen** molecule subunit is a rod made up of three polypeptide strands, each of which is a left-handed helix; these three left-handed helices are twisted together into a clockwise coil, a triple helix, a cooperative quaternary structure stabilized by numerous hydrogen bonds.

colloid structure: dispersion of colloidal particles in a continuous phase (the **dispersion** medium) of a different composition or state; true solutions of materials that have dimensions within the colloidal range (1 nm to 100 nm), e.g., molecules of a very large relative molecular mass such as polymers and proteins, or aggregates of small molecules, as in soaps and detergents under certain conditions (association colloids).

colloid isoelectric point: the pH value of the **dispersion** medium of a colloidal suspension at which the colloidal particles do not move in an electric field.

colon: the first or anterior part of the large intestine; it possesses sacculated walls lined with a smooth **mucous membrane**; contains the lining **glands** that secret **mucus** but no digestive **enzymes**; water is absorbed from the unassimilated liquid material, leaving feces; the colon leads directly into the rectum.

colonic: related to **colon**.

colposcope: an endoscopic instrument that provides *in vivo* examination of the vaginal and cervical mucous with a high magnification.

colposcopy: the *in situ* examination of the vaginal and cervical mucous using the colposcope.

columnar epithelium: **cells** that are taller than they are wide; the **nucleus** is closer to the base of the cell; the small **intestine** is a tubular organ lined with this type of **tissue**; unicellular **glands** called goblet cells are scattered throughout the simple columnar epithelial cells and secrete **mucus**; the free surface of the columnar cell has tiny hairlike projections called microvilli; they increase the surface area for absorption.

computed tomography: the imaging of a selected plane (slice) in the body and 3D computer reconstruction of the image of a whole object on the basis of many individual images (slices) using one of the following methods: x ray, magnetic resonance, positron-emission, or optical imaging (slicing).

conjunctiva: a delicate **mucous membrane** that covers the **cornea**, the front part of the **sclera**, and lining of the eyelids: it protects the cornea.

connective tissue: various body tissues that bind together and support organs and other tissues, e.g., connective tissue surrounds **muscles** and **nerves**, connects **bones** and muscles and underlies the **skin**; **cartilage**, and bone are also connective tissues; typical connective tissue consists of cells scattered in an amorphous mucopolysaccharide matrix in which there are varying amounts of connective tissue fibers (mainly **collagen**, but also **elastin** and **reticulin**).

contrast (ing) agents: see Glossary 1.

copolymer (or **heteropolymer**): a polymer derived from two (or more) monomeric species as opposed to a homopolymer, where only one monomer is used.

coproporphyrin: a **porphyrin** that occurs as several isomers; the III isomer, an intermediate in **heme** biosynthesis, is excreted in the feces and **urine** in such diseases as hereditary coproporphyria and variegate porphyria; the I isomer, a side product, is excreted in congenital erythropoietic porphyria disease; is used as a test to measure **red blood cell** porphyrin levels in evaluating of porphyrin disorders.

cornea: the transparent covering at the front of the **eyeball**; it is the modified continuation of the **sclera**; it refracts light and is the most important element in the refractive system of the eye.

corneocyte: the dead **keratin**-filled **squamous cell** of the **stratum corneum**; synonym: horny cell, keratinised cell (**keratinocyte**).

cornification (keratinization): the conversion of epithelium to the stratified **squamous** type; for instance, the conversion of skin cells into **keratin** or other horny material such as nails or scales.

coronary arteries: the right and left coronary arteries that originate from the beginning (root) of the **aorta**, immediately above the aortic **valve**; the left coronary **artery** originates from the left aortic sinus, while the right coronary artery originates from the right aortic sinus.

cortex (cerebral cortex): the outer layer of the cerebral hemisphere, consisting of **gray matter** and rich in synapses; it is extensive in mammals, with fissures and folds.

cosmetics: the substances used to enhance or protect the appearance or odor of the human body; cosmetics include skincare creams, lotions, powders, perfumes, lipsticks, fingernail polishes, **eye** and facial makeup, permanent waves, **hair** colors, deodorants, baby products, bath **oils**, bubble baths, and many other types of products; the FDA defines cosmetics as: "intended to be applied to the human body for cleansing, beautifying, promoting attractiveness, or altering the appearance without affecting the body's structure or functions."

craniocaudal projection: the projection that shows the direction from the **cranium** to the posterior end of the body.

cranium: a domed, bony case composed of several **bone**s joined by sutures; it encloses and protects the **brain**.

cross-linking: to attach by a cross-link: a bond, atom, or group linking the chains of atoms in a polymer, protein, or other complex organic molecule.

cryogenic: related to treatments using very low temperatures (below $-150°C$, $-238°F$ or 123 K) and the behavior of materials at those temperatures.

cryosection: a frozen sectioning procedure that is used to perform rapid microscopic analysis of a tissue specimen; it is used most often in oncological surgery and tissue research.

cryosurgery (cryotherapy): the application of extreme cold to destroy abnormal or diseased tissue; is used to treat a number of diseases and disorders, especially **skin** conditions; warts, moles, skin tags, solar keratoses, and small skin **cancers** are candidates for cryosurgical treatment; some internal disorders are also treated with cryosurgery, including **liver** cancer, **prostate** cancer, and cervical disorders.

crystalline lens: a transparent structure surrounded by a thin capsule and situated immediately behind the **pupil** of the eye; it has the shape of a biconvex lens; it is attached to the **eyeball** by suspensory ligaments; it refracts light onto the **retina**.

crystallins: the structural proteins that are the components of a mammalian **crystalline lens**.

crystalline: a crystalline material has a regular crystal lattice; it does not necessarily form a single regular crystal, e.g., all metals are crystalline because the atoms have a regular arrangement.

cutaneous: relating to the skin.

cyclophotocoagulation: is a transscleral laser procedure for **glaucoma** treatment; in this procedure, the **ciliary body** is treated with a laser to decrease its production of aqueous—this in turn reduces pressure inside the **eye**; about 20 to 40 laser delivery applications are completed.

cyst: a closed, bladderlike sac formed in tissues; it contains fluid or semifluid matter.

cystitis: an infection of the **bladder**, but the term is often used indiscriminately and covers a range of infections and **irritation**s in the lower urinary system.

cytochrome: the **proteins** that contain iron and act as **coenzymes** in cellular respiratory; they are found abundantly in aerobic organisms; they are oxidized by dissolved **oxygen** in a **cell**, and reduced by oxidizable substances in the cell; they are the main **vehicle** for the use of oxygen in **metabolism**.

cytochrome oxidase: a large transmembrane protein complex found in bacteria and the mitochondrion; it is the last protein in the electron transport chain; it receives an electron from each of four **cytochrome** C molecules, and transfers them to one **oxygen** molecule, converting molecular oxygen to two molecules of **water**; in the process, it translocates four protons, helping to establish a chemiosmotic potential that the **ATP** synthase then uses to synthesize ATP.

cytometry: the methods and instruments for the structural and functional study of **cells** and bacteria; e.g., **flow cytometry** is a technique for automatic measurement and analysis of cells and other small particles suspended in a medium.

cytoplasm: the protoplasm of a **cell** exclusive of the **nucleus**; it is not just a simple, slightly viscous, fluid; in it are situated various structures, called **organelles**, each concerned with different functions of the cell; the **plasma membrane** is part of the cytoplasm.

cytoskeleton: made of **fibrous proteins** (e.g., **microfilaments**, **microtubules**, and intermediate filaments); in many organisms maintains the shape of the **cell**, anchors **organelles**, and controls internal movement of structures.

cytosol: (cf. **cytoplasm**, which also includes the **organelles**) is the internal fluid of the **cell**, and a portion of cell **metabolism** occurs here; **proteins** within the cytosol play an important role in signal transduction pathways and glycolysis; in **prokaryotes**, all chemical reactions take place in the cytosol; in **eukaryotes** the portion of cytosol in the nucleus is called nucleohyaloplasm; the cytosol also surrounds the **cytoskeleton**; the cytosol is a "soup" with free-floating particles, but is highly organized on the molecular level.

decylmethylsulfoxide: a tissue penetrating agent similar to **DMSO**; is also known as n-decylmethylsulfoxide (nDMSO).

dehydration: the removal or loss of water by a **tissue** or a **cell**.

denaturation: the alteration of a **protein** shape through some form of external stress (for example, by applying heat, acid, or alkali), in such a way that it will no longer be able to carry out its cellular function; denatured proteins can exhibit a wide range of characteristics, from loss of solubility to communal **aggregation**.

dental plaque: a biofilm (usually a pale yellow to white color) that builds up on the teeth; if not removed regularly, it can lead to dental cavities (**caries**) or periodontal problems (such as gingivitis).

dentin: a hard, calcified, elastic, yellowish material of the same substance as **bone**, but contains no cells; it is the main structural part of a tooth; dentin is composed of base material that is pierced by mineralized dentinal tubules 1–5 μm in diameter; the tubules' density is in the range of $3-7.5 \times 10^6$ cm^{-2}; they contain organic components and natural **HAP [hydroxyapatite (apatite)]** crystals 2–3.5 nm in diameter and up to 100 nm in length, which intensively scatter light.

deoxyhemoglobin: **hemoglobin** disintegrated with **oxygen**.

dermal papilla: extensions of the **dermis** into the **epidermis**; they sometimes can be perceived at the surface of the **skin**.

dermatitis: **inflammation** of the **skin**.

dermatosis: any disease of the **skin**.

dermis: the inner layer of the **skin**; it is composed of **connective tissue**, **blood** and **lymph vessels**, **muscles**, and **nerves**; **collagen fibers** are abundant in the dermis and run parallel to the surface of the skin; they give the skin elasticity; sweat **glands** and hair follicles are scattered throughout the dermis; the dermis is much thicker than the epidermis and is developed from mesoderm.

desorption: a phenomenon and process opposite of sorption (that is, adsorption or absorption), whereby some of a sorbed substance is released; this occurs in a system being in the state of sorption equilibrium between the bulk phase (fluid, i.e., gas or liquid solution) and an adsorbing surface (solid or boundary separating two fluids); when the concentration (or pressure) of substance in the bulk phase is lowered, some of the sorbed substance changes to the bulk state; in chemistry, especially chromatography, desorption is the ability for a chemical to move with the mobile phase—the more a chemical desorbs, the less likely it will adsorb, thus instead of sticking to the stationary phase, the chemical moves up with the solvent front.

desquamation: the shedding of the outer layers of the **skin**; for example, once the rash of measles fades, there is desquamation.

dextran: a complex, branched polysaccharide made of many **glucose** molecules joined into chains of varying lengths.

diabetes (**diabetes mellitus**): a metabolic disorder characterized by hyperglycemia (high blood sugar); the World Health Organization (WHO) recognizes three main forms of diabetes: type 1, type 2, and gestational diabetes (occurring during pregnancy), which have similar signs, symptoms, and consequences, but different causes and population distributions; type 1 is usually due to autoimmune destruction of the pancreatic beta cells that produce insulin; type 2 is characterized by tissue-wide insulin resistance and varies widely, it sometimes progresses to loss of beta cell function.

diabetic retinopathy: damage to the **retina** caused by complications of **diabetes mellitus**, which could eventually lead to blindness; it is an ocular manifestation of systemic disease, which affects up to 80% of all diabetics who have had diabetes for 15 years or more.

diaphanography: a **noninvasive method** of examining the breast or other human organ by transillumination, using visible or infrared light.

diaphragm: a dome-shaped sheet of tissue, part **muscle**, part **tendon**, separating the thoracic and abdominal cavities.

diastolic: related to diastole, the period of time when the **heart** relaxes after contraction; ventricular diastole is when the ventricles are relaxing, while atrial diastole is when the **atria** are relaxing.

dioley(o)lphosphatidylethanolamine (DOPE): a neutral lipid, used as the carrier system.

dipropylene glycol: $HOC_3H_6OC_3H_6OH$; molecular weight 134.18; refractive index 1.438–1.442; a colorless, viscous, practically nontoxic, and slightly hygroscopic liquid; its melting point is 78°C, boiling point, 231°C; is miscible in water, alcohols, esters, most organic solvents, and various vegetable oils; is used as a solvent, coupling agent, and chemical intermediate in many fields, including **cosmetics**.

disaggregation: the separation of an aggregate body into its component parts.

dissociation: a general process in which ionic compounds (complexes, molecules, or salts) separate or split into smaller molecules, ions, or radicals, usually in a reversible manner; for instance, reversible dissociation of collagen fibers; dissociation is the opposite of association and recombination.

DMSO: dimethyl sulfoxide is the chemical compound $(CH_3)_2SO$; this colorless liquid is an important polar aprotic solvent; it is readily miscible in a wide range of organic solvents as well as water; it has a distinctive property of penetrating the skin very readily, allowing the handler to taste it; its unique capability of to penetrate living tissues without causing significant damage is most probably related to its relatively polar nature, its capacity to accept hydrogen bonds, and its relatively small and compact structure; this combination of properties results in its ability to associate with water, proteins, **carbohydrates**, **nucleic acid**, ionic substances, and other constituents of living systems (www.dmso.org).

DNA (deoxyribonucleic acid): a long-chain compound formed from many **nucleotides** bonded together as units in the chain; a strand of DNA is formed from molecules of deoxyribose (a sugar) and molecules of phosphoric acid attached alternatively in a chain; it is found only in the chromosomes of animals and plants and in the corresponding structures in bacteria and viruses.

dorsal: the term refers to anatomical structures that are either situated toward or grow off the side of an animal or human; in humans, the top of the hand and top of the foot are considered dorsal.

duct: a tube with an outlet, discharging fluids from one system to another; e.g., the **bile** duct discharges bile from the **liver** into the alimentary canal or milk drains through ducts into a cistern in the **mammary gland**.

ductal carcinoma *in situ*: see **duct** and **carcinoma**.

dura mater (pachymeninx): the tough and inflexible outermost of the three layers of the *meninges* surrounding the brain and spinal cord; the *dura mater* itself has two layers: a superficial layer, which is actually the skull's inner *periosteum*, and a deep layer, the *dura mater proper*.

dysfunction: any disturbance in the function of an organ or body part.

dysplasia: an abnormality in the appearance of cells indicative of an early step toward transformation into a **neoplasia**; it is therefore a preneoplastic or precancerous change; this abnormal growth is restricted to the originating system or location, for example, a dysplasia in the epithelial layer will not invade the deeper tissue, or a dysplasia solely in a red blood cell line (refractory anaemia) will stay within the **bone** marrow and cardiovascular systems; the best known form of dysplasia is the precursor lesions to **cervical cancer**, called **cervical intraepithelial neoplasia (CIN)**; this lesion is usually caused by an infection with the **human papilloma virus (HPV)**.

dysplastic: related to **dysplasia**.

ear: the organ that detects sound; the vertebrate ear shows a common biology from fish to humans, with variations in structure according to order and species; it not only acts as a receiver for sound, but plays a major role in the sense of balance and body position; the word "ear" may be used correctly to describe the whole vertebrate ear, or just the visible portion; in most animals, the visible ear is a flap of tissue that is also called the pinna; in humans, the pinna is more often called the auricle; in biomedical optics the lobe of the human ear used as a convenient model for **noninvasive** blood oxygenation and microcirculation studies; in animals—rat, mouse, rabbit, etc., pinna is used as an *in vivo* model for noninvasive blood studies.

ectodermal dysplasia: a hereditary condition characterized by abnormal development of the skin, hair, nails, teeth, and sweat glands.

edema: the effusion of serous fluid into the interstices of cells in tissue spaces or into body cavities.

elastin: an elastic **fibrous** protein resistant to boiling and acetic acid; it forms highly elastic **yellow fibers** in **connective tissue**; elastin is formed and maintained in tissues by **fibroblasts**.

elastosis: the breakdown of elastic **tissue**; for example, the loss of elasticity in the **skin** of elderly people that results from degeneration of **connective tissue**.

electroosmosis: the motion of a polar liquid through a **membrane** or other porous structure under the influence of an applied electric field; (generally, along charged surfaces of any shape and also through nonmacroporous materials, which have ionic sites and allow for water uptake, the latter is sometimes referred to as "chemical **porosity**") also called electroendosmosis.

electroporation: a short pulse of voltage in the range of 5–200 V/cm^2 applied to a biological **membrane** induces its **porosity** that enhances membrane permeation for big molecules.

enamel: a hard, elastic, white material that contains no cells and that is an almost completely inorganic substance; enamel covers the crown of a tooth; dental enamel consists of 87–95% natural **HAP [hydroxyapatite (apatite)]** crystals; they are organized in keyhole-shaped prisms; these prisms are 4–6 μm wide and extend from the dentine-enamel-junction to the outer surface of the tooth; because of their size, number, and refractive index, the prisms are the main light scatterers in enamel.

encapsulated drugs: an encapsulation of some drugs significantly reduce their toxicity (gastrointestinal, **cutaneous**, etc.) especially at the higher dose level and often increase drug efficiency; various encapsulation technologies are used, including **liposomes**.

endocard (endocardium): the serous membrane that lines the cavities of the heart.

endogenous: means "arising from within"; endogenous substances are those that originate from within an organism, **tissue** or **cell**; in biological systems endogeneity refers to the recipient of **DNA** (usually in prokaryotes); however, due to **homeostasis**, discerning between internal and external influences is often difficult.

endoplasmic reticulum (ER): an elaborate series of membranous sacs that communicate with each other in a three-dimensional network and occur in the endoplasm of a cell; the connection between two sacs is an anastomosis; ER is either rough or smooth surfaced; rough ER carries ribosomes on the outside surface of the sacs; smooth ER carries no ribosomes; the functions of ER include the transfer of materials in cells by providing a circulatory system of channels, the formation of **lysosomes**, and lipid **metabolism**.

endothelium: a single layer of squamous cells lining the **heart**, **blood vessels**, and **lymphatic vessels**; the **cells** are tessellated, i.e., they have wavy boundaries that interdigitate or fit together; endothelium is morphologically similar to **epithelium**, but is derived from mesoderm.

enzyme: a protein produced by a living **cell** and that acts as a catalyst in biochemical changes; there are many different types of enzymes, some of which promote a narrow range of chemical reactions on chemically related substances, while most others promote a single chemical reaction; most of the chemical reactions included in **metabolism** are dependent on enzymes to promote them at the rate required for an organism to function properly; a very small quantity of an enzyme is sufficient to convert a large quantity of a substance; each enzyme has optimum conditions for its action, e.g., a temperature of about 35–40°C, a specific pH, the presence of a **coenzyme** for some reactions, and the absence of inhibiting substances.

epicard (epicardium): the inner serous layer of the pericardium, lying directly upon the **heart**.

epidermal stripping sample: a thin slice of **epidermis** obtained with the use of medical glue and a quartz (glass) or metal plate.

epidermis: the outer layer of the skin is a stratified **epithelium** that varies relatively little in thickness over most of the body (between 75 and 150 μm), except on the **palms** and soles, where its thickness may be 0.4–0.6 mm; the epidermis is conventionally subdivided into (1) **stratum basale**, a basal cell layer of keratinocytes, which is the germinative layer of the epidermis, (2) the **stratum spinosum**, which consists of several layers of polyhedral cells lying above the germinal layer, (3) the **stratum granulosum**, which is a layer of flattened cells containing distinctive cytoplasmic inclusions, keratohyalin granules, and (4) the overlying **stratum corneum**, consisting of lamellae of anucleate thin, flat squames that are terminally differentiated **keratinocytes**.

epidurally: means through the *dura mater*.

episclera: the layer of the eye **sclera**.

epithelial tissue: tissue consisting of a sheet of **cells** held together by a minimal amount of cementlike material between the cells; it covers exposed surfaces and lines the cavities and tubes of the body; beneath most epithelial tissue is a thin sheet of **connective tissue**, the **basement membrane**; besides its protective function, epithelial tissue frequently has a secretory function, in which case it is sometimes known as glandular tissue.

epithelium: a sheet of **epithelial tissue**; epithelium is derived from ectoderm and endoderm.

erythema (skin reddening): an abnormal redness of the **skin** due to local congestion, as in **inflammation**; e.g., the skin's response to UV irradiation.

erythematosus lupus: a chronic disease of unknown cause, occasionally affecting internal organs, characterized by red, scaly patches on the **skin**.

erythematous: relating to, or causing, **erythema**.

erythrocyte: a flattened, disk-shaped (circular biconcave disks, about 8 μm in diameter in humans) **cell** that circulates in **blood**; it contains a respiratory pigment, **hemoglobin**; the cell is readily distorted, elastic, and immotile: mammalian erythrocytes have no nuclei, but the erythrocytes of embryos have nuclei; erythrocytes are formed in red **bone** marrow, are destroyed by erythrophages, and have a relatively short life (average 120 days in humans); there are approximately five million per cubic millimeter in normal human blood.

erythrodermia: the general name of the expressed and usually widespread skin redness (**erythema**), often accompanied by scale of the skin.

esophagus: a tube connecting the **pharynx** with the **stomach**, usually about 25 cm for adults; it is divided into three parts: jugular, **chest**, and **abdominal**.

ethanol: see **alcohol**.

ether: the general name for a class of chemical compounds that contain an ether group, which is an **oxygen** atom connected to two (substituted) alkyl groups; a typical example is the solvent and anesthetic diethyl ether, commonly referred to simply as "ether" (ethoxyethane, $CH_3-CH_2-O-CH_2-CH_3$).

ethylene glycol: a colorless, sweet liquid (see **alcohol**) used chiefly as a solvent.

eukaryotes: animals, plants, **fungi**, and protists are eukaryotes, organisms with a complex cell or **cells**, in which the genetic material is organized into a membrane-bound **nucleus** or nuclei; animals, plants, and fungi are mostly multicellular.

evans blue: a biological dye (stain).

evaporation: the process whereby atoms or molecules in a liquid state gain sufficient energy to enter the gaseous state; it is the opposite process of condensation; evaporation is exclusively a surface phenomena and should not be confused with boiling; most notably, for a liquid to boil, its vapor pressure must equal the ambient pressure, whereas for evaporation to occur, this is not the case.

excision: means "to remove as if by cutting"; in surgery, an excision (or resection) is the complete removal of an organ or a tumor, as opposed to a **biopsy**; an "excisional biopsy" (sometimes called a "tumorectomy") is the removal of a tumor with a minimum of healthy tissue; it is therefore an excision rather than a biopsy.

exogenous: an action or object coming from outside a system; the opposite of **endogenous**; for example, an exogenous **contrast agent** in medical imaging refers to a liquid injected in the patient that enhances visibility of a pathology, such as a **tumor**; an exogenous factor is any material present and active in an individual organism or living **cell**, but that originated outside of that organism, as opposed to an endogenous factor, including both pathogens and therapeutics; **DNA** introduced to cells via transfection or viral infection (transduction) is an exogenous factor; carcinogens are exogenous factors.

extracellular: means "outside the **cell**"; outside the plasma **membranes** and occupied by fluid.

ex vivo: taken from the living organism; pertaining to experiments on animal or human organs that are excised from the living body and kept in conditions very close to the natural ones.

eyeball: the spherical structure composed of supporting tissues in which the photoreceptors and refractive media for concentrating light on the **nervous tissues** are situated; the eyeball is divided into anterior and posterior chambers by the **iris**.

facial tissue: refers to a class of soft, absorbant, disposable paper that is suitable for use on the face.

fat: (1) any substance that can be extracted from tissues by ether, hot ethanol, or gasoline (fat solvents); this is a wide definition covering neutral fats, sterols, steroids, carotenes, and terpens; in this sense, lipids, lipins, and lipoids are fats; (2) true fat or neutral fat, as considered in dietetics, is an ester of **glycerol** with one, two, or three different fatty acids replacing the three hydroxyl groups of the trihydric alcohol, glycerol; (3) any substance that is a true fat and solid below 20°C; this is in contrast to an **oil**; (4) see **adipose tissue**.

fat (fatty) acid: a carboxylic acid often with a long unbranched aliphatic tail (chain), which is either saturated or unsaturated; fatty acids derived from natural **fats** and **oils** may be assumed to have at least 8 carbon atoms, e.g., caprylic acid (octanoic acid); most of the natural fatty acids have an even number of carbon atoms, because their biosynthesis involves acetyl-CoA, a **coenzyme** carrying a two-carbon-atom group; fatty acids are produced by the hydrolysis of the ester linkages in a fat or biological oil (both of which are **triglycerides**), with the removal of **glycerol**.

fat cell: a cell in which a food reserve is deposited in the form of droplets of **oil**; the quantity of oil increases until the oil globule formed distends the cell and pushes the nucleus and **cytoplasm** to one side; a collection of fat cells forms **adipose tissue**.

fatty tissue: see **adipose tissue**.

female breast (mamma): the milk-secreting organ of females; it contains a **mammary gland**.

femoral biceps muscle: the muscle pertaining to the thigh or femur.

fibers: the long strands of scleroprotein; they are either **collagen**-forming **white fibers** or **elastin**-forming **yellow fibers**, or **reticulin**-forming **reticular fibers**; fibers form part of a noncellular matrix around and among **cells**; they are formed and maintained in a tissue by **fibroblasts**; a matrix may consist of an amorphous, jellylike polysaccharide together with the three types of fiber; cells and a matrix form a **connective tissue**; different forms of connective tissue possess varying proportions of the constituents of the matrix.

fibril: a fine fiber or filament; e.g., in **muscles** their diameters are in the range of 5–15 nm with a length of about 1–1.5 μm and in the eye cornea, their diameters are in the range of 26–30 nm with a mean of length up to a few millimeters.

fibril D-periodicity: for example, corneal fibrils are D-periodic (axial periodicity of collagen fibrils, where $D \approx 67$ nm), uniformly narrow ($\approx$30–35 nm in diameter), and indeterminate in length, particularly in older animals; the D-periodicity of the fibril arises from side-to-side associations of triple-helical collagen molecules that are $\approx$300 nm in length (i.e., the molecular length $= 4.4 \times D$) and are staggered by D; the D-stagger of collagen molecules produces alternating regions of protein density in the fibril, which explains the characteristic gapping and overlapping appearance of fibrils negatively contrasted for transmission electron microscopy.

fibroadenoma: a **benign tumor** originating in a glandular **epithelium**, having a conspicuous **stroma** consisting of the proliferating **fibroblasts** and other elements of **connective tissue**.

fibroblast: a **cell** that contributes to the formation of **connective tissue**.

fibrocystic: pertaining to the nature of or having a **fibrous cyst** or **cysts**.

fibroglandular tissue: a glandular tissue that has a large number of **fibers**, such as the tissue of the female breast.

fibroid: (1) resembling a **fiber** or **fibrous tissue**; (2) composed of fibers, as a **tumor**; (3) a tumor largely composed of smooth **muscle**.

fibrous: containing, consisting of, or resembling **fibers**.

fibrous cyst: any **cyst** that is surrounded by or situated in a large amount of **fibrous connective tissue**.

fibrous plaque: a small, flat formation or area of **fibrous tissue**.

fibrous tissue: a tissue mainly consisting of conjunctive **collagen** (or **elastin**) **fibers**, often packed in lamellar bundles.

fissure: a groove, natural division, deep furrow, or cleft found in the brain, spinal cord, and liver; a tear in the anus (anal fissure); in dentistry, a break in the tooth enamel.

fixation: in the fields of **histology**, pathology, and **cell** biology, fixation is a chemical process by which biological **tissues** are preserved from decay; fixation terminates any ongoing biochemical reactions, and may also increase the mechanical strength or stability of the treated tissues.

flagelium: a long, fine, threadlike process on a noncellular or unicellular organism, such as bacteria and spermatozoa; usually the organism possesses only one flagellum or possibly two; flagella are used for locomotion; their movements are in 3D and undulate in a wavelike or helical fashion.

flavin: a complex heterocyclic ketone that is common to the nonprotein part of several important yellow **enzymes**, the flavoproteins.

flavin adenine dinucleotide (FAD): formed as the **flavin** moiety is attached with an adenosine diphosphate; a **coenzyme** for many proteins including monoamine **oxidase**, D-amino acid oxidase, glucose oxidase, xanthine oxidase, and Acyl-CoA dehydrogenase.

flavin mononucleotide (FMN): a prosthetic group found in and amongst other **proteins**, **NADH** dehydrogenase and old yellow **enzyme**; a phosphorylated form of riboflavin.

flow cytometry: a technique for automatic measurement and analysis of **cells** and other small particles suspended in a medium; particles flowing at high speed through a narrow opening are measured individually by optical or electrical methods; in this way, methods, where structural and functional characteristics can be determined quickly and with great precision.

fluorescein: a **fluorophore** commonly used in biological microscopy; has an absorption maximum at 494 nm and emission maximum of 521 nm (in water); has an isoabsorptive point (equal absorption for all pH values) at 460 nm.

foot sole: the thickest layers of **skin** on the human body due to the weight that is continually placed on them; contains significantly less pigment than the skin of the rest of the body; one of two areas of the human body that grow no vellus **hair**; houses a denser population of **sweat glands** than most other regions of skin.

forearm: the lower part of the human **arm**, between the elbow and the **wrist**; an arm consists of the upper arm and the forearm.

formalin: trade name of 37% aqueous solution of formaldehyde; in water formaldehyde converts to the hydrate $CH_2(OH)_2$; preserves or fixes **tissue** or **cells** by irreversibly cross-linking primary amine groups in proteins with other nearby nitrogen atoms in protein or DNA through a $-CH_2-$ linkage; can be used as a disinfectant as it kills most **bacteria** and **fungi** (including their **spores**).

freckles: the small tan spots of **melanin** on the **skin** of people with fair complexions.

front chamber (segment) of the human eye: the **cornea**, an anterior chamber of the **eyeball** filled by the **aqueous humor**, and the **iris** and **crystalline lens** in it.

frontal lobe: (see **lobe**).

functional imaging: an imaging technique of the spatial distribution of blood oxygenation, **blood volume**, blood velocity, or any other functional parameter of a living **tissue**.

fundus: the base of an organ, or the part opposite to or remote from an aperture; e.g., fundus (bottom) of the eye.

fungi cells: a division of the subkingdom *Thallophyta*; its members include the yeasts, mushrooms, moulds, and rusts, i.e., fungal organisms; characteristics of this division are that its members have **eucaryotic cells**, they lack chlorophyll, there are unicellular or coenocytic tubular filaments for the main body of the organism, they are saprophytic or parasitic on plants and animals, and they reproduce by forming spores in very large numbers.

fusogenicity: relates to cell membrane fusion by a physical or chemical action.

FVB/N mouse strain: an inbred mouse strain that was established at the National Institute of Health in 1970 from an outbred colony of Swiss mice; is preferable for transgenic analyses.

gall bladder: a small bladder situated between the lobes of the **liver**; it is connected to the liver through the cystic duct and the hepatic duct; its function is to store **bile**, and its capacity in humans is 30–50 cm^3; the liver secretes bile continuously, but the bile only enters the duodenum during periods of digestion, otherwise the liquid is stored in the gall bladder; the walls are contractile and empty the **gall bladder** when food, especially fat, passes through the duodenum; the contractions are probably activated by a hormone secreted by the intestinal walls.

gallstone (biliary calculus): a calculus, or hard stone, formed in the **gall bladder** or a **bile duct**; it contains cholesterol crystals combined with other substances (e.g., calcium salts); the different types of stones are called porcinement, cholesterol, etc.

gastric juice: a secretion from **glands** in the **stomach** wall that contains hydrochloric acid (0.2–0.5%) and digestive **enzymes** (e.g., pepsin), and in young mammals only, rennin.

gastrointestinal (GI): related to the GI tract, also called the digestive tract, alimentary canal, or gut, is the system of organs that takes in food, digests it to extract energy and nutrients, and expels the remaining waste.

gel: an intermediate stage in the **coagulation** of a sol; a mass of intertwining filaments enclose the whole of the dispersion medium to produce a pseudo-solid; a gel is jellylike in appearance and forms a distortable mass.

gelatin: many substances that form lyophilic sols can be obtained in a jellylike condition; the process is called "gelation"; e.g., gelatin mixed with water forms a colloidal solution; when cooled, this becomes a semisolid.

gingival: pertaining to gingiva, or gums, consist of the mucosal tissue that lays over the jawbone; the gingiva are naturally transparent, they are rendered red in color because of the blood flowing through tissue; the gingiva are connected to the teeth and **bone** by way of the periodontal fibers.

gland: an organ manufacturing substances for secretion; it may be large (e.g., the **liver** or **mammary gland**), or it may be small (e.g., a sweat gland); it functions by taking chemical substances and water from the **blood** and synthesizing the compounds for secretion; glands are either exocrine or endocrine, and their methods of secretion are either holocrine, merocrine, or apocrine; they are also described by their shape (e.g., tubular, racemose, flask-shaped).

glandular tissue: a tissue-bearing **gland**.

glaucoma: a disease of the eye characterized by increased pressure within the **eyeball** and a progressive loss of vision.

glioma: a **tumor** of the **brain** arising from and consisting of **neuroglia**.

globulin: one of the two types of **serum proteins**, the other being **albumin**; this generic term encompasses a heterogeneous series of protein families that have larger molecules, are less soluble in pure water, and migrate less during serum electrophoresis than albumin.

α-globulins: a group of globular **proteins** in **blood plasma**, that are highly mobile in alkaline or electrically charged solutions; they inhibit certain blood protease and inhibitor activity.

β-globulins: a group of globular **proteins** in **blood plasma** that are less mobile in alkaline or electricaly charged solutions than **α-globulins**.

glucose: a sugar, $C_6H_{12}O_6$, that have several optically different forms; the common or dextrorotatory form (dextroglucose or D-glucose) occurs in many fruits, animal tissues, and fluids, etc., and has a sweetness about one-half that of ordinary sugar; the levorotatory form (levoglucose or L-glucose) is rare and not naturally occurring; also called "starch syrup," a syrup containing dextrose, maltose, and dextrine, obtained by the incomplete hydrolysis of starch.

glucose clamp experiment: **blood sugar** monitoring at basal (euglycemic) or elevated (hyperglycemic) levels during variable **insulin infusion**; it measures tissue-specific insulin action and **glucose metabolism**.

glucose tolerance test: the **administration** of **glucose** to determine how quickly it is cleared from the **blood**; used to test for **diabetes**, **insulin** resistance, and sometimes reactive hypoglycemia; the glucose is most often given orally so the common test is technically an **oral glucose tolerance test (OGTT)**.

glycation: the result of a **sugar** molecule, such as **glucose**, bonding to a **protein** or **lipid** molecule without the controlling action of an **enzyme**; glycation may occur either inside the body (**endogenous** glycation) or outside the body (**exogenous** glycation); enzyme-controlled addition of sugars to protein or lipid molecules is termed glycosylation; glycation is a haphazard process that impairs the function of biomolecules, while glycosylation occurs at defined sites on the target molecule and is required in order for the molecule to function.

glycerol (glycerin, glycerine): a colorless, odorless, syrupy, sweet liquid (see **alcohol**) usually obtained by the saponification of natural **fats** and **oils**; used in the manufacture of **cosmetics**, perfumes, inks, and certain glues and cements; it also acts as a solvent; and in medicine in suppositories and **skin** emollients.

glycogen: a polysaccharide that is the principal storage form of **glucose** in animal **cells**; it is found in the form of granules in the **cytosol** in many cell types, and plays an important role in the glucose cycle; it forms an energy reserve that can be quickly mobilized to meet a sudden need for glucose, but one that is less compact than the energy reserves of **triglycerides**.

glyco-lipid: any of the class of lipids that comprise the cerebrosides and gangliosides, that upon hydrolysis yield galactose or a similar sugar, a **fatty acid**, and sphingosine or dihydrosphingosine.

Golgi apparatus (complex): a netlike mass of material in the **cytoplasm** of animal **cells** believed to function in cellular secretion.

grandular: pertaining to **tissue** structures that contain granular inclusions; for example, grandular-cystic **hyperplasia**, grandular **tumor**, grandular-**squamous-cell carcinoma**.

gray matter: a **nervous tissue** found in the central nervous system; it contains numerous **cell** bodies (cytons), dendrites, synapses, terminal processes of axons, **blood vessels** and **neuroglia**; it is internal to **white matter** in the spinal cord and some other parts of the **brain**; it is external to white matter in the cerebral hemisphere and in the **cerebellum**; coordination in the central nervous system is effected in gray matter; brain nuclei and nerve centers are composed of gray matter.

green fluorescent protein (GFP): a **protein**, comprised of 238 **amino acids** (27 kDa), from the jellyfish *Aequorea victoria* that fluoresces green when exposed to blue light; GFP has a unique cylindrical shape consisting of an 11-strand β-barrel with a single alpha helical strand containing the **chromophore** that runs through the center; this barrel permits chromophore formation and protects it from quenching by the surrounding microenvironment; in cell and molecular biology, the GFP gene is frequently used as a reporter of expression; in modified forms it has been used to make biosensors.

ground substance: the homogeneous matrix in which the **fibers** and **cells** of **connective tissue** or other particles are embedded.

gyaluronic acid: natural **moisturizing** component in **tissues**; the quantity of gyaluronic acid in an organism decreases with growing older.

hair: a threadlike outgrowth from the **skin**; each hair is a slender rod composed of dead cells strengthened by keratin, but remaining soft and supple; it grows from a hair follicle and its length varies according to species and the part of the body on which it is growing; a follicle surrounds the hair root and **hair shaft**; it penetrates deep into the **dermis**.

hair follicle: a part of the **skin** that grows hair by packing old cells together; attached to the follicle is a **sebaceous gland**.

hair shaft: a mature hair shaft is nonliving biological fiber; it is composed of a central pith (or medulla), surrounded by a more solid cortex, and is enclosed in a thin, hard, cuticle; inside the hair follicle it is surrounded by the inner and outer root sheaths.

hand: is one of the two intricate, prehensile, multifingered body parts normally located at the end of each arm (medically: "terminating each anterior limb/appendage") of a human or other primate.

head: a part of body that comprises the **brain**, eyes, ears, nose, and mouth (all of which aid in various sensory functions, such as sight, hearing, smell, and taste).

heart: a hollow, muscular organ by which rhythmic contractions and relaxations keeps the blood in circulation throughout the body.

heart beats: contractions of the heart; usually the heart rate that describes the frequency of the cardiac cycle, calculated as the number of heart beats in one minute and expressed as "beats per minute" (bpm); the heart beats up to 120 times per minute in childhood; when resting, the adult human heart beats at about 70 bpm (males) and 75 bpm (females).

heart valve leaflet: valves in the **heart** maintain the unidirectional flow of **blood** by opening and closing, depending on the difference in pressure on each side; the mitral valve is the heart valve that prevents the backflow of blood from the left ventricle into the left **atrium**; it is composed of two leaflets (one anterior, one posterior) that close when the left ventricle contracts; each leaflet is composed of three layers of **tissue**: the atrialis, fibrosa, and spongiosa.

hemangioma: a native abnormality caused by proliferation of **endothelial cells**; formed aggregates consisting chiefly of dilated or newly formed **blood vessels** that look like **tumors**; types are capillary hemangioma, **cavernous hemangioma**, senile hemangioma, and verrucous hemangioma.

hematocrit (Hct): the relative volume of the **red blood cells** in a **blood** sample expressed in percentages.

hematoporphyrin: a complex mixture of monomeric and aggregated **porphyrins** used in the **photodynamic therapy** of **tumors**; a purified component of this mixture is known as dihematoporphyrin **ether**.

hematoporphyrin derivative (HPD): a **photosensitizer** with an excitation band around 620 nm; used in **cancer** diagnostics and **photodynamic therapy**.

heme: an iron-containing substance; the basic unit of the **hemoglobin** molecule; mammals have four heme units in their hemoglobin.

hemodynamics: the physiology branch dealing with the forces involved in the circulation of **blood**.

hemoglobin: a red iron-containing respiratory pigment found in the **blood**; it conveys oxygen to the tissues and occurs in reduced form (**deoxyhemoglobin**) in venous blood and in combination with oxygen (**oxyhemoglobin**) in arterial blood; it consists of **heme** combined with globin, a blood protein; it is chemically related to chlorophyll, **cytochrome**, hemocyanin, and **myoglobin** (see **hemoglobin spectrum**, Glossary 1).

hemolysis: the breaking open of **red blood cells** and the release of **hemoglobin** into the surrounding fluid (**plasma**, *in vivo*).

hemolytic disease: hemolytic disease of the newborn is an alloimmune condition that develops in a fetus, when the IgG antibodies that have been produced by the mother and have passed through the placenta, including ones that attack the **red blood cells** in the fetal circulation; the red cells are broken down and the fetus can develop reticulocytosis and anemia; **hemolysis** leads to elevated **bilirubin** levels; after delivery bilirubin is no longer cleared (via the placenta) from the neonate's **blood** and the symptoms of yellowish skin and yellow discoloration of the whites of the eyes increase within 24 hours after birth.

hemorheological status: the status determined on the basis of **blood** rheology parameters, such as whole blood and plasma viscosity, cell transit time, cell deformability, clogging rate, clogging particle changes.

hemorrhage: a discharge of **blood**, as from ruptured **blood vessels**.

histology: the study of **tissue** sectioned as a thin slice, using a microtome; it can be described as microscopic anatomy.

H_2O_2: hydrogen peroxide is a very pale blue liquid that appears colorless in a dilute solution, slightly more viscous than water; it is a weak acid; it has strong oxidizing properties and is therefore a powerful bleaching agent that has found use as a disinfectant.

homologous series: in chemistry, this is a series of organic compounds with a similar general formula, possessing similar chemical properties due to the presence of the same functional group, and shows a gradation in physical properties as a result of increase in molecular size and mass.

horny-skin layer: the same as **stratum corneum**.

human epidermal membrane (HEM): a skin flap containing epidermis and used to perform *in vitro* permeability experiments under varied experimental conditions for different deliverable agents and drugs; two-chamber diffusion cells are typically used.

human papilloma virus (HPV): a diverse group of **DNA**-based viruses that infect the **skin** and **mucous membranes** of humans and a variety of animals; more than 100 different HPV types have been characterized; some HPV types cause benign skin warts, or papillomas, for which the virus family is named; HPVs associated with the development of such "common warts" are transmitted environmentally or by casual skin-to-skin contact.

humidity: see Glossary 1.

hydration: the absorption of water by tissues and cells; the organic hydration reaction, a reaction in which water is added across a double bond; mineral (component of tooth or **bone** tissue) hydration, a reaction in which water is combined into the crystalline structure of a mineral.

hydraulic conductivity: a property of material that describes the ease with which water can move through pore spaces or fractures; it depends on the intrinsic permeability of the material and on the degree of saturation; saturated hydraulic conductivity describes water movement through saturated media.

hydrocephalus: an abnormal increase in the amount of **cerebospinal fluid (CSF)**.

hydrocortisone: a corticosteroid that is similar to a natural hormone produced by adrenal **glands**.

hydrophilic: that which has an affinity with water.

hydrophobic: that which has little or no affinity with water.

α-hydroxy acids (AHAs): such as glycolic, lactic, or fruit acids are the mildest of the peel formulas and produce light peels.

hydroxyapatite (HAP): see **apatite**.

hydroxyethyl cellulose: is a nonionic, water-soluble polymer that can thicken, suspend, bind, emulsify, and form films.

3-hydroxy-L-kynurenine-0-β-glucoside (3-HKG): an important age-related chromophore of the human-eye lens, protecting it from UVA radiation.

hygroscopic: the ability of a substance to attract water molecules from the surrounding environment through either absorption or adsorption; hygroscopic substances include **glycerol**, **ethanol**, methanol, concentrated sulfuric acid, and concentrated sodium hydroxide; calcium chloride is so hygroscopic that it eventually dissolves in the water it absorbs.

hypaque: a commonly used x-ray contrast medium; as diatrizoate meglumine and as diatrizoate sodium, it is used for gastrointestinal studies, angiography, and urography.

hyperchromaticity: the increase in optical density of **DNA** molecules in solution, which increase upon nuclease digestion due to the release of nucleotides that absorb more UV light; such a chromic shift is also seen during the process of denaturation due to temperature of DNA separation.

hyperdermal: refers exclusively to the skin, when, for instance, intradermal injection is provided; see **cutaneous**.

hyperglycemia: a condition in which an excessive amount of **glucose** circulates in the **blood plasma**; it is primarily a symptom of **diabetes** in which there are elevated levels of blood sugar, or glucose, in the bloodstream; in type I diabetes, hyperglycemia results from malfunctioning in the supply of insulin, the chemical that enables cells to receive energy from glucose; type II diabetes is due to a combination of defective insulin secretion and defective responsiveness to insulin, often termed "reduced insulin sensitivity."

hyperinsulinemic-hypoglycemic clamp: a procedure that suppresses endogenous insulin secretion by hyperinsulinemia- and hypoglycemia-mediated feedback inhibition of beta-cells.

hyperosmotic: see Glossary 1.

hyperplasia: the enlargement of a part due to an abnormal increase in the number of its **cells**.

hyperthermia: an acute condition that occurs when the body produces or absorbs more heat than it can dissipate; it is usually due to excessive exposure to heat; it can be created artificially by drugs or medical devices (based on acoustics, microwaves, light, etc.), in these instances it may be used to treat cancer and other conditions.

hypertonic: a solution that has a higher concentration of solutes than that in a cell is said to be hypertonic; this solution has more solute particles and, therefore, relatively less water than the cell contents.

hypodermic: situated or lying under the **skin**, as **a tissue**; performed or introduced under the skin, e.g., injection by a syringe, etc.

hypoosmotic: describes a cell or other membrane-bound object that has a lower concentration of solutes than its surroundings; for example, a cell in a high-salt-concentration medium is hypoosmotic; water is more likely to move out of the cell by osmosis as a result; this is the opposite of **hyperosmotic**.

hypothesis of Roy and Sherrington: the hypothesis widely accepted to account for the phenomenon of increased neuronal metabolic activity giving rise to the accumulation of vasoactive catabolites, which decrease vascular resistance and thereby increase blood flow until normal homeostasis is reestablished.

hypotonic: conditions or bathing media owing to the osmotic flow of water into the **cell cytoplasm**.

hypoxia: lack of **oxygen** in air, **blood**, or **tissue**.

hysterectomy: the excision of the **uterus**.

immobilize: to deprive mobility.

implant: a material grafted (implanted) or introduced into a **tissue**.

incision: a surgical cut of a **tissue**; the separation of soft tissues using a scalpel.

India ink: used in preparation of phantoms as an absorbing medium.

indocyanine green: a tricarbocyanine type of dye (stain) having a high absorption in NIR (800 nm) and little or no absorption in the visible range; it is used in diagnostics for **blood volume** determination, **hemodynamic**, cardiac output, or hepatic function studies.

infiltrate: to perform infiltration, i.e., to penetrate a cell or tissue with a substance; also refers to the substance infiltrated.

infiltrating: the process of percolation and impregnation of material, cell or tissue by gas, liquid, or solution; also related to **cell** migration; examples: adipose infiltration-appearance of **fat** cells in the places where they are normally absent, calcareous infiltration—see **calcification**, fatty infiltration—a pathological storage of fat drops in cell **cytoplasm**.

inflammation: traditionally Western medicine has recognized the four signs of inflammation as *tumor, rubor, calor, and dolor*—swelling, redness, heat, and pain; besides these physical changes, there are also important psychological ones, including lethargy, apathy, loss of appetite and increasing sensitivity to pain; in response to acute damage or entrance of foreign material, monocytes enlarge and synthesis increases the amount of **enzymes** that help to break down the material; in doing so they are transformed to more active **phagocytes** called **macrophages**; http://freespace.virgin.net/ahcare.qua/index4.html.

Infracyanine25 (IC25): an NIR contrasting agent (see **indocyanine green**).

infusion: the administration of a drug parenterally by the intravenous route, **subcutaneous** or intramuscle injection.

injury: tissue damage, **wound**, **trauma**.

in situ: pertaining to experiments on an object in its original place.

insulin: a polypeptide hormone that regulates **carbohydrate metabolism**; it is produced in the islets of Langerhans in the pancreas; it has effects on fat metabolism and it changes the **liver's** activity in storing or releasing **glucose** and in processing **blood lipids**, and in other **tissues** such as **fat** and **muscle**; insulin is used medically to treat some forms of **diabetes mellitus**.

intact: not changed or diminished; not influenced or swayed.

intercellular: between **cells**, as in an intercellular bridge.

interferons (**IFNs**): the natural **proteins** produced by the **cells** of the immune system of most vertebrates in response to challenges by foreign agents such as viruses, **bacteria**, parasites, and **tumor** cells; they belong to the large class of glycoproteins known as cytokines and assist the immune response by inhibiting viral replication within other cells of the body.

interfibrillar spacing: the spacing between fibrils.

intermolecular spacing: the spacing between molecules.

interstitial fluid: a solution that bathes and surrounds the **cells** of multicellular animals; it is the main component of the **extracellular** fluid, which also includes **plasma** and transcellular fluid; on average, a subject has about 11 liters of interstitial fluid, providing the cells of the body with nutrients and a means of waste removal.

interstitial space: space where **interstitial fluid** is circulating.

intestine: a part of the alimentary canal, in the shape of a long tube, which is concerned with the digestion and absorption of nutrients and the reabsorption of water from feces; most of the digestion and almost all the absorption takes place in the intestine; the internal surface area of the intestine is increased by folds in the lining and projections on the lining; the intestine is coiled in the abdominal cavity; its length is greater than the length of the body; the anterior part of the intestine that contains **glands** for secreting digestive **enzymes** and receives **ducts** from the large digestive glands.

intima: the innermost **membrane** or lining of some organ or part, especially that of an **artery**, **vein**, or **lymphatic vessel**.

intracellular fluid: see **cytoplasm**.

intracellular motility: the motility of **cytoplasm** components.

intralipid, nutralipid, liposyn: intravenously administered nutrients that are fat emulsions containing soybean oil, egg **phospholipids**, and **glycerol**.

invasive: characterized by invasion; denoting: (1) a procedure that requires insertion of an instrument or device into the body through the **skin** or a body orifice for diagnosis or treatment, (2) a diffusion of **malignant tumor** by its growing into or destruction of the adjacent tissue, (3) spread of infection.

in vitro: in medicine, pertaining to experiments on dead tissue.

in vivo: in medicine, pertaining to experiments on living animals and humans.

ionic strength: characterizes a solution, the concentration of all ions present in a solution; generally multivalent ions contribute strongly to the ionic strength.

iontophoresis: a **noninvasive method** of propelling high concentrations of a charged substance, normally medication or bioactive-agents, transdermally by repulsive electromotive force using a small electrical charge applied to an iontophoretic chamber that contains a similarly charged active agent and its **vehicle**; one or two chambers are filled with a solution that contains an active ingredient and its solvent, termed the vehicle; the positively charged chamber, termed the anode will repel a positively charged chemical, while the negatively charged chamber, termed the cathode, will repel a negatively charged chemical into the skin.

iris: the thin, circular, colored sheet of muscular tissue at the front of the **eyeball**, forming the colored part of the eye; the central opening, the **pupil**, allows light to enter the **eyeball**; the iris controls the amount of light that enters the eyeball and assists in accommodating for near objects.

irritation: the enhanced **inflammatory** reaction of **tissue** on its **injury**.

ischemia: a restriction in **blood** supply, generally due to factors in the blood **vessels**, with resultant damage or dysfunction of **tissue**.

isoosmotic: pertaining to solutions that exert the same **osmotic pressure**.

isopropyl laurate: a synthetic compound derived from **fatty acids**; emollient, **moisturizer**.

isopropyl myristate: $C_{17}H_{34}O_2$; refractive index $n_D = 1.435–1.438$ at 20°C; used in **cosmetic** and **topical** medicinal preparations where good absorption through the **skin** is desired; binding agent, emollient, **moisturizer**, and solvent.

isopropyl palmitate: an ester of palmitic acid from coconut oil used to impart silkiness to the **skin** and **hair**; a synthetic antistatic agent, binding agent, emollient, **moisturizer**, and solvent.

isotonic: (1) noting or pertaining to solutions characterized by equal **osmotic pressure**; (2) noting or pertaining to a solution containing just enough salt to prevent the destruction of the **erythrocytes** when added to the **blood**.

keloid: a kind of **fibrous tumor** that forms hard, irregular, clawlike excrescences upon the **skin**, especially postburn.

keratin: a family of **fibrous** structural proteins; tough and insoluble, they form the hard but nonmineralized structures found in animals and other living objects.

keratinocyte: the principal cell type of **epidermis**; so named because of the family of filamentous proteins, the keratins, that comprise its distinctive **cytoskeleton**.

keratectomy: the **incision** of part of the **cornea**.

keratotomy: the **incision** of the **cornea**.

kidney: either of a pair of bean-shaped glandular organs in the back part of the abdominal cavity that excrete urine; a kidney contains numerous nephrons and their associated blood supply; it consists of two zones, a cortex and **medulla**, encased in a fatty protective capsule.

knee: the lower extremity joint that connects the femur and the tibia; since in humans the knee supports nearly the entire weight of the body, it is vulnerable both to acute **injury** and to the development of osteoarthritis.

labeling: the specific marking of **cells** or cell compartments to track probes and measure functional parameters from molecular- and cellular-based studies to *in vivo* systems to understand how marked components impact human physiology and disease; for instance, in cellular transplantation technology, labeling provides information about location, tracking, and quantifying of implanted cells in *in vivo* systems; a wide variety of labeling probes and systems are available; they are mostly based on fluorescing molecules and **nanoparticles** with a possibility of specific binding to cell and **tissue** compartments; CW and time-resolved **fluorescence** techniques are typically used to monitor these markers.

lamella: a platelike structure, appearing in multiples, that occurs in various situations, such as biology (**connective tissue** structures) or materials sciences.

lamina fusca: the layer of the eye **sclera**.

lamina propia: a thin vascular layer of **areolar connective tissue** beneath the **epithelium** and is part of the **mucous membrane**.

Langer's skin tension lines: the local **skin** tension directed lines caused by bundles of fibroconnective **tissues** within the *reticular dermis*.

larynx: a muscular and cartilaginous structure lined with **mucous membrane** at the upper part of the trachea, in which the **vocal cords** are located.

laser coagulation: a **coagulation** of **tissue** caused by laser heating.

laser cyclophotocoagulation: see **cyclophotocoagulation**.

laser interferential retinometer: see Glossary 1.

laser refractive surgery: a special laser (typically UV **excimer laser**, see Glossary 1) reshapes the **cornea** by the precise and controled removal of corneal **tissue** and therefore changes corneal focusing power.

lecithin: technical lecithin contains 60% natural **phospholipids** (major phosphatidylcholine), 30–35% plant oil, glycerol, et al.; it is a basis for many **nourishing (nutritive) creams** due to its ability to penetrate deep into the **skin**.

lens: see **crystalline lens**.

lens syneresis: water is released from the bound state in the hydration layers of lens proteins and becomes bulk water; this increases the difference in refractive index between the lens proteins and the surrounding fluid.

lesion: an **injury** or an alteration of an organ or **tissue**.

leukoplakia: a condition in which thickened, white patches form on a subject's gums, on the inside of his/her cheeks and sometimes on his/her tongue; these patches can't easily be scraped off; the cause of leukoplakia is unknown, but it's considered to result from chronic **irritation**, caused by tobacco or long-term alcohol use; it is the most common of all chronic mouth lesions, more frequently appears in older men.

ligament: a band of **tissue**, usually white and **fibrous**, serving to connect **bones**, hold organs in place, etc.

limbus: the border, edge, or fringe of a part.

lipid: any substance occurring in plants or animals that is soluble in **ether**, hot **ethanol**, and gasoline (i.e., **fat** solvents); the term includes true **fats**, waxes, sterols, steroids, **phospholipids**, etc.

lipid bilayer or **bilayer lipid membrane**: a membrane or zone of a membrane composed of lipid molecules (usually **phospholipids**); the lipid bilayer is a critical component of all biological membranes, including **cell** membranes; the structure of a bilayer explains its function as a barrier; lipids are **amphiphilic** molecules since they consist of polar head groups and nonpolar **acyl** tails; the bilayer is composed of two opposing layers of lipid molecules arranged so that their hydrocarbon tails face one another to form an oily core, while their charged heads face the aqueous solutions on either side of the membrane; thus, the bilayer consists of the **hydrophobic** core region formed by the acyl chains of the lipids, and membrane interfacial regions that are formed by the polar head groups of lipids; the **hydrophilic** interfacial regions are saturated with water, whereas the hydrophobic core region contains almost no water; because of the oily core, a pure lipid bilayer is permeable only to small hydrophobic solutes, but has a very low permeability to polar inorganic compounds and ionic molecules.

lipophilic: that which has affinity with **lipids**.

liposomes: microscopic spherical vesicles prepared by adding a water solution to a **phospholipid** gel; a liposome is a good model of a cell **organelle**; liposome diameters are usually in the range of 20–100 nm; they are used for drug delivery in medicine and **cosmetics**.

lips: a visible organ at the mouth of humans and many animals; both lips are soft, protruding, movable, and serve primarily for food intake, as a tactile sensory organ, and in articulation of speech.

liquid-crystalline phase: a liquid that has certain crystalline characteristics, especially different optical properties in different directions.

liver: a large, reddish-brown, glandular organ located in the upper right side of the abdominal cavity, divided by fissures into five lobes, and functioning in the secretion of **bile** and various **metabolic** processes.

lobe (cerebral lobe): each cerebral hemisphere is divided, more or less arbitrarily, into different regions, each region being a lobe; the deeper fissures are used to distinguish the lobes; each lobe is named from the part of the **skull** near which it is situated; they are terms of convenience, not of anatomical or physiological significance: frontal lobe, temporal lobe, occipital lobe, etc.

lung: either of the two saclike respiratory organs in the **thorax** of humans; the lungs connect to the larynx via trachea, bronchi, and ramifications of bronchial tubes.

lymph: an alkaline colorless liquid obtained from **blood** by filtration through **capillary** walls; it contains a smaller amount of soluble blood proteins and **white blood cells** than blood, but more **lymphocytes**; it contains no **red blood cells**.

lymph nodes (lymphoid tissues): the tissues that produce **lymphocytes** by division of some **cells**; they are found in lymph **glands**, which are formed from a network of **reticular fibers** that enclose **lymphocytes**, lymphoblasts, and **macrophages**; they also occur in the **spleen**, tonsils, and thymus.

lymphatic (lymph) vessels: the thin-walled tubular vessels resembling **veins** in structure but with thinner walls and more valves; the walls are enclosed by smooth **muscle** and **connective tissue**; lymph vessels drain into lymph **ducts**; lymph vessels act as channels along which pathogens are conducted from infected areas of the body, the pathogens being unable to enter the **blood capillaries**; **lymph nodes** are distributed along the lymph vessels; the lymph flow is maintained by peristaltic contractility of the lymph vessels, aided by the squeezing of the vessels by skeletal muscles, with the **valves** maintaining a flow in one direction only.

lymphocyte: a spherical **white blood cell** with one large nucleus and relatively little **cytoplasm**; two types exist, small and large lymphocytes; they are produced continually in **lymphoid tissues**, such as **lymph nodes**, by cell division; the cells are nonphagocytic, exhibit amoeboid movement, and produce antibodies in the **blood**; they constitute about 25% of all leukocytes in the human body.

lymphotropic agent: an agent that influences the functioning of **lymph vessels**.

lyophilized: related to **tissues**, **blood**, **serum**, or other biological substances that are dried by freezing in a high vacuum; the samples preserved to prevent decay, spoilage, and prepared for future use.

lysed blood: the product of **blood** at its **hemolysis**.

lysosomes: the membrane-bound particles that are smaller than **mitochondrion**, occurring in large numbers in the **cytoplasm** of **cells**; they contain hydrolytic **enzymes** that are released when the cell is damaged; these enzymes assist in the digestion and removal of dead cells, the digestion of food and other substances, and the destruction of redundant **organelles**.

macromolecules: pertaining to conventional polymers and biopolymers (such as **DNA**) as well as nonpolymeric molecules with large molecular mass such as **lipids** or macrocycles.

macrophage: a large phagocytic **cell** with one nucleus; movement is by membranelike pseudopodia; these cells are found in contact with **blood** and **lymph** at the sites of corpuscle formation (e.g., in **bone** marrow, **lymph nodes**, and **spleen**); their function is to remove foreign particles from blood and lymph; macrophages are also found in all loose **connective tissue**, but they only become active when the tissue is damaged; their function is to remove the debris from damaged tissues; they form the reticuloendothelial system; in the inactive state (i.e., in undamaged tissue), the resting form of the cell is called a "histiocyte"; macrophages are closely related to **monocytes**.

magnetic resonance imaging (MRI): see Glossary 1, **magnetic resonance imaging**.

malignant: a clinical term that means to be severe and become progressively worse, as in malignant hypertension; in the context of **cancer**, a malignant **tumor** can invade and destroy nearby tissue and may also spread (metastasize) to other parts of the body.

malignant tissue: an abnormally growing **tissue** that has the tendency to spread to other parts of the body, even when the original growth is removed by surgery; eventually it causes death.

mammary gland: a large **gland** on the ventral surface of the mature female; it is thought to be a modified sweat gland; it consists of clusters of gland cells that can extract the necessary substances from **blood** to produce milk; the milk drains through **ducts** into a cistern; a canal leads from the cistern to a mammary papilla; the growth and activity of the gland is under the control of gonadal hormones and the state of the gland is influenced by the estrous cycle; milk production is stimulated by the pituitary lactogenic hormone.

mammography: x-ray **imaging**, **magnetic resonance imaging (MRI)**, **ultrasound (US)**, and positron-emission imaging of a **female breast**, especially for screening or early detection of **cancer**.

mammogram: an image of a **female breast** obtained by **mammography**.

mannitol (or hexan-1,2,3,4,5,6-hexol $[C_6H_8(OH_6)]$): an osmotic diuretic agent and a weak renal vasodilator; it is a sorbitol isomer; it is used clinically to reduce acutely raised intracranial pressure.

mastopathy: any disease of the **female breast**.

meal tolerance test (MTT): the complete nutrient test (**carbohydrate**, **fat**, and **protein** containing meals) that induces both **glucose** and **insulin** responses; the MTT is a more potent insulin stimulator than glucose alone (see **OGTT**).

media: the middle layer of an **artery** or **lymphatic vessel** wall.

mediolateral projection: the medial and lateral planes of the body.

medulla: the inner, paler-colored region of a **kidney**, surrounded by the cortex; it contains the collecting tubules leading from the uriniferous tubules to the pyramid.

medulloblastomas: most **brain tumors** are named after the type of cells from which they develop; medulloblastomas are **malignant** tumors formed from poorly developed cells at a very early stage of their life; they develop in the **cerebellum**, in a part of the brain called the posterior fossa, but may spread to other parts of the brain; very rarely, medulloblastomas may spread to other parts of the body; if they do spread to other parts of the brain, or to the spinal cord, this is usually through the **cerebrospinal fluid (CSF)**; they are more common in children.

melanin: a dark-brown or black pigment; melanin in **melanosomes** of normal **skin** is an extremely dense, virtually insoluble polymer of high molecular weight and is always attached to a structural protein; mammalian melanin pigments have one of two chemical compositions: eumelanin, a brown polymer, and pheomelanin, a yellow-reddish alkali-soluble pigment.

melanin granular (melanosome): the cytoplasmic **organelles** on which melanin pigments are synthesized and deposited; normal human skin color is primarily related to the size, type, color, and distribution of melanosomes; melanosomes are the product of specialized exocrine glands: **melanocytes**.

melanocytes: the components of the **melanin** pigmentary system, which is made up of melanocytes distributed in various sites: the eye (retinal pigment **epithelium**, uveal tract), the ear (in the stria vascularis), the central nervous system (in leptomeninges), the hair (in the hair matrix), the **mucous membranes**, and the **skin** (at the dermal-epidermal interface, where they rest on the basement membrane); in the skin melanocytes project their dendrites into the **epidermis**, where they transfer melanosomes to **keratinocytes**.

melanoma: a darkly pigmented **tumor**, especially of the **skin** or eye, of **cells** containing **melanin**.

melanoma maligna: a special kind of **melanoma** *in situ* that occurs on the sun damaged **skin** of the face or neck may be described as lentigo maligna **melanoma**.

membrane: (1) a very thin layer of **connective tissue** covering an organ: (2) connective tissue that divides **cells**; (3) a thin layer of cells.

meningiomas: the most common **benign tumors** of the **brain** (95% of benign tumors); however, they can also be **malignant**; they arise from the arachnoidal cap cells of the meninges and represent about 15% of all primary brain tumors; they are more common in females than in males (2:1) and has a peak incidence in the sixth and seventh decades.

meniscus: a disk of **cartilage** between the articulating ends of the **bones** in a joint.

mesentery: (1) sheets of thin **connective tissue** by which the **stomach** and **intestines** are suspended from the dorsal wall of the abdominal cavity; (2) the **tissue** supporting the intestines; the mesenteries carry **blood**, **lymph vessels**, and **nerves** to the organs of the alimentary canal.

metabolism (metabolic processes): the chemical processes that take place in a living organism or within part of a living organism (e.g., **cell**) are collectively known as metabolism; metabolism consists of **catabolism** and **anabolism**.

metabolite: a substance that takes part in a **metabolic process**; those metabolites that the organism cannot manufacture have to be obtained from the environment; some metabolites are supplied partly by the environment and partly by the organism; the majority of the metabolites in an organism are manufactured by the organism.

methylene blue: a biological dye (stain) showing a phototoxic effect; its absorption bands are at 609 and 668 nm; it is used as a stain in bacteriology and as an oxidation-reduction indicator; it can be activated by light to an excited state, which in turn activates oxygen to yield oxidizing radicals, such radicals can cause **crosslinking** of **amino acid** residues on proteins and achieve some degree of cross-linking.

micelles: an aggregate of **surfactant** molecules dispersed in a liquid **colloid**; a typical micelle in aqueous solution forms an **aggregate** with the **hydrophilic** "head" regions in contact with surrounding solvent, sequestering the **hydrophobic** tail regions in the micelle center; this type of micelle is know as a normal phase micelle (oil-in-water micelle); inverse micelles have the headgroups at the center with the tails extending out (water-in-oil micelle); micelles are approximately spherical in shape; other phases, including shapes such as ellipsoids, cylinders, and bilayers are also possible; the shape and size of a micelle is a function of the molecular geometry of its surfactant molecules and solution conditions such as surfactant concentration, temperature, **pH**, and **ionic strength**; the process of forming micelles is known as micellization and forms part of the phase behaviour of many **lipids** according to their **polymorphism**.

microcirculation: the flow of **blood** from **arterioles** to capillaries or sinusoids to **venules**; blood flows freely between an arteriole and a venule through a vessel channel called a thoroughfare channel; capillaries extend from this channel to structures called precapillary sphincters, which control the flow of blood between the arteriole and capillaries; the precapillary sphincters contain **muscle** fibers that allow them to contract; when the sphincters are open, blood flows freely to the capillary beds, where gases and waste can be exchanged with body **tissue**; when the sphincters are closed, blood is not allowed to flow through the capillary beds and must flow directly from the arteriole to the venule through the thoroughfare channel; it is important to note that blood is supplied to all parts of the body at all times but all capillary beds do not contain blood at all times (http://biology.about.com/library/organs/heart/blmicrocirc.htm).

microfibril: a very fine **fibril**, or fiber like strand, consisting of glycoproteins; its most frequently observed structural pattern is $9+2$ in which two central protofibrils are surrounded by nine others; the cellulose inside plants is one of the examples of nonprotein compounds that are using this term with the same purpose.

microfilaments: the fine, threadlike protein fibers, 3–6 nm in diameter; they are composed predominantly of a contractile protein called actin, which is the most abundant cellular protein; microfilaments' association with the protein myosin is responsible for **muscle** contraction; microfilaments can also carry out cellular movements including gliding, contraction, and cytokinesis.

microtubules: cylindrical tubes, 20–25 nm in diameter; they are composed of subunits of the **protein** tubulin; they act as a scaffold to determine **cell** shape, and provide a set of "tracks" for cell **organelles** and vesicles to move on; microtubules also form the spindle fibers for separating chromosomes during **mitosis**; when arranged in geometric patterns inside **flagella** and cilia, they are used for locomotion.

microvessels: see **capillary**.

mineralization: the process where a substance is converted from an organic substance to an inorganic substance, thereby becoming mineralized.

mitochondrion (*pl.* mitochondria): a threadlike, or rodlike, granular **organelle** in the **cytoplasm** of **cells**, about 0.5 μm in width, and up to 10 μm in length for threadlike mitochondria; mitochondria are bounded by a double **membrane**; the inner membrane is folded inward at a number of places to form cristae; mitochondria contain phosphates and numerous **enzymes** that vary in different tissues; their function is cellular respiration and the release of chemical energy in the form of **ATP** for use in most of the cell's biological functions; the cells of all organisms, except bacteria and blue-green algae, contain mitochondria in varying numbers mitochondria are especially numerous in cells involved in significant metabolic activity, such as **liver** cells; mitochondria are self-replicating.

mitosis: the process by which a **cell** nucleus usually divides into two; the process takes place in four phases: prophase, metaphase, anaphase, and telophase; the daughter nuclei are genetically identical to each other and to the parent nucleus.

modified amino resin (MAR): material used in preparation of tissue **phantoms** (see Glossary 1).

monocyte: a spherical **white blood cell** with an oval nucleus; monocytes are the largest of the white blood cells; the cells are voraciously phagocytic and exhibit amoeboid movement; they are produced in **lymphoid tissues** and constitute about 5% of all leukocytes.

moisturizers: complex mixtures of chemical agents specially designed to make the external layers of the **skin (epidermis)** softer and more pliable, by increasing its **hydration**; naturally occurring skin lipids and sterols as well as artificial or natural **oils**, humectants, emollients, lubricants, etc., may be part of the composition of commercial skin moisturizers; they usually are available as commercial products for **cosmetic** and therapeutic uses.

molar mass: the mass of one mole of a chemical element or chemical compound.

monomer (monomeric form): the original compound from which a polymer is formed, e.g., ethylene is the monomer from which polyethylene is formed.

mononucleotide (nucleotide): a unit in a long-chain molecule of **nucleic acid**; it is a chemical compound formed from one molecule of a sugar (ribose or deoxyribose), one molecule of phosphoric acid, and one molecule of a base (containing an amino group); nucleotides are also found free in **cells** (see **DNA**).

monounsaturated fatty acid: a **fatty acid** with one double-bonded carbon in the molecule, with all of the others single-bonded carbons, in contrast to polyunsaturated fatty acids, which have more than one double bond.

mucin: a mucoprotein that forms **mucus** in solution.

mucinous: pertaining to or containing **mucin**.

mucopolysaccharides (or glycosaminoglycans): are long unbranched polysaccharides that consist of a repeating disaccharide unit; they are synthesized in **endoplasmic reticulum** and **Golgi apparatus**; they form an important component of **connective tissues**; their chains may be covalently linked to a **protein** to form **proteoglycans**.

mucous membrane (mucosa): a **membrane** consisting of moist **epithelium** and the **connective tissue** immediately beneath it; it usually consists of simple epithelium, but is stratified near openings to the exterior; it is often ciliated and often contains goblet **cells**; mucosa is found in the lining of the gut and in the urinogenital **ducts**.

mucus: a thin, slimy, viscous liquid secreted by epithelial **cells** in **tissues** or **glands**; it protects and lubricates the surface of structures, e.g., the internal surfaces of the greater part of the alimentary canal are lubricated with mucus.

muscle: an organ of movement which is highly contractile, extensible and elastic; it is composed of **muscular tissue**; a muscle contracts and relaxes; it can also be stretched beyond its normal length, and return to its original length and shape when the stretching force is removed.

muscular tissue: a **tissue** characterized by its ability to contract on being stimulated by a motor nerve; there are three main types of muscular tissue forming three types of muscle: skeletal, smooth, and cardiac.

myelin: an electrically insulating phospholipid layer that surrounds the axons of many neurons; it is an outgrowth of glial cells.

myocard (myocardium): the muscular substance of the **heart**.

myocardial infarction: commonly known as a heart attack, is a disease state that occurs when the blood supply to a part of the **heart** is interrupted; the resulting **ischemia** or **oxygen** shortage causes damage and potential death of heart tissue.

myofibrils: cylindrical **organelles**, found within **muscle cells**; they are bundles of filaments that run from one end of the cell to the other and are attached to the cell surface membrane at each end.

myofilaments: the filaments of **myofibrils** constructed from **proteins**; they consist of two types, thick and thin; thin filaments consist primarily of the protein actin; thick filaments consist primarily of the protein myosin; in striated **muscle**, such as skeletal and cardiac muscle, the actin and myosin filaments each have a specific and constant length.

myoglobin: a variety of **hemoglobin** found in voluntary **muscle fibers**; it has a higher affinity for oxygen than hemoglobin, and thus assists in the transfer of oxygen to **muscles**.

NAD, NAD+, NADH: NAD (nicotinamide adenine dinucleotide); an important **coenzyme** found in **cells**; it plays key roles as carriers of electrons in the transfer of reduction potential; cells produce NAD from niacin, and use it to transport electrons in redox reactions; during this process NAD picks up a pair of electrons and a proton and is thus reduced to NADH, releasing one proton (H^+): $MH_2 + NAD+ \rightarrow NADH + H^+ + M$ + energy, where M is a metabolite; two hydrogen atoms (a hydride ion and a proton H^+) are removed from the metabolite and the proton is released into solution; from the hydride electron pair, one electron is transferred to the positively-charged nitrogen, and one hydrogen attaches to the carbon atom opposite to the nitrogen; the reducing potential stored in NADH can be converted to **ATP** through the aerobic electron transport chain or used for anabolic **metabolism**; ATP is the universal energy currency of cells, and the contribution of NADH to the synthesis of ATP under aerobic conditions is substantial; however, under certain conditions (e.g., **hypoxia**) the aerobic regeneration of oxidized NAD+ is unable to meet the cell's immediate demand for ATP; in contrast, glycolysis does not require **oxygen**, but it does require the anaerobic regeneration of NAD+; the oxidation of NADH to NAD+ in the absence of oxygen is called fermentation.

nanospheres: the **fat** particles used for transportation of biologically active substances to the deep layers of **epidermis** and **hair** follicles.

neck: supports the weight of the **head** and protects the **nerves** that travel from the **brain** down to the rest of the body; the cervical portion of the human spine comprises seven bony segments, typically referred to as C-1 to C-7, with cartilaginous disks between each vertebral body; In addition, the neck is highly flexible and allows the head to turn and flex in all directions.

necrosis: the death or decay of **tissue**.

necrotic: pertaining to **necrosis**.

needle-free injection gun: a device creating the required pressure to ensure the medicine penetrates skin tissue directly without needle and correctly being distributed; traditional systems use compressed gas or a spring device to create the pressure that triggers the injection; a novel gas generator system that produces a few milliseconds-gas sparks with a predetermined pressure profile at the moment the injection is made.

neoplasia: (1) **tumor** growth; (2) the formation and growth of new **tissue**.

neoplasm: a new growth of different or abnormal **tissue**; **tumor**.

neoplastic: pertaining to **neoplasia**, **neoplasm**.

nerve: a bundle of parallel funiculi with associated **connective tissue** and **blood vessels**, enclosed in a sheath of connective tissue that forms a tough external coat called the "epineurium."

nervous tissue: **tissue** that consists of nerve cells and their fibers or of nerve fibers alone, together with accessory cells surrounding the **cells** or **fibers**, and **connective tissues** with **blood vessels**.

neurofibroma: a **tumor** that incorporates all sorts of **cells** and structural elements, **infiltrate** the **nerve** and splay apart the individual nerve fibers; although usually benign, they can sometimes degenerate into **cancer**; single neurofibromas often occur in middle and old age and grow at the margins of the peripheral nerves, displacing the nerve's main body; the vestibulocochlear (acoustic) nerve is the most commonly affected; other cranial nerves and spinal nerves are less commonly involved.

neuroglia: the delicate **connective tissue** elements of **nerve tissue** in the central nervous system.

neutral polymer: one that has no electrical charge or ionizable groups such as polyethylene oxide, cellulose, **sugar**, **dextrans**, polyvinyl alcohol, or polystyrene, there are many other examples; some neutral polymers are water soluble, others are not.

nevus: a general term that refers to a number of different, usually benign, pigmented lesions of the **skin**; most birthmarks and moles fall into the category of nevi.

nonionic: not converted into ions.

noninvasive method: a diagnostic method that avoids **trauma** to the **skin** or insertion of an instrument through a body orifice.

nourishing (nutritive) creams: used in **skin cosmetics** for preventing **transepidermal water loss (TEWL)**; they easy penetrate to the deep layers of **epidermis**; skin **hydration** can be provided by two mechanisms—**osmotic** or physiological; as the hydrating substances **sodium lactate**, **pyrrolidonecarboxylic acid**, derivatives of **amino acids** and **sugars**, **proteins**, **mucopolysaccharides** are usually used; as a **hygroscopic** component **glycerol** often use (usually less than 10% in composition), at present glycerol usually replaced by a **propylene glycol**.

nuclear envelope: the main structural elements of the **nucleus**; it is a double **membrane** that encloses the entire **organelle** and keeps its contents separated from the cellular **cytoplasm**.

nuclear pores: because the **nuclear envelope** is impermeable to most molecules, pores are required to allow movement of molecules across the envelope; these pores cross both membranes of the envelope, providing a channel that allows free movement of small molecules and ions; movement through the pores is required for both gene expression and chromosomal maintenance.

nucleic acid: a complex, high-molecular-weight biochemical **macromolecule** composed of nucleotide chains that convey genetic information; the most common nucleic acids are **DNA** and **RNA**; nucleic acids are found in all living **cells** and viruses.

nucleolus: a "sub-organelle" of the **cell nucleus**, which itself is an **organelle**; a main function of the nucleolus is the production and assembly of ribosome components; the nucleolus is roughly spherical, and is surrounded by a layer of condensed **chromatin**; no **membrane** separates the nucleolus from the nucleoplasm.

nucleus: a membrane-enclosed **organelle** found in most eukaryotic cells; it contains most of the cell's genetic material, organized as multiple long linear **DNA** molecules in complex with a large variety of **proteins** such as histones to form chromosomes; the genes within these chromosomes make up the cell's nuclear genome; the function of the nucleus is to maintain the integrity of these genes and to control the activities of the cell by regulating gene expression.

occlusion: a term indicating that the state of something, which is normally open, is now totally closed; in medicine, the term is often used to refer to **blood vessels**, **arteries**, or **veins** that have become totally blocked to any **blood flow**; for issues of artery occlusion, see **stenosis** and **atheroma**; in dentistry, occlusion refers to the manner in which the teeth from upper and lower arches come together when the mouth is closed.

occlusion spectroscopy: the most important **blood** parameters such as **hemoglobin**, **glucose**, **oxygen** saturation, etc., influence the optical transmission growth over systolic **occlusion** and, therefore, may be extracted from the detailed analysis of the time evolution of optical transmission; this forms a basis for a kind of **noninvasive** measurements, i.e., occlusion spectroscopy.

oil: a neutral liquid, soluble in ether, hot **ethanol**, and gasoline, but not in water; it contains carbon and hydrogen, is capable of combustion, and has a marked viscosity; the main types of oils are essential oils, fixed oils, mineral oils; oils are esters of **glycerol** with unsaturated **fatty acids**, of which the most usually occurring are **oleic acid**, linoleic acid, and linolenic acid; oleic acid has one double bond, linoleic has two double bonds, and linolenic has three double bonds; a neutral **fat**, liquid below 20°C, is usually called an oil; it contains a higher proportion of **unsaturated fatty acids** than a solid fat.

ointment: a viscous semisolid preparation used **topically** on a variety of body surfaces; these include the **skin** and the **mucous membrane**.

oleic acid: a **monounsaturated fatty acid** found in various animal and vegetable sources; it has the formula $C_{18}H_{34}O_2$ (or $CH_3(CH_2)_7CH=CH(CH_2)_7COOH$); it comprises 55–80% of olive **oil**.

olfactory tract: a narrow white band, triangular on the coronal section, the apex being directed upward, that lies in the olfactory sulcus on the inferior surface of the **frontal lobe** of the **brain**, and divides posteriorly into a medial and lateral striae.

optic (optical) nerve: the second cranial **nerve**; it is connected to the **retina** and simulated by light; the sensory nerve of sight.

optical biopsy: a measurement of the localized optical properties of **tissues** for diagnostic purpose.

optical clearing: making a **tissue** more translucent by reducing light scattering through matching the refractive index of the scatterers and ground substances; immersion liquids (**osmolytes**) with the appropriate refractive index and the rate of diffusion are usually used (see **immersion technique**, Glossary 1, and **optical immersion technique**).

optical clearing agent (OCA): chemical agents used for controlling the optical properties of **cells** and **tissues** with the result of the increase of their optical transmittance and reduction of the backreflectance.

optical immersion technique: based on impregnation of a **tissue** by a biocompatible chemical agent with a refractive index higher than an interstitial refractive index or **topical** application of a **hyperosmotic** agent inducing tissue **dehydration**; both processes cause an increase of refractive index of interstitial space relating to other tissue compartments and make up tissue more optically transparent (less scattering); for **cell** systems, such as **blood**, an addition to **plasma** a biocompatible chemical agent with a refractive index higher than plasma causes an increase of blood optical transmittance.

oral glucose tolerance test (OGTT): see **glucose tolerance test**.

organelle: a part of a **cell** that is a structural and functional unit, e.g., a **mitochondrion** is a respiratory organelle; organelles in a cell correspond to organs in an organism.

osmolality: a measure of the **osmoles** of solute per kilogram of solvent.

osmolarity: a measure of the **osmoles** of solute per liter of solution; if the concentration is very low, osmolarity and **osmolality** are considered equivalent; in calculations for these two measurements, salts are presumed to dissociate into their component ions; for example, a mole of **glucose** in solution is one osmole, whereas a mole of sodium chloride in solution is two osmoles (one mole of sodium and one mole of chloride), both sodium and chloride ions affect the osmotic pressure of the solution.

osmole (Osm): a unit of measurement that defines the number of moles of a chemical compound that contribute to a solution's **osmotic stress** (pressure).

osmolyte: an osmotically active liquid (molecules) (see **osmotic phenomenon**, **osmotic stress**, Glossary 1).

osmolytic: pertaining to **osmolyte**.

ovalbumin: the main **protein** found in egg white, making up 60–65% of the total protein; is made up of 385 **amino acids**, and its relative molecular mass is 45 kD; it is a glycoprotein with 4 sites of glycosylation.

oxidase: a type of dehydrogenase; the hydrogen removed from the substrate combines with molecular oxygen.

oxidative stress: caused by an imbalance between the production of reactive **oxygen** and a biological system's ability to readily detoxify the reactive intermediates or easily repair the resulting damage; all forms of life maintain a reducing environment within their **cells**; the cellular redox environment is preserved by **enzymes** that maintain the reduced state through a constant input of metabolic energy; disturbances in this normal redox state can cause toxic effects through the production of peroxides and free radicals that damage all components of the cell, including **proteins**, **lipids**, and **DNA**; in humans, oxidative stress is involved in many diseases, such as atherosclerosis, Parkinson's disease and **Alzheimer's disease** and it may also be important in aging; however, reactive oxygen species can be beneficial, as they are used by the immune system as a way to attack and kill pathogens and as a form of cell signaling.

oxygen: chemical element with the chemical symbol O and atomic number 8; it is usually bonded to other elements covalently or ionically; an important example of common oxygen-containing compound is water (H_2O); dioxygen (O_2) is the second most common component of the atmosphere (about 21% by volume) and produced predominantly through photolysis (light-driven splitting of water) during photosynthesis in cyanobacteria, green algae, and plants; oxygen is essential for cellular respiration in all aerobic organisms; triatomic oxygen (ozone, O_3) forms through radiation in the upper layers of the atmosphere and acts as a shield against UV radiation.

oxygenated blood: **blood** saturated by oxygen.

oxyhemoglobin: **hemoglobin** combined with oxygen.

pacemaker: a local rhythm driver; the region of the **heart** or the skeletal **muscles** around a **vessel** where the nervous impulse that starts the contraction of the heart or **blood vessel** muscles is sent out.

pain: a subjective experience; the system that carries information about **inflammation**, damage or near-damage in **tissue**, to the spinal cord and **brain**.

palm: flat of the **hand**.

papillary: related to papilla; papillary **dermis** is the part of the dermis that lies immediately below the **epidermis**, it has vertically oriented **connective tissue** fibers and a rich supply of **blood vessels**; papillary **muscles** of the **heart** serve to limit the movements of the mitral and tricuspid **valves**; papillary **tumors** are the tumors shaped like a small mushroom, with its stem attached to the **epithelial** layer (inner lining) of an organ; papillary tumors are the most common of all **thyroid cancers** ($>70\%$); papillary thyroid cancer forms in cells in the thyroid and grows in small fingerlike shapes, it grows slowly, is more common in women than in men, and often occurs before age 40; papillary **carcinoma** typically exhibits as an irregular, solid or cystic mass that arises from otherwise normal thyroid tissue; papillary serous carcinoma is an aggressive cancer that usually affects the **uterus**/endometrium, peritoneum, or ovary.

parakeratosis: a disorder of the **horny layer** of the **skin epidermis** manifested by the appearance of **cell** nuclei in this layer; it can be seen in chronic **dermatitis**, such as **psoriasis**.

parakeratotic: pertaining to **parakeratosis** (e.g., parakeratotic focus); the **cell** structure near such a focus is substantially disordered, or consists of parakeratotic scales.

pars conv.: pars (partes) convalescent; a part of convalescent tissue.

partially permeable membrane: also termed a semipermeable **membrane**, a selectively permeable membrane or a differentially permeable membrane, is a membrane that allows certain molecules or ions to pass through it by **diffusion** and occasionally specialized "facilitated diffusion"; the rate of passage depends on the pressure, concentration, and temperature of the molecules or solutes on either side, as well as the permeability of the membrane to each solute; depending on the membrane and the solute, permeability may depend on solute size, solubility, properties, or chemistry; an example of a semipermeable membrane is a **lipid bilayer**, on which is based the plasma membrane that surrounds all biological **cells**.

pathological: pertaining to pathology, the study and diagnosis of disease through examination of organs, **tissues**, **cells** and body fluids; the term encompasses both the medical specialty which uses tissues and body fluids to obtain clinically useful information, as well as the related scientific study of disease processes.

peeling: a body treatment technique used to improve and smooth the texture of the facial **skin** using physical (mechanical, acoustical, laser, etc.) or chemical action; for instance chemical solution causes the skin to blister and eventually peel off; the regenerated skin is usually smoother and less wrinkled than the old skin; **α-hydroxy acids (AHAs)** are naturally occurring organic carboxylic acids such as glycolic acid, a natural constituent of sugar cane juice and lactic acid and found in sour milk and tomato juice, is the mildest of the peel formulas and produce light peels for treatment of fine wrinkles, areas of dryness, uneven pigmentation and acne.

percutaneous: pertains to any medical procedure where access to inner organs or other tissue is done through the skin, for instance via needle-puncture of the skin, rather than by using an "open" approach where inner organs or tissue are exposed; phototherapy is another example of percutaneous treatment .

perfusion: **oxygen perfusion** see **blood perfusion**.

peripapillary: surrounding a papilla; papilla is a projection occurring in various animal tissues and organs.

perivascular: around the blood vessels; for instance, perivascular lymphatics.

peroxisome: a specialized **organelle** containing the oxidizing **enzymes** that degrade peroxides.

petrolatum: a semisolid mixture of hydrocarbons obtained from petroleum; used in medicinal **ointments** and for lubrication.

pH: a measure of the acidity or alkalinity of a solution; solutions with a pH less than 7 are considered acidic, while those with a pH greater than 7 are considered basic (alkaline); pH 7 is defined as neutral because it is the pH of pure water at $25°C$; pH is formally dependent upon the activity of hydrogen ions (H^+), but for very pure dilute solutions.

phagocyte: a **white blood cell** that engulfs foreign bodies, particularly pathogens, by enclosing the body in **cytoplasm** through a process of extending pseudopodia around it (the amoeboid movement for engulfing); in mammals, **polymorphs**, **monocytes**, and **macrophages** are phagocytes; macrophages can be phagocytes of other **WBCs**; phagocytes are an important part of the defense mechanism of most animals against invading pathogens.

phenols: a class of chemical compounds consisting of a hydroxyl group ($-OH$) attached to an aromatic hydrocarbon group; the simplest of the class is phenol (C_6H_5OH).

phenylalanine: an essential alpha-**amino acid**; it exists in two forms, a D and an L form, which are enantiomers (mirror-image molecules) of each other; it has a benzyl side chain; its name comes from its chemical structures consisting of a phenyl group substituted for one of the hydrogens in the side chain of alanine; because of its phenyl group, phenylalanine is an aromatic compound.

Philly mice: a new model for genetic **cataracts**, in which there is an apparent defect in **lens membrane** permeability.

phonophoresis: the use of **ultrasound** to enhance the delivery of topically applied drugs.

phosphate-buffered saline (PBS): a **saline** solution with phosphates added to keep the **pH** approximately constant.

phospholipids: a class of **lipids**, and a major component of all biological **membranes**, along with glycolipids, **cholesterol**, and **proteins**; understanding of the **aggregation** properties of these molecules is known as lipid **polymorphism** and forms part of current academic research.

photo-aging: this condition is most noticeable in women who have spent hours in the sun without the benefit of sunscreen; the most obvious symptoms of photo-aging are: dark age spots on the face and décolleté; deep wrinkles around the eyes; fine lines; leathery skin; a gradual thickening of the skin; uneven complexion.

photobleaching: removal of color from a sample by irradiating it with light of a certain wavelength and intensity; in **photochemotherapy**, the **photosensitizer** can be photobleached, either permanently and/or transiently, by the treatment light; the term "photobleaching" is variously used to denote actual photochemical destruction of the photosensitizer or simply decreased optical absorbance and/or fluorescence, which may not be equal and which does not necessarily involve molecular decomposition.

photochemical therapy (photochemotherapy): the branch of therapy that deals with the biochemical action of light on a tissue photosensitized by the appropriate chemical or a chemical that induces the photosensitive agent in **tissue**, e.g., **photodynamic therapy**, **PUVA therapy**.

photocoagulation: the **coagulation** (clotting) of **tissue** using a laser (or lamp) that produces light in the visible (green) wavelength that is selectively absorbed by **hemoglobin**, which is the pigment in **red blood cells**, in order to seal off bleeding **blood vessels**; photocoagulation has diverse uses such as in **cancer** treatment it is used to treat tumors to destroy blood vessels entering a **tumor** and deprive it of nutrients; in the treatment of a detached **retina** it is used to destroy abnormal blood vessels in the retina and to treat tumors in the eye, etc.; NIR light that is selectively absorbed by not very intensive water and fat bands also is used for tissue coagulation, for instance in the treatment of gastric **ulcers**.

photodestruction: intensive or focused laser beams used in the destruction of a **tissue**, **cell**, or their part (**coagulation**, **ablation**), or photochemically by light with a moderate intensity (**necrosis**).

photodisruption: a localized breakdown of semitransparent biological **tissues** that do not strongly absorb light in the visible range by an intensive tightly focused femtosecond laser pulse; the nonlinearity of the process ensures absorption and, therefore, material alterations are confined to the extremely small focal volume; typically submicrometer-sized photodisrupted regions can be produced inside single **cells**; femtosecond pulses deposit very little energy but still causing breakdown, therefore producing surgical photodisruption, while minimizing collateral damage.

photodynamic therapy (PDT): therapy of **malignant lesions** that involves administration to the patient of a **photosensitizer**, a time delay to allow adequate concentration of the drug in the **tumor**, followed by irradiation of the target **tissue** volume by light of a wavelength appropriate to activate the **photosensitizer** effectively; the consequent photochemical damage results in **tissue necrosis** by directly killing tumor cells and/or by vascular damage leading to ischemic necrosis; for most photosensitizers, it is believed that the PDT effect is mediated by the production of highly active singlet oxygen, $^1O_2$, formed by energy transfer from the excited-state photosensitizer to molecular oxygen in the tissue.

Photofrin II: the effective biological photodynamic dye for red light [see, **hematoporphyrin derivative (HPD)**]; its molecular weight is about 500; it has an extinction coefficient of about 5000 cm^{-1} M^{-1} at 626 nm when dissolved in dextrose.

photorefractive surgery: see **laser refractive surgery**.

photothermal therapy: therapy based on the thermal reaction of a living **tissue** and a **cell** to light that is intensive enough to produce thermal effects; laser-induced thermotherapy (LITT) includes laser-induced hyperthermia (LIHT) using temperatures from 42 to 60°C, high-temperature laser-induced **coagulation** (LIC) for temperatures above 60°C, and laser-induced interstitial thermotherapy (LIITT) for coagulation of deep tissues using special fiber optic probes; LIITT is a safe procedure with minimum physical strain for the patient.

physiological solution: there are a number of physiological solutions that provide safety and normal functioning of biological cells, tissues, and organs; they contain electrolytes and organic acids at concentrations similar to that found in animal or human **serum**; **saline** is a physiological solution of sodium chloride (NaCl) in sterile water, used frequently for intravenous infusion, rinsing contact lenses, and nasal irrigation; saline solutions are available in various concentrations for different purposes; normal saline is the solution of 0.9% w/v of NaCl; it has a slightly higher degree of **osmolality** compared to **blood** (hence, though it is referred to as being **isotonic** with blood in clinical contexts, this is a technical inaccuracy), about 300 mOsm/L.

pigment: any substance whose presence in the **tissues** or **cells** colors them.

pigmentary glaucoma: a form of **glaucoma** that usually presents in young males, 20 to 50 years old; in fact, all patients with pigmentary glaucoma will necessarily have pigmentary dispersion syndrome prior to the onset of glaucoma (i.e., actual **optic nerve** damage and peripheral vision loss); the mechanism of glaucoma development in this syndrome is the deposition of **pigment** from the **iris** into the trabecular meshwork (primary site of fluid egress), essentially "plugging" the microscopic spaces through which fluid escapes.

pigmentation: many **tissues** and organs, such as **skin, iris, retina**, etc., as well as blood contain **pigments**, such as **melanin** or **hemoglobin**, in specialized cells called **melanocyte**s or **erythrocytes**; many conditions affect the levels or nature of pigments in **cells**; for instance, albinism is a disorder affecting the low level of melanin production in animals and humans; pigment color differs from structural color in that it is the same for all viewing angles, whereas structural color is the result of selective reflection or interference, usually because of multilayer structures and light scattering.

plague: an infectious, epidemic disease of high mortality caused by the **bacterium** *Pasteurella pestis*.

plasma: the clear, waterlike, colorless liquid of **blood** and other body liquids.

plasma-membrane: see **cell membrane**.

plasmid: a **DNA** molecule separated from the chromosomal DNA and capable of autonomous replication; it is typically circular and double-stranded, occurs in **bacteria**, sometimes in eukaryotic organisms.

plastic surgery: surgical techniques that change the appearance and function of a person's body; some of these operations are called "**cosmetic**," and others are called "reconstructive."

platelets: the very small, nonnucleated, round or oval disks that are fragments of **cells** from red **bone** marrow; they are found only in mammalian **blood**; there are approximately 200,000–400,000 per mm^3 in human **blood**; they initiate blood clotting by disintegrating and releasing thrombokinase.

polar aprotic solvent: a solvent that does not contain an O$-$H or N$-$H bond; acetone ($CH_3-C(=O)-CH_3$) is the polar aprotic solvent.

polyethylene glycol (PEG): any of a series of polymers that have the general formula $HOCH_2(CH_2OCH_2)_nCH_2OH$ or $H(OCH_2CH_2)_nOH$ and a molecular weight of from about 200 to 20,000; they are obtained by condensation of **ethylene glycol** or ethylene oxide and water and used as an emulsifying agent and lubricant in ointments, creams, etc., PGs with a high molecular weight are used as effective **osmolytes**.

polyglycerylmethacrylate: used in drugs and **cosmetic** preparations for personal care/use.

polymorph (polymorphonuclear leukocyte): a polynucleated, irregularly shaped **white blood cell** that exhibits amoeboid movement; the nucleus consists of two or more lobes (in humans up to five) joined by threads; the number of lobes increases with the age of the cell; in a healthy person, the distribution of polymorphs by the number of lodes remains constant; any variation indicates a diseased condition; the cells are all active **phagocytes**; they are produced continually in **bone** marrow and constitute about 70% of all **leukocytes** in humans; the **cytoplasm** of polymorphs in humans is granular; some granulations stain with acid dyes (eosinophils), some with basic dyes (basophils), and some with neutral dyes (neutrophils); all three types increase in number during infection.

polymorphism (pleomorphism): the multiple possible states for a single property, for instance, the property of amphiphiles that gives rise to various **aggregations** of **lipids**; it is also defined as the occurrence of two or more structural forms.

polyorganosiloxane (POS, silicone): the host media for a solid-state **tissue phantom**.

polyp: an abnormal growth of **tissue** (**tumor**) projecting from a **mucous membrane**; if it is attached to the surface by a narrow elongated stalk it is said to be pedunculated; if no stalk is present it is said to be sessile; polyps are commonly found in the **colon, stomach**, nose, **urinary bladder**, and **uterus**; they may also occur elsewhere in the body where mucous membranes exist like the **cervix** and small **intestine**.

polypropylene glycol (PPG): the polymer of **propylene glycol**; chemically it is a polyether; the term polypropylene glycol or PPG is reserved for low to medium range molar mass polymer when the nature of the end-group, which is usually a hydroxyl group, still matters.

pons: a structure located on the **brain** stem; in humans it is above the medulla, below the midbrain, and anterior to the **cerebellum**.

porosity: a measure of the void spaces in a material, and is measured as a fraction, between 0–1, or as a percent between 0–100%; the term porosity is used in multiple fields including biology, for instance, porosity of biological **membranes**.

porphyrin: a heterocyclic macrocycle derived from four pyrrolelike subunits interconnected via their α-carbon atoms via methine bridges ($=CH-$); the macrocycle, therefore, is highly conjugated, and consequently deeply colored; the macrocycle has 26 π-electrons; many porphyrins occur in nature, they are pigments found in both animals and plants.

Porphyromonas gingivalis: is a gram-negative oral anaerobe found in periodontal lesions and associated with adult periodontal disease.

port wine stain (cavernous hemangioma): a **skin** discoloration characterized by a deep red to purple color (see **hemangioma**).

postmenopausal: after menopause; the period of permanent cessation of menstruation, usually occurring between the ages of 45 and 50.

post mortem: occurring after death; colloquial expression for an examination of the body after death (autopsy).

precancerous: a **tissue** lesion that carries the risk of turning into **cancer**; it is a preliminary stage of cancer; these precancerous lesions can have several causes: UV radiation, genetics, exposure to such cancer-causing substances as arsenic, tar or x-ray radiation.

preembryos: a stage of the human life cycle in in-vitro fertilization; an oocyte (egg) is fertilized (but not yet implanted) and forms a preembryo, also called a blastocyte, usually on day five or six.

premenopausal: prior to menopause, the period of permanent cessation of menstruation, usually occurring between the ages of 45 and 50.

presbyopia: the **eye**'s diminished ability to focus that occurs with aging; the most widely held theory is that it arises from the loss of elasticity of the **crystalline lens**, although changes in the lens's curvature from continual growth and loss of power of the **ciliary muscles** have also been postulated as its cause.

Prevotella intermedia: an obligatory anaerobic, black-pigmented, gram-negative rod that is frequently associated with periodontal disease: adult periodontitis, acute necrotizing ulcerative gingivitis, and pregnancy gingivitis; this organism is also involved in extraoral infections such as nasopharyngeal infection and intraabdominal infection; it coaggregates with *Porphyromonas gingivalis*.

Prevotella nigrescens: a genospecie that is very close to *Prevotella intermedia*; *Prevotella intermedia* is likely to be more associated with periodontal sites, whereas *Prevotella nigrescens* seems to be more frequently recovered from healthy gingivae.

prickle cells layer: the layer between the **stratum granulosum** and **stratum basale**, characterized by the presence of prickle **cells**—cells with delicate radiating processes connecting with similar cells, being a dividing **keratinocyte** of the **stratum spinosum** of the **epidermis**.

prokaryotes: are organisms without a **cell nucleus**, or any other membrane-bound **organelles**; most are unicellular, but some prokaryotes are multicellular; the **bacteria** are the prokaryotes.

proliferative disorder: the growth or production of **cells** by multiplication of parts.

Propionibacterium acnes: a relatively slow growing, (typically) aerotolerant anaerobe gram-positive bacterium that is linked to the skin condition acne; this bacteria is largely commensal and thus present on most people's skin; and lives on

fatty acids in the **sebaceous glands** and the **sebum** secreted by them; it is named after its ability to generate propionic acid.

propylene glycol (PG): a colorless, viscous, hygroscopic liquid, $CH_3CHOHCH_2OH$, used chiefly as a lubricant and as a solvent for **fats**, **oils**, and waxes.

prostate (prostate gland): the muscular, **glandular** organ that surrounds the ure-thra of males at the base of the **bladder**.

protein: the result of **amino acids** joined together in a chain by peptide bonds between their amino and carboxylate groups; an amino acid residue is one amino acid that is joined to another by a peptide bond; each different protein has a unique sequence of amino acid residues, this is its primary structure; just as the letters of the alphabet can be combined to form an almost endless variety of words, amino acids can be linked in varying sequences to form a huge variety of proteins.

proteoglycans: a special class of glycoproteins that are heavily glycosylated; they consist of a core **protein** with one or more covalently attached glycosaminogly-can chain(s); these chains are long, linear **carbohydrate** polymers that are nega-tively charged under physiological conditions, due to the occurrence of sulphate and uronic acid groups.

protoporphyrin IX (Pp IX): pertaining to carbonic acids: its absorbing bands are the same as for other porphyrins (see **hemoglobin spectra**, Glossary l), e.g., in di-ethyl ester solution, Pp IX has the following peaks: 403 nm (Soret band), and 504, 535, 575, and 633 (Q bands); Pp IX is the immediate **heme** precursor; exogenous administration of 5-aminolaevulinic acid (ALA), an early precursor in heme syn-thesis, induces accumulation of endogenous photoactive porphyrins, particularly Pp IX; induced Pp IX is used in **cancer** diagnostics and **photodynamic therapy**.

psoralene: pertaining to furocumarins; its absorbing peaks lie in the UV at 295 and 335 nm; used in **photochemical therapy** of **psoriasis** and other **dermatosis** (**PUVA**).

psoriasis: a common chronic **skin** disease characterized by scaly patches (psoriasis focus or psoriatic plaque).

psoriatic: pertaining to **psoriasis**.

pulse wave: a wave of pressure sent down the **arteries** by every contraction of the ventricle; the increased pressure can be felt if the **artery** is pressed against a **bone** (usually in the wrist); the pressure wave travels much faster than the flow of **blood** through the artery; the pulse becomes fainter the farther it is from the **heart**; in the **capillaries** it completely disappears.

pupil: the opening of the **iris** of the **eyeball**; radiating muscles dilate the pupil and a ring of **muscle** (a sphincter) a round it constricts the pupil; the regulation of the size of the pupil is a reflex action caused by the stimulus of light on the **optic nerve**.

PUVA therapy: a **photochemical therapy** method based on impregnation of the diseased **skin** by a **psoralen** as a **photosensitizer** and on use of UVA radiation to provide the phototoxic effect (reduction of **cell** abnormal proliferation due to **cross-linking** of **DNA** molecules in the cell nuclei); used for treatment of **psoriasis** and other **skin** disease.

pyloric: pertaining to the region of the **stomach** that connects to the duodenum.

pyrrolidonecarboxylic acid: 2-pyrrolidone-5-carboxylic acid (PCA) is a cyclic derivative of glutamic acid, physiologically present in mammalian **tissues**.

radial artery: the main **artery** of the lateral aspect of the **forearm**.

raffinose: a complex **carbohydrate**, a trisaccharide composed of galactose, fructose, and **glucose**.

raft tissue: the organotypic **tissue** culture systems permitting the growth of differentiated **keratinocytes** *in vitro* or even creating of skin-equivalent tissue model composed of dermis with type I **collagen** and **fibroblast cells** and **epidermis** of differentiated keratinocytes.

red blood cell (RBC): see **erythrocyte**.

rehydration: the replenishment of water and electrolytes lost through **dehydration**.

ren: see **kidney**.

respiratory chain: in aerobic respiration, electrons are transferred from metabolites to molecular **oxygen** through a series of redox reactions mediated by an electron transport chain; the resulting free energy is used for the formation of **ATP** and **NAD**; in anaerobic respiration, analogous reactions take place with an inorganic compound other than oxygen as an ultimate electron acceptor.

reticular fibers: very thin, almost inextensible threads of **reticulin**; they form a network of intercellular **fibers** around and among the **cells** of many **tissues**, e.g., in many large organs such as the **liver** and **kidney**, and also in tissues, such as **nerves** and **muscles**; they especially support and unite **reticular tissue**.

reticular layer of dermis (*reticular dermis*): the lower layer of the **dermis**; it is made primarily of coarse **collagen** and elastic fibers, it is denser than the **papillary dermis**; it strengthens the **skin**, providing structure and elasticity; it also supports other skin components such as **sweat glands** and **hair** follicles.

reticular tissue: a **tissue** consisting of a network of **reticular fibers** around and among **cells**, with **lymph** in the intercellular spaces; it occurs in **muscles, nerves**, and the larger **glands**.

reticulin: a tough **fibrous** protein, similar to **collagen**, but more resistant to higher temperatures and chemical reagents; it occurs in vertebrate **connective tissue** as

reticular fibers; reticulin is formed in embryos and also in **wounds**; it often changes to collagen.

retina: the innermost coat of the posterior part of the **eyeball** that receives the image produced by the **crystalline lens**; it is continuous with the **optic nerve** and consists of several layers, one of which contains the rods and cones sensitive to light.

retinal: pertaining to **retina**.

retinal nerve fiber layer (RNFL): formed by the expansion of the fibers of the **optic nerve**.

retinol (vitamin A): a fat-soluble, yellow **oil** stored in the **liver**; it is not excreted and can accumulate in the body to produce toxic effects; retinol is used in the body to produce visual purple, the pigment in rods of the **retina**; a deficiency impairs vision and also causes **epithelial cells** to become flattened and heaped up on one another; this leads to xerophthalmia and also to the formation of hard, rough **skin**.

rhodamine: a family of related chemical compounds; examples are Rhodamine 6G, Rhodamine B, and Rhodamine 123; they are used as dyes; they fluoresce and can thus be measured easily and inexpensively; they are generally toxic, and are soluble in water, methanol, and ethanol; Rhodamine B (excitation 510 nm/fluorescence 580 nm) is used in biology as a staining fluorescent dye; Rhodamine 123 (excitation 480 nm/fluorescence 540 nm) is used in biochemistry to inhibit mitochondrion function.

rheumatic arthritis: a chronic disease marked by **inflammation** of the joints; it is frequently accompanied by marked deformities and is ordinarily associated with manifestations of a general or systemic affliction.

rib: one of a series of curved **bones** that are articulated with the vertebrae and occur in pairs (12 in humans) on each side of the vertebrate body; certain pairs are connected with the sternum and form the thoracic wall.

ribosome: in the **cytoplasm** of a **cell** any of several minute, angular, or spherical particles composed of protein and **RNA**.

Ringer's solution: an aqueous solution of the chlorides of sodium, potassium, and calcium that is **isotonic** to animal **tissue** and is used topically as a **physiological solution** and, in experiments, to bathe animal tissues.

RNA (ribonucleic acid): a **nucleic acid** polymer consisting of nucleotide monomers that acts as a messenger between **DNA** and **ribosomes**, and is also responsible for making proteins out of amino acids; RNA polynucleotides contain ribose sugars and are predominantly uracil unlike DNA; it is transcribed (synthesized) from DNA by **enzymes** called RNA polymerases and further processed by other enzymes; RNA serves as the template for translation of genes into proteins, transferring amino acids to the ribosome to form proteins, and also translating the transcript into proteins.

rouleaux: for a condition wherein the **blood cells** clump together to form what looks like stacks of coins.

saccharose (sucrose): a disaccharide (**glucose** + fructose) with the molecular formula $C_{12}H_{22}O_{11}$; it is best known for its role in human nutrition.

saline: a salt solution that is used for medical treatment; this solution is designed to have the same **osmotic pressure** as **blood**.

saliva: a viscid, colorless, watery fluid secreted into the mouth by the salivary **glands**; it functions in tasting, chewing, and swallowing food; it keeps the mouth moist and starts the digestion of starches.

sarcoma: a malignant growth of abnormal **cells** in **connective tissue**.

scalp: the anatomical area bordered by the face anteriorly and the neck to the sides and posteriorly.

scar: the areas of **fibrous tissue** that replace normal **skin** (or other tissue) after **injury**; a scar results from the biologic process of **wound** repair; thus, scarring is a natural part of the healing process.

Scheimpflug camera: a camera that is based on the Scheimpflug principle; this technique allows the assessment of the anterior segment of the **eye**, from the front of the **cornea** to the back of the **lens**, in a sagittal plane; the Scheimpflug principle allows for quantification of the light scatter.

sciatic nerve: a large **nerve** that runs down the lower limbs; it is the longest single nerve in the body.

sclera (sclerotic): a dense, white, **fibrous membrane**, which, with the **cornea**, forms the external covering of the **eyeball**; scleral regions: limbal, equatorial, and posterior pole region.

scleroderma: a disease in which all the layers of the **skin** become hardened and rigid.

sclerotomy: surgical **incision** of the sclerotic coat of the eye.

sea collagen: used in **cosmetics**; sea **collagen** is a better choice than bovine collagen because of its affinity to the **skin** and its richness in trace elements; deep skin **moisturizer**, prevents and reduces **wrinkles** and lines by redensifying and restructuring the skin.

sebaceous gland: a tiny sebum-producing **gland** found everywhere except on the palms, lips, and soles of the feet; the thicker density of **hair**, the more sebaceous glands are found; they are classified as holocrine glands.

sebum: an oily substance secreted by the **sebaceous glands**; it is made of **fat (lipids)** and the debris of dead fat-producing cells; in the **glands**, sebum is produced within specialized **cells** and is released as these cells burst; sebum is odorless, but its bacterial breakdown can produce odors.

sedimentation: the motion of molecules in solutions or particles in suspensions in response to an external force such as gravity, centrifugal force or electric force; sedimentation may pertain to objects of various sizes, ranging from suspensions of dust and pollen particles to cellular suspensions and solutions of single molecules such as **proteins** and peptides.

semen: the viscid, whitish fluid produced in the male reproductive organ; it contains **spermatozoa**.

sensory nerve: **nerves** that receive sensory stimuli, such as how something feels and if it is painful; they are made up of nerve fibers, called sensory fibers (mechanoreceptor fibers sense body movement and pressure placed against the body, and nociceptor fibers sense tissue **injury**).

serum: **blood plasma** with clotting factors removed.

sheath: a protective covering fitting closely to a structure or a part of an organism, especially an elongated structure or part.

silicon oils: polymerized siloxanes are silicon analogues of carbon based organic compounds, and can form (relatively) long and complex molecules based on silicon rather than carbon; chains of alternating silicon are formed: oxygen atoms or siloxane, rather than carbon atoms.

silicon waxes: a semicrystalline with a melting point of 53°C to 75°C to liquid; a clear, light-straw to white and off-white colored flake; they are semiocclusive or occlusive formulations, lubricants, emollients, water-repellents, replacement for petrolatum, thickeners, **moisturizers**, emulsifiers for water-in-oil and water-in-silicone emulsions (up to 80% water content).

skeletal muscle: a type of striated **muscle**, usually attached to the skeleton; skeletal muscles are used to create movement, by applying force to **bone**s and joints via contraction; they generally contract voluntarily (via somatic nerve stimulation), although they can contract involuntarily through reflexes.

skin: the external protective covering on a body, joined by **connective tissue** to the **muscles**; it consists of an inner **dermis** and an outer **epidermis**.

skin appendages: structures associated with the skin such as **hair**, **sweat glands**, **sebaceous glands** and nails; they have their roots in the **dermis** or even in the hypodermis.

skin barrier function: protects internal organs from the environment, resides in the uppermost thin heterogeneous layer called **stratum corneum**, which is composed of dead protein-rich cells and intercellular lipid domains; this two-compartment structure is renewed continuously and when the barrier function is damaged, it is repaired immediately; under low humidity, the stratum corneum becomes thick, the lipid content in it increases and water impermeability is enhanced; the heterogeneous field in the epidermis induced by ions, such as calcium and potassium, regulates the self-referential, self-organizing system to protect the living organism, http://www.scipress.org/journals/forma/pdf/1503/15030227.pdf.

skin flap: a tear of the skin away from the body, which leaves one side of the skin still attached.

skin flap window: a special window surgically made in the skin area void of blood vessels and an implanted glass plate allows for prolonged *in vivo* studies of skin vessels seen through this window.

skin irritation: some physical and chemical exposures to the **skin** may cause **dermatitis** with redness, itching, and discomfort.

skin microdermabrasion: a quick, **noninvasive** procedure used to resurface the **skin**; it gently removes only the very top layers of damaged skin by "sand blasting" them with tiny crystals; the technique exfoliates and gently resurfaces the skin, promoting the formation of new smoother, clearer skin.

skin stripping: a mechanical disruption and reduction of the **stratum corneum** and partially of the living **epidermis**, which are the outermost layers of the **skin**; medical adhesive tapes and medical glues with such substrates as glass, quartz, and metal plates are typically used; the technique is used for study of drug delivery and protective filters distribution in the skin.

skull: the part of the skeleton consisting of the **cranium** and the facial skeleton; the latter includes the sense capsules, the jaws, the hyoid **bone**, and the **cartilage** of the larynx; the skull may also be divided into the neurocranium and the viscerocranium.

slit-lamp: an optical device that consists of a high-intensity light source that can be focused to shine as a slit; it is used in conjunction with a microscope; the lamp facilitates an examination, which looks at the anterior segment, or frontal structures, of the human **eye**, which including the eyelid, **sclera**, **conjunctiva**, **iris**, natural **crystalline lens**, and **cornea**; the binocular slit-lamp examination provides a stereoscopic magnified view of the eye structures in striking detail, enabling exact anatomical diagnoses to be made for a variety of eye conditions.

sodium fluorescein: probably the most commonly used dye in the biological world; it is a **protein** dye, so some caution is required in its use within the living body; sodium **fluorescein** has its peak absorption at 450 nm and produces a yellow/green emission when stimulated by light in the blue region; fluorescent angiography is a technique for examining the **blood** circulation of the **retina** using the dye tracing method that involves the injection of sodium fluorescein into the systemic circulation, and then an angiogram is obtained by photographing the fluorescence emitted after illumination of the retina with blue light.

sodium lactate: natural salt that is derived from a naturally fermented product, lactic acid that is produced naturally in foods such as cheese, yogurt, hard salami, pepperoni, sourdough bread and many others by the action of lactic acid starter cultures (also known as a "good" bacteria); sodium lactate can correct the normal acid-base balance in patients whose **blood** has become too acidic; it can also help treat overdoses of certain medications by increasing removal of the drug from the body.

sodium lauryl sulfate: a detergent and **surfactant** found in many personal care products (soaps, shampoos, toothpaste, etc.).

somatosensory: pertaining to the sense of touch that is mediated by the somatosensory system; touch may simply be considered one of the five human senses; however, when a person touches something or somebody this gives rise to various feelings: the perception of pressure (shape, softness, texture, vibration, etc.), relative temperature and sometimes pain; thus the term "touch" is actually the combined term for several senses; in medicine, the colloquial term "touch" is usually replaced with somatic senses, to better reflect the variety of mechanisms involved; the somatosenses include: **cutaneous** (**skin**), kinesthesia (movement) and visceral (internal) senses; visceral senses have to do with sensory information from within the body, such as **stomach** aches.

sonophoresis: see **phonophoresis**.

spherocytes: the sphere-shaped **erythrocytes**; an auto-hemolytic anemia (a disease of the blood) characterized by the production of spherocytes; it is caused by a molecular defect in one or more of the **proteins** of the erythrocyte **cytoskeleton** (usually ankyrin, sometimes spectrin); because the **cell** skeleton has a defect, the **blood cell** contracts to its most surface-tension efficient and least flexible configuration: a sphere, rather than the more flexible donut-shape.

spleen: a highly vascular, glandular, ductless organ situated in humans at the cardiac end of the **stomach**; it serves chiefly in the formation of **lymphocytes**, in the destruction of worn-out **erythrocytes**, and as a reservoir for red corpuscles and platelets.

spore: a reproductive structure that is adapted for dispersion and surviving for extended periods of time in unfavorable conditions; spores form part of the life cycles of many plants, algae, fungi, and some protozoans.

squames: flat, keratinized, dead **cells** shed from the outermost layer of a stratified **squamous epithelium**.

squamous cell carcinomas: a form of **carcinoma cancer** that may occur in many different organs, including the **skin**, mouth, **esophagus**, prostate, **lungs**, and **cervix**; it is a **malignant tumor** of **epithelium** that shows **squamous cell** differentiation.

squamous epithelium: an **epithelium** characterized by its most superficial layer consisting of flat, scalelike cells called squamous cells; it may possess only one layer of these cells, in which case it is referred to as simple squamous epithelium, or it may possess multiple layers, referred to then as stratified squamous epithelium; both types perform differing functions, ranging from nutrient exchange to protection.

Staphylococcus: any of several spherical bacteria of the genus *Staphylococcus*, occurring in pairs, tetrads, and irregular clusters, certain species of which, such as *S. aureus* can be pathogenic for man.

Staphylococcus toxin: some **Staphylococci** produce a toxin, which is responsible for food poisoning; this toxin is heat stable; bacteria can be killed during processing but the toxin may remain behind.

stasis: stopping of **blood flow** or flows of other biological fluids in the organism.

stenosis: a narrowing of a canal (as in the walls of **arteries** or a cardiac **valve**).

steroid: a terpenoid **lipid** characterized by a carbon skeleton with four fused rings; different steroids vary in the functional groups attached to these rings; hundreds of distinct steroids are found in plants, animals, and fungi; all steroids are derived either from the sterol lanosterol (animals and fungi) or the sterol cycloartenol (plants); both sterols are derived from the cyclization of the triterpene squalene.

stomach: a saclike enlargement of the alimentary canal that forms an organ for storing, diluting, and digesting food; it is situated between the **esophagus** and duodenum and divided into a **fundus**, cardiac portion, pyloric portion, and pylorus.

stratum basale: see **epidermis**.

stratum corneum: see **epidermis**.

stratum granulosum: see **epidermis**.

stratum spinosum: see **epidermis**.

stromal layer (stroma): the supporting framework of an organ, as distinguished from the parenchyma, e.g., the main layer of the eye **sclera** and **cornea**.

subcutaneous: see **hypodermic**.

subdermis (subdermal): the **skin** layer that primarily consists of globular **fat cells**.

subepidermal: situated immediately below the **epidermis**.

submucous: the **tissue** layer under the **mucous membrane**.

sucrose: see **saccharose**.

sugar: a white, crystalline **carbohydrate**, soluble in water and sweet to the taste; sugars are classified as reducing or nonreducing, according to their reaction with Fehling's solution, and also as monosaccharides or disaccharides, according to their structure; **glucose**, fructose, galactose, and mannose are monosaccharides, and sucrose is a disaccharide.

sulci: the superficial fissures that increase the surface area of the **cerebral cortex**; the *pia mater* dips down into the fissures.

sulfonated tetraphenyl porphines (TPPSn): are photosensitizing dyes that localize in cell **lysosomes**; extralysosomal location of hydrophylic $TPPS_3$ and $TPPS_4$ in close proximity to the plasma membrane is also found; the Soret band of the dye is 400–440 nm, the emission maxima at 655 nm (dye) and 610 nm (photoproduct).

sunscreen cream: a cream for protecting the **skin** from the sunlight, especially for blocking UVC and UVB rays.

superior sagittal sinus: also known as the superior longitudinal sinus, occupies the attached or convex margin of the falx cerebri.

surfactant: a wetting agent that lowers the surface tension of a liquid that allows easier spreading and lowers the interfacial tension between two liquids.

sweat glands: in humans, there are two kinds of sweat **glands**, which differ greatly in both the composition of the sweat and its purpose; eccrine sweat glands are distributed over the entire body surface but are particularly abundant on the **palms** of **hands**, soles of feet, and on the forehead; these glands produce sweat that is composed chiefly of water with various salts and are used for body temperature regulation; apocrine sweat glands develop during the early to middle puberty ages approximately around the age of 15, releasing more than normal amounts of sweat for approximately a month and subsequently regulate and release normal amounts of sweat after a certain period of time; they are located wherever there is body **hair**.

swelling: the enlargement of organs caused by the accumulation of excess fluid in **tissues**, called **edema**; it can occur throughout the body (generalized), or only in some part or organ that is affected (localized); it is considered one of the five characteristics of **inflammation**.

synergetic (synergistic): refers to the phenomenon in which two or more discrete influences or agents acting together create an effect greater than that predicted by knowing only the separate effects of the individual agents.

T84: the carcinoma cell line.

tendon: a cord of **white fibrous tissue**; it usually attaches **muscle** to **bones**.

Tenon's capsule: adherent to **episcleral tissue**.

thalamus: a pair and symmetric part of the brain; it constitutes the main part of the diencephalons; in the caudal (tail) to oral (mouth) sequence of neuromeres, the diencephalons is located between the mesencephalon (cerebral peduncule, belonging to the brain stem) and the telencephalon.

thermodynamic activity: ions in solution are in constant motion; this movement is temperature dependent, i.e., as water becomes hotter the particles within it move faster and, conversely, as it becomes cooler they slow down.

thigh: in humans the thigh is the area between the pelvis and buttocks and the **knee**; anatomically, it is part of the lower limb; the single **bone** in the thigh is called the femur; this bone is very thick and strong (due to the high proportion of cortical bone), and forms a ball and socket joint at the hip, and a condylar joint at the knee.

thorax: see **chest**.

thrombocyte: a small, spindle-shaped, nucleated **cell**; the cell readily disintegrates, releasing thrombokinase and initiating **blood** clotting.

thyroid: one of the larger endocrine **glands** in the body; this gland is found in the **neck** just below the Adam's apple; the thyroid controls how quickly the body burns energy, makes **proteins** and how sensitive the body should be to other hormones.

tissue: an aggregate of similar **cells** and cell products forming a definite kind of structural material.

tissue phantom: see Glossary 1, **phantom**.

tissue shrinkage: the loss of volume and weight caused, for instance, by **tissue dehydration**.

titanium dioxide (TiO_2) **particles**: an insoluble white powder, is used extensively in many commercial products, including paint, cosmetics, plastics, paper, and food as an anticaking or whitening agent; it is produced and used in the workplace in varying particle-size fractions, including fine and ultrafine sizes.

TMP (**trimethylolpropanol**): ester of methyl phosphoric acid; a lubricator.

tonometer: an instrument for determining pressure or tension, particularly that for measuring tension within the **eyeball**.

tooth: the hard body composed of **dentin** surrounding a sensitive pulp and covered on the crown with **enamel**.

topical: an application of medication to body surfaces such as the **skin** or **mucous membranes**; some **hydrophobic** chemicals such as steroid hormones can be absorbed into the body after being applied to the skin in the form of a cream, gel, or lotion; **transdermal** patches have become a popular means of administering some drugs for birth control, hormone replacement therapy, and prevention of motion sickness; in dentistry, a topical medication may also mean one that is applied to the surface of teeth.

total hemoglobin: the sum concentration of the **oxy-** and **deoxyhemoglobin**.

toxin: any of a group of poisonous, usually unstable compounds generated by microorganisms or plants, or of animal origin; certain toxins are produced by specific pathogenic microorganisms and are the causative agents in various diseases; some are capable of inducing the production of antibodies in certain animals.

trabeculae: the rodlike **cells** or a row of cells forming supporting structures lying across spaces or lumina, e.g., outgrowths of the cell wall across the lumen of tracheids, the supporting meshwork in spongy **bone**.

trachea: a common biological term for an airway through which respiratory air transport takes place in organisms.

transdermal: means through **skin**.

transepidermal water lost (TEWL): describes the total amount of water lost through the **skin**, a loss that occurs constantly by passive diffusion through the **epidermis**; although TEWL is a normal physiological phenomenon, if it rises too high, the skin can become **dehydrated**, which disrupts form and function, and potentially leads to infection or transepidural passage of deleterious agents.

transpupillary: through the **eye pupil**.

transscleral: through the **eye sclera**.

trauma: an often serious and body-altering physical **injury**, such as the removal of a limb.

trazograph: a derivative of 2,4,6-triiodobenzene acid, molecular weight of about 500; a water-soluble colorless liquid usually used at concentrations of 60 or 76% as an intravenous x-ray contrasting agent; very good agent for **optical clearing** of **fibrous tissue** owing to its high **osmolarity** and high index of refraction.

tremor: an unintentional, somewhat rhythmic, to-and-fro **muscle** movement (oscillations) involving one or more parts of the body; it is the most common of all involuntary movements and can affect the **hands**, **arms**, **head**, face, vocal cords, trunk, and legs; most tremors occur in the hands; in some people, tremor is a symptom of another neurological disorder.

triphenylmethane: a biological dye (stain).

trypan blue: a vital stain that is used to color dead **tissues** or **cells** blue; it is a diazo dye; live cells or tissues with intact cell **membranes** will not be colored; since cells are very selective in the compounds that pass through the membrane, in a viable cell trypan blue is not absorbed it traverses the membrane in a dead cell, hence, dead cells are shown as a distinctive blue color under a microscope.

tryptophan: a colorless, crystalline, aromatic essential amino acid that occurs in the seeds of some leguminous plants; it is released from proteins by tryptic digestion and is important in the nutrition of animals.

tubular structure: a structure consisting of tubes.

tumor: an abnormal or diseased swelling in any part of the body, especially a more or less circumscribed overgrowth of new **tissue** that is autonomous, differs more or less in structure from the part in which it grows, and serves no useful purpose; **neoplasm**.

tympanic membrane: a **membrane** separating the tympanium or middle ear from the passage of the external ear; eardrum.

type I collagen: the most abundant **collagen** of the human body; it is present in **scar tissue**, the end product when tissue heals by repair; it is found in **tendons**, the endomysium of **myofibrils** and the organic part of **bone**.

tyrosine: a crystalline amino acid that results from the hydrolysis of proteins.

ulcer: a sore open either to the surface of the body or to a natural cavity, and accompanied by the disintegration of **tissue**, the formation of pus, etc.

ultrasound gel: a viscous gel for medical ultrasound transmission used for diagnostic and therapeutic ultrasound applications; acoustically corrects US energy transmission for frequencies used.

unsaturated fatty acid: a **fatty acid** in which there are one or more double bonds in the fatty acid chain; a fat molecule is **monounsaturated** if it contains one double bond, and polyunsaturated if it contains more than one double bond; where double bonds are formed, hydrogen atoms are eliminated; thus, a saturated fat is "saturated" with hydrogen atoms; the greater the degree of unsaturation in a fatty acid (i.e., the more double bonds in the fatty acid), the more vulnerable it is to lipid peroxidation (rancidity); antioxidants can protect unsaturated fat from lipid peroxidation; unsaturated fats also have a more enlarged shape than saturated fats.

urea: an organic compound of carbon, nitrogen, **oxygen**, and hydrogen, with the formula CON_2H_4 or $(NH_2)_2CO$; urea is also known as carbamide.

urinary bladder: a sac for storing **urine**; it is a diverticulum of the hindgut; urine is conducted to the **bladder** by a ureter; the exit to the bladder is closed by a sphincter **muscle**.

urine: a yellowish, slightly acid, watery fluid; waste matter excreted by the **kidneys**.

urocanic acid: 4-imidazoleacrylic acid, found in the **skin epidermis**; it has high absorption in the UV range with a peak at 260 nm.

uroporphyrin: any of several porphyrins produced by oxidation of uroporphyrinogen; one or more are excreted in excess in the **urine** in several of the porphyries.

uterine: of or to do with the **uterus**, e.g., the uterine walls.

uterus: an organ in which an embryo develops and is nourished; it has walls of unstriated **muscle** that increase greatly in thickness during pregnancy and whose contractions expel the fetus at birth; the uterus is lined with endometrium, which undergoes modification during pregnancy and is also modified under control of sex hormones during the estrus cycle; the uterus is connected through the **cervix** to the vagina.

UV skin filter: a **sunscreen cream** or lotion based on chemical formulations and/or reflecting nanoparticles that filter broadband UV radiation and helps shield the **skin** from the UVA and UVB rays that do skin damage every day.

vacuole: (1) a cavity within a **cell**, often containing a watery liquid or secretion; (2) a minute cavity or vesicle in organic **tissue**.

valve: a structure that allows fluids to flow through it in one direction only; this is done by closing the vessel, or canal, to stop backward flow [see **lymphatic (lymph) vessels**].

vascularization: consisting of or containing **vessels** that conduct **blood** and **lymph**.

vascular: related to **vessels**.

vasculature: arrangement of **blood vessels** in the body or in an organ or body part; the **vascular** network of an organ.

vasoconstriction: constriction of the **blood** or **lymph vessels**, as by the action of a **nerve**.

vasodilation: where **blood vessels** in the body become wider following the relaxation of the smooth **muscle** in the vessel wall; this reduces blood pressure—since there is more room for the blood; vasodilation also occurs in superficial blood vessels of warm-blooded animals when their ambient environment is hot; this process diverts the flow of heated blood to the skin of the animal, where heat can be more easily released into the atmosphere; the opposite physiological process is **vasoconstriction**.

vehicles: the nonliving means of transportation; pertaining to transportation of drugs by various solutes.

vein: a **blood vessel** that conducts **blood** from the **tissues** and organs back to the **heart**; the vein is lined with **endothelium** (smooth flat cells) and surrounded by muscular and **fibrous tissue**; the walls are thin and the diameter large compared with an **artery**; the vein contains **valves** that allow blood to flow only toward the heart.

vein *femoralis* **(femoral)**: the vein pertaining to the thigh or femur.

ventricle: a chamber in **brain** or **heart**; in a heart a chamber that collects **blood** from an **atrium** (another heart chamber that is smaller than a ventricle) and pumps it out of the heart.

venule: a small **vein** that collects **blood** from **capillaries**; it joins other venules to form a vein; a venule has more **connective tissue** than a capillary **muscle**; the permeability of the venule wall to blood is similar to that of a capillary wall.

verografin: a water-soluble colorless liquid usually used at concentrations of 60% or 76% as an intravenous x-ray contrasting agent; very good agent for **optical clearing** of **fibrous tissue**, owing to its high **osmolarity** and high index of refraction; an analog of **trazograph**.

vesicle: a relatively small and enclosed compartment, separated from the **cytosol** by at least one **lipid bilayer**; if there is only one lipid bilayer, they are called unilamellar vesicles; otherwise they are called multilamellar; vesicles store, transport, or digest cellular products and waste.

vessel: a tube or duct, such as an **artery** or **vein**, containing or conveying **blood** or some other body fluid.

videokeratoscope: a keratoscope fitted with a video camera; keratoscope is an instrument marked with lines or circles by means of which the corneal reflex can be observed.

visual acuity: acuteness or clearness of vision, especially form vision, which is dependent on the sharpness of the **retinal** focus within the **eye**, the sensitivity of the nervous elements, and the interpretative faculty of the **brain**.

vital activity (function): any function of the body, **tissue**, **cell**, etc., that is essential for life.

vitality: the peculiarity distinguishing the living from the nonliving; capacity to live and develop.

vitreopathy: a pathology of vitreous body.

vitreous humor: the transparent gelatinous substance filling the **eyeball** behind the **crystalline lens**, called vitreous body.

vocal chord: either of the two pairs of folds of **mucous membrane** projecting into the cavity of the **larynx**.

volar: pertaining to both the **palm** and sole of feet.

volatile solvents: liquids that vaporize at room temperature; these organic solvents can be inhaled for psychoactive effects and are present in many domestic and industrial products such as glue, aerosol, paints, industrial solvents, lacquer thinners, gasoline, and cleaning fluids; some substances are directly toxic to the **liver**, **kidney**, or **heart**, and some produce peripheral neuropathy (nerve damage usually affecting the feet and legs) or progressive **brain** degeneration.

white blood cell (WBC) (leukocyte): a nucleated, motile, colorless cell found in the **blood** and **lymph** of animals; it contains no respiratory pigments; it is either a **lymphocyte**, a **polymorph**, or a **monocyte**; in humans, there are approximately 8000 WBCs per cubic millimeter.

white fibers: see **white fibrous tissue**.

white fibrous tissue: a **connective tissue** that consists of a matrix of very fine, white wavy fibers arranged parallel to each other, in bundles and unbranching, with **fibroblasts** embedded in the bundle; the tissue is tough and inelastic; it is found pure in **tendons**; the **white fibers** are composed of **collagen**.

white matter: the **nervous tissue** found in the central nervous system; it consists of tracts of medullated **nerve** fibers in the **brain** and spinal cord; it also contains **blood vessels** and **neuroglia**; it is mainly external to **gray matter**, but is internal to **gray matter** in the cerebral hemispheres and in the **cerebellum**; the medullated fibers give the tissue its shiny white appearance.

WHO: World Health Organization.

whole blood: **blood** that contains all its natural components: **blood, cells**, and **plasma**.

Wister rat: an animal strain widely used in experimental studies.

wound: a type of physical **trauma** wherein the **skin** is torn, cut, or punctured (an open wound), or where blunt force trauma causes a contusion (a closed wound); in pathology, it specifically refers to a sharp **injury** that damages the **dermis** of the skin.

wrinkle: a ridge or crease of a surface; usually refers to the skin of an organism; in skin a wrinkle or fold may be permanent; skin wrinkles typically appear as a result of the aging processes such as glycation or, temporarily, as the result of prolonged (more than a few minutes) immersion in water; wrinkling in skin is caused by habitual facial expressions, aging, sun damage, smoking, poor **hydration**, and various other factors.

wrist: in human anatomy, the wrist is the flexible and is a narrowing connection between the **forearm** and the **hand**; the wrist is essentially a double row of small short **bone**s, called carpals, intertwined to form a malleable hinge.

yellow fibers: see **yellow elastic tissue**.

yellow elastic tissue: a **connective tissue** that consists of a matrix of coarse **yellow elastic fibers** that branch regularly and anastomosely with **fibroblasts** in the matrix; **elastin** is the principal constituent of this tissue; the tissue rarely occurs in a pure form, usually containing **white fibers**; yellow elastic fibers are numerous in the **lungs** and in the walls of **arteries**, where elastic supporting tissues are required; yellow elastic tissue occurs in ligaments, where an extensible tissue is required.

Yucatan micropigs: strain of hairless small animals widely used in experimental studies.

Sources

This glossary was compiled using mostly Refs. 2, 7, 26, 57, and the following

1. *Webster's New Universal Unabridged Dictionary*, Barnes & Noble Books, New York, 1994.
2. A. Godman and E. M. F. Payne, *Longman Dictionary of Scientific Usage*, reprint edition, Longman Group, Harlow, UK, 1979.

3. *Stedman's Medical Dictionary*, Williams & Wilkins, Baltimore, MD, 1995.
4. K. Jimbow, W. C. Quevedo, T. B. Fitzpatrick, and G. Szabo, "Biology of Melaninocytes," in *Dermatology in General Medicine*, T. B. Fitzpatrick, A. Z. Eisen, K. Wolff, I. M. Freedberg, and K. F. Austen (eds.), McGraw-Hill, New York, 1993, pp. 261–288.
5. http://en.wikipedia.org
6. http://www.medterms.com
7. http://www.disabled-world.com/artman/publish/glossary.shtml
8. http://cancerweb.ncl.ac.uk/cgi-bin/omd?action=Home&query=
9. http://www.online-medical-dictionary.org/link.asp
10. http://www.stedmans.com/

References

1. G. Müller, B. Chance, R. Alfano, et al. (eds.), *Medical Optical Tomography: Functional Imaging and Monitoring*, vol. IS11, SPIE Press, Bellingham, WA, 1993.
2. G. Müller and A. Roggan (eds.), *Laser-Induced Interstitial Thermotherapy*, vol. PM25, SPIE Press, Bellingham, WA, 1995.
3. V. V. Tuchin (ed.), *Selected Papers on Tissue Optics: Applications in Medical Diagnostics and Therapy*, vol. MS102, SPIE Press, Bellingham, WA, 1994.
4. B. Chance, M. Cope, E. Gratton, N. Ramanujam, and B. Tromberg, "Phase Measurement of Light Absorption and Scatter in Human Tissue," *Rev. Sci. Instrum.*, vol. 69, no. 10, 1998, pp. 3457–3481.
5. A. V. Priezzhev, V. V. Tuchin, L. P. Shubochkin, *Laser Diagnostics in Biology and Medicine*, Nauka, Moscow, 1989.
6. V. V. Tuchin, *Lasers and Fiber Optics in Biomedical Science*, Saratov Univ. Press, Saratov, 1998.
7. A. Katzir, *Lasers and Optical Fibers in Medicine*, Academic Press, San Diego, 1993.
8. V. V. Tuchin and J. A. Izatt (eds.), *Coherence Domain Optical Methods in Biomedical Science and Clinical Applications II, III, Proc. SPIE* 3251, 1998; 3598 (1999).
9. V. V. Tuchin, "Lasers and Fiber Optics in Biomedicine," *Laser Physics*, vol. 3, 1993, pp. 767–820; 925–950.
10. V. V. Tuchin, "Lasers Light Scattering in Biomedical Diagnostics and Therapy," *J. Laser Appl.*, vol. 5, nos. 2, 3, 1993, pp. 43–60.
11. D. H. Sliney and S. L. Trokel, *Medical Lasers and Their Safe Use*, Academic Press, New York, 1993.
12. L. E. Preuss and A. E. Profio (eds.), "Special Section on Optical Properties of Mammalian Tissue," *Appl. Opt.*, vol. 28, no. 12, 1989, pp. 2207–2357.
13. J. M. Shmitt, A. Knüttel, and R. F. Bonnar, "Measurement of Optical Properties of Biological Tissues by Low-Coherence Reflectometry," *Appl. Opt.*, vol. 32, 1993, pp. 6032–6042.
14. "Special Section on Lasers in Biology and Medicine," *IEEE J. Quantum Electr.*, vol. 20, no. 12, 1984, pp. 1342–1532; vol. 23, no. 10, 1987, pp. 1701–1855; vol. 26, no. 12, 1990, pp. 2146–2308.

15. O. Minet, G. Mueller, and J. Beuthan (eds.), *Selected Papers on Optical Tomography, Fundamentals and Applications in Medicine*, vol. MS 147, SPIE Press, Bellingham, 1998.

16. A. Katzir (ed.), "Special Section on Biomedical Optics," *Opt. Eng.*, vol. 31, no. 7, 1992, pp. 1399–1486; vol. 32, no. 2, 1993, pp. 216–367.

17. H. Podbielska, C. K. Hitzenberger, and V. V. Tuchin (eds.), "Special Section on Interferometry in Biomedicine," *J. Biomed. Opt.*, vol. 3, no. 1, 1998, pp. 5–79; no. 3, 1998, pp. 225–266.

18. V. V. Tuchin, H. Podbielska, C. K. Hitzenberger (eds.), "Special Section on Coherence Domain Optical Methods in Biomedical Science and Clinics," *J. Biomed. Opt.*, vol. 4, no. 1, 1999, pp. 94–190.

19. T. J. Dougherty (ed.), "Special Issue on Photodynamic Therapy," *J. Clin. Laser Med. Surg.*, vol. 14, 1996, pp. 219–348.

20. V. S. Letokhov, "Laser Biology and Medicine," *Nature*, vol. 316, no. 6026, 1985, pp. 325–328.

21. J. M. Brunetaud, V. Maunoury, and D. Cochelard, "Lasers in Digestive Endoscopy," *J. Biomed. Opt.*, vol. 2, no. 1, 1997, pp. 42–52.

22. A. V. Priezzhev, T. Asakura, and J. D. Briers, *Optical Diagnostics of Biological Fluids III, Proc. SPIE* 3252, 1998.

23. J. G. Fujimoto and M. S. Patterson (eds.), *Advances in Optical Imaging and Photon Migration, OSA Trends in Optics and Photonics*, vol. 21, Optical Society of America, Washington, DC, 1998.

24. V. V. Tuchin, "Light Scattering Study of Tissues," *Physics—Uspekhi*, vol. 40, no. 5, 1997, pp. 495–515.

25. V. P. Zharov and V. S. Letokhov, *Laser Opto-Acoustic Spectroscopy*, Springer-Verlag, New York, 1989.

26. S. Bown, G. Buonaccorsi (eds.), "Special Issue on VI Biennial Meeting of the International Photodynamic Association," *Lasers Med. Sci.*, vol. 12, no. 3, 1997, pp. 180–284.

27. H. J. Geschwind, "Recent Developments in Laser Cardiac Surgery," *J. Biomed. Opt.*, vol. 1, no. 1, 1996, pp. 28–30.

28. W. Rudolph and M. Kempe, "Topical Review: Trends in Optical Biomedical Imaging," *J. Modern Opt.*, vol. 44, no. 9, 1997, pp. 1617–1642.

29. G. J. Mueller and D. H. Sliney (eds.), *Dosimetry of Laser Radiation in Medicine and Biology*, vol. IS5, SPIE Press, Bellingham, WA, 1989.

30. A. Mahadevan-Jansen and R. Richards-Kortum, "Raman Spectroscopy for Detection of Cancers and Precancers," *J. Biomed. Opt.*, vol. 1, no. 1, 1996, pp. 31–70.

31. B. B. Das, F. Liu, and R R. Alfano, "Time-Resolved Fluorescence and Photon Migration Studies in Biomedical and Random Media," *Rep. Prog. Phys.*, vol. 60, 1997, pp. 227–292.

32. M. Motamedi (ed.), "Special Section on Photon Migration in Tissue and Biomedical Applications of Lasers," *Appl. Opt.*, vol. 32, 1993, pp. 367–434.

33. A. Yodh, B. Tromberg, E. Sevick-Muraca, and D. Pine (eds.), "Special Section on Diffusing Photons in Turbid Media," *Appl. Opt.*, vol. 36, 1997, pp. 9–231.

34. T. Durduran, A. G. Yodh, B. Chance, and D. A. Boas, "Does the Photon-Diffusion Coefficient Depend on Absorption?" *J. Opt. Soc. Am. A*, vol. 14, no. 12, 1997, pp. 3358–3365.

35. D. Kessel (ed.), *Selected Papers on Photodynamic Therapy*, vol. MS82, SPIE Press, Bellingham, WA, 1993.

36. S. L. Jacques, "Strengths and Weaknesses of Various Optical Imaging Techniques," Saratov Fall Meeting'01, Internet Plenary Lecture, Saratov, Russia, 2001, http://optics.sgu.ru/SFM.

37. M. J. C. van Gemert, S. L. Jacques, H. J. C. M. Sterenborg, and W. M. Star, "Skin Optics," *IEEE Tranc. Biomed. Eng.*, vol. 36, no. 12, 1989, pp. 1146–1154.

38. M. J. C. van Gemert, J. S. Nelson, T. E. Milner, et al., "Non-Invasive Determination of Port Wine Stain Anatomy and Physiology for Optimal Laser Treatment Strategies," *Phys. Med. Biol.*, vol. 42, 1997, pp. 937–949.

39. H. J. C. M. Sterenborg, M. J. C. van Gemert, W. Kamphorst, et al., "The Spectral Dependence of the Optical Properties of Human Brain," *Lasers Med. Sci.*, vol. 4, 1989, pp. 221–227.

40. W.-F. Cheong, S. A. Prahl, and A. J. Welch, "A Review of the Optical Properties of Biological Tissues," *IEEE J. Quantum Electr.*, vol. 26, no. 12. 1990, pp. 2166–2185. Updated by W.-F. Cheong, further additions by L. Wang and S. L. Jacques, August 6, 1993.

41. S. L. Jacques, "Monte Carlo Modeling of Light Transport in Tissues," in *Tissue Optics*, A. J. Welch and M. C. J. van Gemert (eds.), Academic Press, New York, 1992.

42. H. Niemz, *Laser-Tissue Interactions. Fundamentals and Applications*, Springer-Verlag, Berlin, 1996.

43. R. G. Johnston, S. B. Singham, and G. C. Salzman, "Polarized Light Scattering," *Comments Mol. Cell. Biophys.*, vol. 5, no. 3, 1988, pp. 171–192.

44. J. R. Lakowicz (ed.), *Time-Resolved Laser Spectroscopy in Biochemistry*, *Proc. SPIE* 1204, Pt. 1-2, 1990.

45. V. V. Tuchin, H. Podbielska, and B. Ovryn (eds.), *Coherence-Domain Methods in Biomedical Science and Clinical Applications, Proc. SPIE* 2981, 1997.

46. A. Kienle, L. Lilge, M. S. Patterson, R. Hibst, R. Steiner, and B. C. Wilson, "Spatially Resolved Absolute Diffuse Reflectance Measurements for Noninvasive Determination of the Optical Scattering and Absorption Coefficients of Biological Tissue," *Appl. Opt.*, vol. 35, no. 13, 1996, pp. 2304–2314.

47. J. R. Mourant, I. J. Bigio, J. Boyer, et al., "Elastic Scattering Spectroscopy as a Diagnostic Tool for Differentiating Pathologies in the Gastrointestinal Tract: Preliminary Testing," *J. Biomed. Opt.*, vol. 1, no. 2, 1996, pp. 192–199.

48. A. Roggan, M. Friebel, K. Dorschel, A. Hahn, and G. Mueller, "Optical Properties of Circulating Human Blood in the Wavelength Range 400–2500 nm," *J. Biomed. Opt.*, vol. 4, no. 1, 1999, pp. 36–46.

49. A. M. K. Nilsson, G. W. Lucassen, W. Verkruysse, S. Andersson-Engels, and M. J. C. van Gemert, "Changes in Optical Properties of Human Whole Blood *In Vitro* Due to Slow Heating," *Photochem. Photobiology*, vol. 65, no. 2, 1997, pp. 366–373.

50. F. A. Marks, "Optical Determination of the Hemoglobin Oxygenation State of Breast Biopsies and Human Breast Cancer Xenografts in Unde Mice," *Proc. SPIE* 1641, 1992, pp. 227–237.

51. M. Bassani, F. Martelli, G. Zaccanti, and D. Contini, "Independence of the Diffusion Coefficient from Absorption: Experimental and Numerical Evidence," *Opt. Lett.*, vol. 22, 1997, pp. 853–855.

52. M. S. Patterson, "Noninvasive Measurement of Tissue Optical Properties: Current Status and Future Prospects," *Comments Mol. Cell. Biophys.*, vol. 8, 1995, pp. 387–417.

53. B. Chance, K. Kang, L. He, H. Liu, and S. Zhou, "Precision Localization of Hidden Absorbers in Body Tissues with Phased-Array Optical Systems," *Rev. Sci, Instrum.*, vol. 67, 1996, pp. 4324–4332.

54. H. J. S. M. Sterenborg and J. C. Van der Leun, "Change in Epidermal Transmission Due to UV-Induced Hyperplasia in Hairless Mice: a First Approximation of the Action Spectrum," *Photodermatology*, vol. 5, 1988, pp. 71–82.

55. M. Ferrari, D. Delpy, and D. A. Benaron (eds.), "Special Section on Clinical Near Infrared Spectroscopy/Imaging," *J. Biomed. Opt.*, vol. 1, no. 4, 1996, pp. 361–434; vol. 2, no. 1, 1997, pp. 7–41; no. 2, pp. 147–175.

56. G. Yoon, A. J. Welch, M. Motamedi, et al., "Development and Application of Three-Dimensional Light Distribution Model for Laser Irradiated Tissue," *IEEE J. Quantum Electr.*, vol. 23, no. 10, 1987, pp. 1721–1733.

57. R. R. Anderson and J. A. Parrish, "Optical Properties of Human Skin," in *The Science of Photomedicine*, J. D. Regan and J. A. Parrish (eds.), Plenum Press, New York, 1982, pp. 147–194.

58. A. Dunn, C. Smithpeter, A. J. Welch, and R. Richards-Kortum, "Finite-Difference Time-Domain Simulation of Light Scattering from Single Cells," *J. Biomed. Opt.*, vol. 2, no. 3, 1997, pp. 262–266.

59. J. M. Schmitt, A. H. Gandjbakhche, and R. F. Bonnar, "Use of Polarized Light to Discriminate Short-Photons in a Multiply Scattering Medium," *Appl. Opt.*, vol. 31, 1992, pp. 6535–6546.

60. B. V. Bronk, W. P. van de Merwe, and M. Stanley, "*In Vivo* Measure of Average Bacterial Cell Size From a Polarized Light Scattering Function," *Cytometry*, vol. 13, 1992, pp. 155–162.

61. V. V. Bakutkin, I. L. Maksimova, P. I. Saprykin, V. V. Tuchin, and L. P. Shubochkin, "Light Scattering by the Human Eye Sclera," *J. Appl. Spectrosc. (USSR)*, vol. 46, no. 1, 1987, pp. 104–107.

62. I. L. Maksimova, V. V. Tuchin, and L. P. Shubochkin, "Light Propagation in Anisotropic Biological Objects," in *Laser Beams*, Khabarovsk Technical Inst. Press, Khabarovsk, USSR, 1985, pp. 91–95.

63. I. L. Maksimova, V. V. Tuchin, L. P. Shubochkin, "Polarization Features of Eye's Cornea," *Opt. Spectrosc. (USSR)*, vol. 60, no. 4, 1986, pp. 801–807.

64. I. L. Maksimova, V. V. Tuchin, and L. P. Shubochkin, "Light Scattering Matrix of Crystalline Lens," *Opt. Spectrosc. (USSR)*, vol. 65, no. 3, 1988, pp. 615–619.

65. G. B. Altshuler and V. N. Grisimov, "Effect of Waveguide Transport of Light in Human Tooth," *USSR Acad. Sci. Reports*, vol. 310, no. 5, 1990, pp. 1245–1248.

66. J. R. Zijp and J. J. ten Bosch, "Angular Dependence of He-Ne-Laser Light Scattering by Bovine Human Dentine," *Archs Oral Biol.*, vol. 36, no. 4, 1991, pp. 283–289.

67. A. V. Priezzhev, V. V. Tuchin, and L. P. Shubochkin, "Laser Microdiagnostics of Eye Optical Tissues and Form Elements of Blood," *Bullet. USSR Acad. Sci., Phys. Ser.*, vol. 53, no. 8, 1989, pp. 1490–1495.

68. I. L. Maksimova, A. P. Mironychev, S. V. Romanov, et al., "Methods and Equipment for Laser Diagnostics in Ophthalmology," *Bullet. USSR Acad. Sci., Phys. Ser.*, vol. 54, no. 10, 1990, pp. 1918–1923.

69. A. N. Korolevich, A. Ya. Khairulina, and L. P. Shubochkin, "Scattering Matrix of a Monolayer of Optically "Soft" Particles at their Dense Package," *Opt. Spectrosc. (USSR)*, vol. 68, 1990, pp. 403–409.

70. E. E. Gorodnichev and D. B. Rogozkin, "Small Angle Multiple Scattering in Random Inhomogeneous Media, *JETP*, vol. 107, 1995, pp. 209–235.

71. H. Rinneberg, "Scattering of Laser Light in Turbid media, Optical Tomography for Medical Diagnostics," in *The Inverse Problem*, H. Lübbig (ed.), Akademie Verlag, Berlin, 1995, pp. 107–141.

72. A. J. Welch and M. C. J. van Gemert (eds.), *Tissue Optics*, Academic Press, New York, 1992.

73. V. L. Kuzmin and V. P. Romanov, "Coherent Effects at Light Scattering in Disordered Systems," *Physics—Uspekhi*, vol. 166, no. 3, 1996, pp. 247–278.

74. K. M. Yoo, F. Liu, and R R. Alfano, "Biological Materials Probed by the Temporal and Angular Profiles of the Backscattered Ultrafast Laser Pulses," *J. Opt. Soc. Am. B.*, vol. 7, 1990, pp. 1685–1693.

75. S. M. Rytov, Yu. A. Kravtsov, and V. I. Tatarskii, *Wave Propagation through Random Media*, vol. 4 of *Principles of Statistical Radiophysics*, Springer-Verlag, Berlin, 1989.

76. V. V. Tuchin (ed.), SPIE CIS Selected Papers: *Coherence-Domain Methods in Biomedical Optics*, vol. 2732, SPIE Press, 1996.

77. V. V. Tuchin, "Coherence-Domain Methods in Tissue and Cell Optics," *Laser Physics*, vol. 8, no. 4, 1998, pp. 807–849.

78. H. Z. Cummins and E. R. Pike (eds.), *Photon Correlation and Light Beating Spectroscopy*, Plenum Press, New York, 1974.

79. H. Z. Cummins and E. R. Pike (eds.), *Photon Correlation Spectroscopy and Velocimetry*, Plenum Press, New York, 1977.

80. M. H. Kao, A. G. Yodh, and D. J. Pine, "Observation of Brownian Motion on the Time Scale of Hydrodynamic Interactions," *Phys. Rev. Lett.*, vol. 70, 1993, pp. 242–245.

81. P. D. Kaplan, A. D. Dinsmore, A. G. Yodh, and D. J. Pine, "Diffuse-Transmission Spectroscopy: a Structural Probe of Opaque Colloidal Mixtures," *Phys. Rev. E.*, vol. 50, 1994, pp. 4827–4835.

82. J. D. Briers, "Laser Doppler and Time-Varying Speckle: a Reconciliation," *J. Opt. Soc. Am. A.*, vol. 13, 1996, pp. 345–350.

83. J. D. Briers and S. Webster, "Laser Speckle Contrast Analysis (LASCA): a Nonscanning, Full-Field Technique for Monitoring Capillary Blood Flow," *J. Biomed. Opt.*, vol. 1, 1996, pp. 174–179.

84. A. F. Fercher, "Optical Coherence Tomography," *J. Biomed. Opt.*, vol. 1, 1996, pp. 157–173.

85. B. Beauvoit, T. Kitai, H. Liu, and B. Chance, "Time-Resolved Spectroscopy of Mitochondria, Cells, and Rat Tissues under Normal and Pathological Conditions," *Proc. SPIE* 2326, 1994, pp. 127–136.

86. A. R. Young, "Chromophores in Human Skin," *Phys. Med. Biol.*, vol. 42, 1997, pp. 789–802.

87. F. A. Duck, *Physical Properties of Tissue: a Comprehensive Reference Book*, Academic Press, London, 1990.

88. B. Chance (ed.), *Photon Migration in Tissue*, Plenum Press, New York, 1989.

89. K. Frank and M. Kessler (eds.), *Quantitative Spectroscopy in Tissue*, PMI Verlag, Frankfurt am Main, 1992.

90. B. W. Henderson and T. J. Dougherty (eds.), *Photodynamic Therapy: Basic Principles and Clinical Applications*, Marcel-Dekker, New York, 1992.

91. H.-P. Berlien and G. J. Mueller (eds.), *Applied Laser Medicine*, Springer-Verlag, Berlin, 2003.

92. N. N. Zhadin and R. R. Alfano, "Correction of the Internal Absorption Effect in Fluorescence Emission and Excitation Spectra from Absorbing and Highly Scattering Media: Theory and Experiment," *J. Biomed. Opt.*, vol. 3, no. 2, 1998, pp. 171–186.

93. A. Kienle, M. S. Patterson, N. Dognitz, R. Bays, G. Wagnieres, and H. van de Bergh, "Noninvasive Determination of the Optical Properties of Two-Layered Turbid Medium," *Appl. Opt.*, vol. 37, 1998, pp. 779–791.

94. J. R. Mourant, J. Boyer, A. H. Hielscher, and I. J. Bigio, "Influence of the Scattering Phase Function on Light Transport Measurements in Turbid Media Performed with Small Source-Detector Separations," *Opt. Lett.*, vol. 21, no. 7, 1996, pp. 546–548.

95. J. R. Mourant, J. P. Freyer, A. H. Hielscher, A. A. Eick, D. Shen, and T. M. Johnson, "Mechanisms of Light Scattering from Biological Cells Relevant to Noninvasive Optical-Tissue Diagnostics," *Appl. Opt.*, vol. 37, 1998, pp. 3586–3593.

96. R. Drezek, A. Dunn, and R. Richards-Kortum, "Light Scattering from Cells: Finite-Difference Time-Domain Simulations and Goniometric Measurements," *Appl. Opt.*, vol. 38, no. 16, 1999, pp. 3651–3661.

97. J. R. Zijp and J. J. ten Bosch, "Anisotropy of Volume-Backscattered Light," *Appl. Opt.*, vol. 36, 1997, pp. 1671–1680.

98. H. Moseley (ed.), "Special Issue on Optical Radiation Technique in Medicine and Biology," *Phys. Med. Biol.*, vol. 24, 1997, pp. 759–996.

99. M. D. Morris (ed.), "Special Section on Biomedical Applications of Vibrational Spectroscopic Imaging," *J. Biomed. Opt.*, vol. 4, 1999, pp. 6–34.

100. K. Okada and T. Hamaoka, "Special Section on Medical Near-Infrared Spectroscopy," *J. Biomed. Opt.*, vol. 4, 1999, pp. 391–428.

101. A. V. Priezzhev and T. Asakura (eds.), "Special Section on Optical Diagnostics of Biological Fluids," *J. Biomed. Opt.*, vol. 4, 1999, pp. 35–93.

102. J. M. Schmitt, "Optical Coherence Tomography (OCT): a Review," *IEEE J. Select. Tops Quant. Electr.*, vol. 5, 1999, pp. 1205–1215.

103. D. Benaron, I. Bigio, E. Sevick-Muraca, and A. G. Yodh (eds.), "Special Issue Honoring Professor Britton Chance," *J. Biomed. Opt.*, vol. 5, 2000, pp. 115–248; pp. 269–282.

104. A. Carden and M. D. Morris, "Application of Vibration Spectroscopy to the Study of Mineralized Tissues (review)," *J. Biomed. Opt.*, vol. 5, 2000, pp. 259–268.

105. R. J. McNichols and G. L. Coté, "Optical Glucose Sensing in Biological Fluids: an Overview," *J. Biomed. Opt.*, vol. 5, 2000, pp. 5–16.

106. L. Beloussov, F.-A. Popp, V. L. Voeikov, and R. van Wijk (eds.), *Biophotonics and Coherent Systems*, Moscow Univ. Press, Moscow, 2000.

107. R. K. Wang, "Modelling Optical Properties of Soft tissue by Fractal Distribution of Scatterers," *J. Modern Opt.*, vol. 47, no. 1, 2000, pp. 103–120.

108. J. G. Fujimoto, W. Drexler, U. Morgner, F. Kartner, and E. Ippen, "Optical Coherence Tomography: High Resolution Imaging Using Echoes of Light," *Optics & Photonics News*, January, 2000, pp. 24–31.

109. J. G. Fujimoto and M. E. Brezinski, "Optical Coherence Tomography Imaging," in *Biomedical Photonics Handbook*, Tuan Vo-Dinh (ed.), CRC Press, Boca Rotan, Florida, 2003, pp. 13-1–29.

110. J. Welzel, "Optical Coherence Tomography in Dermatology: a Review," *Skin Res. Technol.*, vol. 7, 2001, pp. 1–9.

111. A. M. Sergeev, L. S. Dolin, and D. H. Reitze, "Optical Tomography of Biotissues: Past, Present, and Future," *Optics & Photonics News*, July, 2001, pp. 28–35.

112. J. D. Briers, "Laser Doppler, Speckle and Related Techniques for Blood Perfusion Mapping and Imaging," *Physiol. Meas.*, vol. 22, 2001, pp. R35–R66.

113. W. V. Meyer, A. E. Smart, and R. G. W. Brown (eds.), "Special Issue on Photon Correlation and Scattering," *Appl. Opt.*, vol. 40, no. 24, 2001, pp. 3965–4242.

114. P. T. C. So, C. Y. Dong, B. R. Masters, and K. M. Berland, "Two-Photon Excitation Fluorescence Microscopy," in *Annual Review of Biomedical Engineering*, Annual Reviews, Palo Alto, CA. 2000.

115. R. K. Wang, J. C. Hebden, and V. V. Tuchin, "Special Issue on Recent Developments in Biomedical Optics," *Phys. Med. Biol.*, vol. 49, no. 7, 2004, pp. 1085–1368.

116. V. V. Tuchin, J. A. Izatt, and J. G. Fujimoto (eds.), *Coherence Domain Optical Methods in Biomedical Science and Clinical Applications IV-VIII, Proc. SPIE* 3915, 2000; 4251, 2001; 4619, 2002; 4956, 2003; 5316, 2004; *Coherence Domain Optical Methods and Optical Coherence Tomography in Biomedicine IX, X, Proc. SPIE* 5690, 2005; 6079, 2006.

117. V. V. Tuchin, and D. A. Zimnyakov, and A. B. Pravdin (eds.), *Saratov Fall Meeting, Optical Technologies in Biophysics and Medicine, Proc. SPIE* 4001, 2000.

118. V. V. Tuchin (ed.), *Saratov Fall Meeting, Optical Technologies in Biophysics and Medicine II-VII, Proc. SPIE* 4241, 2001; 4707, 2002; 5068, 2003; 5474, 2004; 5771, 2005; 6163, 2006; 6535, 2007.

119. A. Periasamy (ed.), *Methods in Cellular Imaging*, Oxford University Press, New York, 2001.

120. D. B. Murphy, *Fundamentals of Light Microscopy and Electronic Imaging*, Wiley-Liss, New York, 2001.

121. P. Sebbah (ed.), *Waves and Imaging through Complex Media*, Kluwer Academic Publishers, New York, 2001.

122. A. Diaspro (ed.), *Confocal and Two-Photon Microscopy: Foundations, Applications, and Advances*, Wiley-Liss, New York, 2002.

123. J. M. Chalmers and P. R. Grifiths (eds.), *Handbook of Vibrational Spectroscopy*, John Wiley & Sons, Chichester, 2002.

124. Q. Luo, B. Chance, and V. V. Tuchin (eds.), *Photonics and Imaging in Biology and Medicine, Proc. SPIE* 4536, Bellingham, WA, 2002.

125. Q. Luo, V. V. Tuchin, M. Gu, L. V. Wang (eds.), *Photonics and Imaging in Biology and Medicine, Proc. SPIE* 5254, Bellingham, WA, 2003.

126. B. Masters (ed.), *Selected Papers on Optical Low-Coherence Reflectometry and Tomography*, vol. MS165, SPIE Press, Bellingham, WA, 2002.

127. B. E. Bouma and G. J. Tearney (eds.), *Handbook of Optical Coherence Tomography*, Marcel-Dekker, New York, 2002.

128. D. R. Vij and K. Mahesh (eds.) *Lasers in Medicine*, Kluwer Academic Publishers, Boston, Dordrecht, and London, 2002.

129. V. V. Tuchin (ed.), *Handbook of Optical Biomedical Diagnostics*, vol. PM107, SPIE Press, Bellingham, WA, 2002.

130. Tuan Vo-Dinh (ed.), *Biomedical Photonics Handbook*, CRC Press, Boca Raton, 2003.

131. B. R. Masters (ed.), *Selected Papers on Multiphoton Excitation Microscopy*, vol. MS175, SPIE Press, Bellingham, WA. 2003.

132. R. G. Driggers (ed.), *Encyclopedia of Optical Engineering*, Marcel-Dekker, New York, 2003. www.dekker.com/servlet/product/productid/E-EOE/

133. U. Utzinger and R. Richards-Kortum, "Fiber-Optic Probes for Biomedical Optical Spectroscopy," *J. Biomed. Opt.*, vol. 8, 2003, pp. 121–147.

134. R. R. Alfano and B. R. Masters (eds.), *Biomedical Optical Biopsy and Optical Imaging: Classic Reprints on CD-ROM Series*, Optical Society of America, Washington, DC, 2004.

135. V. V. Tuchin, L. V. Wang, and D. A. Zimnyakov, *Optical Polarization in Biomedical Applications*, Springer-Verlag, New York, 2006.

136. V. V. Tuchin (ed.), *Coherent-Domain Optical Methods: Biomedical Diagnostics, Environmental and Material Science*, Kluwer Academic Publishers, Boston, vol. 1 & 2, 2004.

137. B. R. Masters and T. P. C. So, *Handbook of Multiphoton Excitation Microscopy and other Nonlinear Microscopies*, Oxford University Press, New York, 2004.

138. L. V. Wang, G. L. Coté, and S. L. Jacques (eds.), "Special Section on Tissue Polarimetry," *J. Biomed. Opt.*, vol. 7, no. 3, 2002, pp. 278–397.

139. D. A. Zimnyakov and V. V. Tuchin, "Optical Tomography of Tissues (overview)," *Quantum Electron.*, vol. 32, no. 10, 2002, pp. 849–867.

140. W. R. Chen, V. V. Tuchin, Q. Luo, and S. L. Jacques, "Special Issue on Biophotonics," *J. X-Ray Sci. and Technol.*, vol. 10, nos. 3-4, 2002, pp. 139–243.

141. P. French and A. I. Ferguson (eds.), "Special Issue on Biophotonics," *J. Phys. D: Appl. Phys.*, vol. 36, no. 14, 2003, pp. R207–R258; 1655–1757.

142. A. F. Fercher, W. Drexler, C. K. Hitzenberger, and T. Lasser, "Optical Coherence Tomography—Principles and Applications," *Rep. Progr. Phys.*, vol. 66, 2003, pp. 239–303.

143. E. D. Hanlon, R. Manoharan, T.-W. Koo, K. E. Shafer, J. T. Motz, M. Fitzmaurice, J. R. Kramer, I. Itzkan, R. R. Dasari, and M. S. Feld, "Prospects for *In Vivo* Raman Spectroscopy," *Phys. Med. Biol*, vol. 45, 2000, pp. R1–R59.

144. R. R. Ansari and J. Sebag (eds.), "Ophthalmic Diagnostics," *J. Biomed. Opt.*, vol. 9, no. 1, 2004, pp. 8–179.

145. M. I. Mishchenko, J. W. Hovenier, and L. D. Travis (eds.), *Light Scattering by Nonspherical Particles*, Academic Press, San Diego, 2000.

146. M. I. Mishchenko, L. D. Travis, and A. A. Lacis, *Scattering, Absorption, and Emission of Light by Small Particles*, Cambridge Univ. Press, Cambridge, 2002.

147. D. A. Zimnyakov (ed.), *Saratov Fall Meeting, Coherent Optics of Ordered and Random Media*, Proc. SPIE 4242, 2001; 4705, 2002; 5067, 2003; 5475, 2004.

148. C. F. Bohren and D. R. Huffman, *Absorption and Scattering of Light by Small Particles*, Wiley, New York, 1983.

149. G. C. Salzmann, S. B. Singham, R. G. Johnston, and C. F. Bohren, "Light Scattering and Cytometry," in *Flow Cytometry and Sorting*, 2nd ed.,

M. R. Melamed, T. Lindmo, and M. L. Mendelsohn (eds.), Wiley-Liss, New York, 1990, pp. 81–107.

150. V. Backman, R. Gurjar, K. Badizadegan, R. Dasari, I. Itzkan, L. T. Perelman, and M. S. Feld, "Polarized Light Scattering Spectroscopy for Quantitative Measurement of Epithelial Cellular Structures *In Situ*," *IEEE J. Sel. Top. Quant. Elect.*, vol. 5, 1999, pp. 1019–1027.

151. J. R. Mourant, M. Canpolat, C. Brocker, O. Esponda-Ramos, T. M. Johnson, A. Matanock, K. Stetter, and J. P. Freyer, "Light Scattering from Cell: the Contribution of the Nucleus and the Effects of Proliferative Status," *J. Biomed. Opt.*, vol. 5, no. 2, 2000, pp. 131–137.

152. J. R. Mourant, R. R. Gibson, T. M. Johnson, S. Carpenter, K. W. Short, Y. R. Yamada, and J. P. Freyer, "Methods for Measuring the Infrared Spectra of Biological Cells," *Phys. Med. Biol.*, vol. 48, 2003, pp. 243–257.

153. R. Drezek, M. Guillaud, T. Collier, I. Boiko, A. Malpica, C. Macaulay, M. Follen, and R. Richards-Kortum, "Light Scattering from Cervical Cells throughout Neoplastic Progression: Influence of Nuclear Morphology, DNA Content, and Chromatin Texture," *J. Biomed. Opt*, vol. 8, 2003, pp. 7–16.

154. J. M. Schmitt and G. Kumar, "Turbulent Nature of Refractive-Index Variations in Biological Tissue," *Opt. Lett.*, vol. 21, 1996, pp. 1310–1312.

155. D. A. Zimnyakov, V. V. Tuchin, and A. A. Mishin, "Spatial Speckle Correlometry in Applications to Tissue Structure Monitoring," *Appl. Opt.*, vol. 36, 1997, pp. 5594–5607.

156. J. M. Schmitt and G. Kumar, "Optical Scattering Properties of Soft Tissue: a Discrete Particle Model," *Appl. Opt.*, vol. 37, no. 13, 1998, pp. 2788–2797.

157. J. W. Goodman, *Statistical Optics*, Wiley-Interscience Publication, New York, 1985.

158. S. Ya. Sid'ko, V. N. Lopatin, and L. E. Paramonov, *Polarization Characteristics of Solutions of Biological Particles*, Nauka, Novosibirsk, 1990.

159. A. G. Borovoi, E. I. Naats, and U. G. Oppel, "Scattering of Light by a Red Blood Cell," *J. Biomed. Opt.*, vol. 3, no. 3, 1998, pp. 364–372.

160. M. Born and E. Wolf, *Principles of Optics*, 7th ed., Cambridge Univ., Cambridge, 1999.

161. R. D. Dyson, *Cell Biology: a Molecular Approach*, Allyn and Bacon, Boston, 1974.

162. P. Latimer, "Light Scattering and Absorption as Methods of Studying Cell Population Parameters," *Ann. Rev. Biophys. Bioeng.*, vol. 11, no. 1, 1982, pp. 129–150.

163. K. Sokolov, R. Drezek, K. Gossagee, and R. Richards-Kortum, "Reflectance Spectroscopy with Polarized Light: is it Sensitive to Cellular and Nuclear Morphology," *Opt. Express*, vol. 5, 1999, pp. 302–317.

164. A. N. Yaroslavsky, I. V. Yaroslavsky, T. Goldbach, and H.-J. Schwarzmaier, "Influence of the Scattering Phase Function Approximation on the Optical Properties of Blood Determined from the Integrating Sphere Measurements," *J. Biomed. Opt.*, vol. 4, no. 1, 1999, pp. 47–53.

165. G. Kumar and J. M. Schmitt, "Micro-Optical Properties of Tissue," *Proc. SPIE* 2679, 1996, pp. 106–116.

166. J. R. Mourant, T. M. Johnson, S. Carpenter, A. Guerra, T. Aida, and J. P. Freyer, "Polarized Angular Dependent Spectroscopy of Epithelial Cells and Epithelial Cell Nuclei to Determine the Size Scale of Scattering Structures," *J. Biomed. Opt.*, vol. 7, no. 3, 2002, pp. 378–387.

167. M. J. Hogan, J. A. Alvardo, and J. Weddel, *Histology of the Human Eye*, W. B. Sanders Co., Philadelphia, 1971.

168. Q. Zhou and R. W. Knighton, "Light Scattering and Form Birefringence of Parallel Cylindrical Arrays that Represent Cellular Organelles of the Retinal Nerve Fiber Layer," *Appl. Opt.*, vol. 36, no. 10, 1997, pp. 2273–2285.

169. G. Videen and D. Ngo, "Light Scattering Multipole Solution for a Cell," *J. Biomed. Opt.*, vol. 3, 1998, pp. 212–220.

170. J. R. Mourant, T. M. Johnson, V. Doddi, and J. P. Freyer, "Angular Dependent Light Scattering from Multicellular Spheroids," *J. Biomed. Opt.*, vol. 7, no. 1, 2002, pp. 93–99.

171. K. S. Shifrin, *Physical Optics of Ocean Water*, American Institute of Physics, New York, 1988.

172. V. V. Tuchin, I. L. Maksimova, D. A. Zimnyakov, I. L. Kon, A. H. Mavlutov, and A. A. Mishin, "Light Propagation in Tissues with Controlled Optical Properties," *J. Biomed. Opt.*, vol. 2, 1997, pp. 401–417.

173. V. V. Tuchin and D. M. Zhestkov, "Tissue Structure and Eye Lens Transmission and Scattering Spectra," *Proc. SPIE* 3053, 1997, pp. 123–128.

174. A. Brunsting and P. F. Mullaney, "Differential Light Scattering from Spherical Mammalian Cells," *Biophys. J.*, vol. 10, 1974, pp. 439–453.

175. J. Beuthan, O. Minet, J. Helfmann, M. Herring, and G. Mueller, "The Spatial Variation of the Refractive Index in Biological Cells," *Phys. Med. Biol.*, vol. 41, no. 3, 1996, pp. 369–382.

176. F. H. Silver, *Biological Materials: Structure, Mechanical Properties, and Modeling of Soft Tissues*, New York Univ. Press, New York, 1987.

177. R. G. Kessel, *Basic Medical Histology: The Biology of Cells, Tissues, and Organs*, Oxford Univ. Press, New York, 1998.

178. F. P. Bolin, L. E. Preuss, R. C. Taylor, and R. J. Ference, "Refractive Index of Some Mammalian Tissues using a Fiber Optic Cladding Method," *Appl. Opt.*, vol. 28, 1989, pp. 2297–2303.

179. R. Graaff, J. G. Aarnoudse, J. R. Zijp, P. M. A. Sloot, F. F. M. de Mul, J. Greve, and M. H. Koelink, "Reduced Light Scattering Properties for Mixtures of Spherical Particles: a Simple Approximation Derived from Mie Calculations," *Appl. Opt.*, vol. 31, 1992, pp. 1370–1376.

180. L. T. Perelman, V. Backman, M. Wallace, G. Zonios, R. Manoharan, A. Nusrat, S. Shields, M. Seiler, C. Lima, T. Hamano, I. Itzkan, J. Van Dam, J. M. Crawford, and M. S. Feld, "Observation of Periodic Fine Structure in Reflectance from Biological Tissue: a New Technique for Measuring Nuclear Size Distribution," *Phys. Rev. Lett.*, vol. 80, 1998, pp. 627–630.

181. H. C. van de Hulst, *Light Scattering by Small Particles*, Wiley, New York, 1957 [reprint, Dover, New York, 1981].

182. H. C. van de Hulst, *Multiple Light Scattering. Tables, Formulas and Applications*, Academic Press, New York, 1980.

183. A. Ishimaru, *Wave Propagation and Scattering in Random Media*, IEEE Press, New York, 1997.

184. C. Chandrasekhar, *Radiative Transfer*, Dover, Toronto, Ontario, 1960.

185. V. V. Sobolev, *Light Scattering in Planetary Atmospheres*, Pergamon Press, Oxford, 1974.

186. E. P. Zege, A. P. Ivanov, and I. L. Katsev, *Image Transfer through a Scattering Medium*, Springer-Verlag, New York, 1991.

187. A. Z. Dolginov, Yu. N. Gnedin, and N. A. Silant'ev, *Propagation and Polarization of Radiation in Cosmic Media*, Gordon and Breach, Basel, 1995.

188. M. I. Mishchenko, L. D. Travis, and A. A. Lacis, *Multiple Sacattering of Light by Particles*: *Radiative Transfer and Coherent Backscattering*, Cambridge University Press, New York, 2006.

189. E. J. Yanovitskij, *Light Scattering in Inhomogeneous Atmospheres*, Springer-Verlag, Berlin, 1997.

190. G. E. Thomas and K. Stamnes, *Radiative Transfer in the Atmosphere and Ocean*, Cambridge University Press, New York, 1999.

191. D. J. Durian, "The Diffusion Coefficient Depends on Absorption," *Opt. Lett.*, vol. 23, 1998, pp. 1502–1504.

192. A. Ishimaru, "Diffusion of Light in Turbid Material," *Appl. Opt.*, vol. 28, 1989, pp. 2210–2215.

193. T. J. Farrell, M. S. Patterson, and B. C. Wilson, "A diffusion Theory model of Spatially Resolved, Steady-State Diffuse Reflectance for the Noninvasive Determination of Tissue Optical Properties *In Vivo*," *Med. Phys.*, vol. 19, 1992, pp. 881–888.

194. M. Keijzer, W. M. Star, and P. R. M. Storchi, "Optical Diffusion in Layered Media," *Appl. Opt.*, vol. 27, pp. 1988, pp. 1820–1824.

195. G. Yoon, S. A. Prahl, and A. J. Welch, Accuracies of the Diffusion Approximation and its Similarity Relations for Laser Irradiated Biological Media," *Appl. Opt.*, vol. 28, 1989, pp. 2250–2255.

196. K. M. Yoo, F. Liu, and R. R. Alfano, "When Does the Diffusion Approximation Fail to Describe Photon Transport in Random Media?" *Phys. Rev. Lett.*, vol. 64, no. 22, 1990, pp. 2647–2650.

197. I. Dayan, S. Halvin, and G. H. Weiss, "Photon Migration in a 2-Layer Turbid Medium—a Diffusion Analysis," *J. Modern Opt.*, vol. 39, no. 7, 1992, pp. 1567–1582.

198. L. V. Wang and S. L. Jacques, "Source of Error in Calculation of Optical Diffuse Reflectance from Turbid Media Using Diffusion Theory," *Comput. Meth. Progr. Biomed.*, vol. 61, 2000, pp. 163–170.

199. F. Martelli, M. Bassani, L. Alianelli, L. Zangheri, and G. Zaccanti, "Accuracy of the Diffusion Equation to Describe Photon Migration through an

Infinite Medium: Numerical and Experimental Investigation," *Phys. Med. Biol.*, vol. 45, 2000, pp. 1359–1373.

200. S. Del Bianko, F. Martelli, and G. Zaccanti, "Penetration Depth of Light Re-emitted by a Diffusive Medium: Theoretical and Experimental Investigation," *Phys. Med. Biol.*, vol. 47, 2002, pp. 4131–4144.

201. R. Graaff and K. Rinzema, "Practical Improvements on Photon Diffusion Theory: Application to Isotropic Scattering," *Phys. Med. Biol.*, vol. 46, 2001, pp. 3043–3050.

202. D. C. Sahni, E. B. Dahl, and N. G. Sjostrand, "Diffusion Coefficient for Photon Transport in Turbid Media," *Phys. Med. Biol.*, vol. 48, 2003, pp. 3969–3976.

203. T. Khan and H. Jiang, "A New Diffusion Approximation to the Radiative Transfer Equation for Scattering Media with Spatially Varying Refractive Indices," *J. Opt. A: Pure Appl. Opt.*, vol. 5, 2003, pp. 137–141.

204. R. C. Haskell, L. O. Svaasand, T.-T. Tsay, T. C. Feng, M. N. McAdams, and B. J. Tromberg, "Boundary Conditions for the Diffusion Equation in Radiative Transfer," *J. Opt. Soc. Am. A*, vol. 11, no. 10, 1994, pp. 2727–2741.

205. A. Kienle and M. S. Patterson, "Improved Solutions of the Steady-State and Time-Resolved Diffusion Equations for Reflectance from a Semi-Infinite Turbid Media," *J. Opt. Soc. Am. A*, vol. 14, 1997, pp. 246–254.

206. T. J. Farrell and M. S. Patterson, "Experimental Verification of the Effect of Refractive Index Mismatch on the Light Fluence in a Turbid Medium," *J. Biomed. Opt.*, vol. 6, no. 4, 2001, pp. 468–473.

207. M. Motamedi, S. Rastegar, G. LeCarpentier, and A. J. Welch, "Light and Temperature Distribution in Laser Irradiated Tissue: The Influence of Anisotropic Scattering and Refractive Index," *Appl. Opt.*, vol. 28, 1989, pp. 2230–2237.

208. S. R. Arridge, M. Schweiger, M. Hiraoka, and D. T. Delpy, "A Finite Element Approach for Modelling Photon Transport in Tissue," *Med. Phys.*, vol. 20, 1993, pp. 299–309.

209. W. M. Star, "Comparing the P3-approximation with diffusion theory and with Monte Carlo Calculations of Light Propagation in a Slab Geometry," in *Dosimetry of Laser Radiation in Medicine and Biology*, G. J. Mueller and D. H. Sliney (eds.), vol. IS5, SPIE Press, Bellingham, WA, 1989, pp. 146–154.

210. W. M. Star, "Light Dosimetry *In Vivo*," *Phys. Med. Biol.*, vol. 42, 1997, pp. 763–787.

211. D. J. Dickey, R. B. Moore, D. C. Rayner, and J. Tulip, "Light Dosimetry Using the P3 Approximation," *Phys. Med. Biol.*, vol. 46, 2001, pp. 2359–2370.

212. B. Phylips-Invernizzi, D. Dupont, and C. Caze, "Bibliographical Review for Reflectance of Diffusing Media, *Opt. Eng.*, vol. 40, no. 6, 2001, pp. 1082–1092.

213. V. V. Tuchin, S. R. Utz, and I. V. Yaroslavsky, "Tissue Optics, Light Distribution, and Spectroscopy," *Opt. Eng.*, vol. 33, 1994, pp. 3178–3188.

214. S. M. Ermakov and G. A. Mikhailov, *Course on Siatistical Modeling*, Nauka, Moscow, 1982.

215. B. C. Wilson and G. A. Adam, "Monte Carlo Model for the Absorption and Flux Distributions of Light in Tissue," *Med. Phys.*, vol. 10, 1983, pp. 824–830.

216. S. A. Prahl, M. Keijzer, S. L. Jacques, and A. J. Welch, "Monte Carlo Model of Light Propagation in Tissues," in *Dosimetry of Laser Radiation in Medicine and Biology*, G. J. Mueller and D. H. Sliney (eds.), vol. IS5, SPIE Press, Bellingham, WA, 1989, pp. 102–111.

217. M. Keijzer, S. L. Jacques, S. A. Prahl, and A. J. Welch, "Light Distribution in Artery Tissue: Monte Carlo Simulation for Finite-Diameter Laser Beams," *Lasers Surg. Med.*, vol. 9, 1989, pp. 148–154.

218. M. Keijzer, R. R. Richards-Kortum, S. L. Jacques, and M. S. Feld, "Fluorescence Spectroscopy of Turbid Media: Autofluorescence of the Human Aorta," *Appl. Opt.*, vol. 28, no. 20, 1989, pp. 4286–4292.

219. S. T. Flock, B. C. Wilson, D. R. Wyman, and M. S. Patterson, "Monte Carlo Modeling of Light–Propagation in Highly Scattering Tissues I: Model Predictions and Comparison with Diffusion Theory," *IEEE Trans. Biomed. Eng.*, vol. 36, no. 12, 1989, pp. 1162–1168.

220. S. T. Flock, B. C. Wilson, and M. S. Patterson, "Monte Carlo Modeling of Light-Propagation in Highly Scattering Tissues II: Comparison with Measurements in Phantoms," *IEEE Trans. Biomed. Eng.*, vol. 36, no. 12, 1989, pp. 1169–1173.

221. J. M. Schmitt, G. X. Zhou, E. C. Walker, and R. T. Wall, "Multilayer Model of Photon Diffusion in Skin," *J. Opt. Soc. Am. A*, vol. 7, 1990, pp. 2141–2153.

222. S. L. Jacques, "The Role of Skin Optics in Diagnostic and Therapeutic Uses of Lasers," in *Lasers in Dermatology*, Springer-Verlag, Berlin, 1991, pp. 1–21.

223. H. Key, E. R. Davies, P. C. Jackson, and P. N. T. Wells, "Optical Attenuation Characteristics of Breast Tissues at Visible and Near-Infrared Wavelengths," *Phys. Med. Biol.*, vol. 36, no. 5, 1991, pp. 579–590.

224. I. V. Yaroslavsky and V. V. Tuchin, "Light Propagation in Multilayer Scattering Media. Modeling by the Monte Carlo Method," *Opt. Spectrosc.*, vol. 72, 1992, pp. 505–509.

225. L.-H. Wang and S. L. Jacques, "Hybrid Model of the Monte Carlo Simulation and Diffusion Theory for Light Reflectance by Turbid Media," *J. Opt. Soc. Am. A.*, vol. 10, 1993, pp. 1746–1752.

226. R. Graaff, M. H. Koelink, F. F. M. de Mul, et al., "Condensed Monte Carlo Simulations for the Description of Light Transport," *Appl. Opt.*, vol. 32, no. 4, 1993, pp. 426–434.

227. V. V. Tuchin, S. R. Utz, and I. V. Yaroslavsky, "Skin Optics: Modeling of Light Transport and Measuring of Optical Parameters," in *Medical Optical Tomography: Functional Imaging and Monitoring*, G. Müller, B. Chance, R. Alfano, et al. (eds.), vol. IS11, SPIE Press, Bellingham, WA, 1993, pp. 234–258.

228. R. Graaff, A. C. M. Dassel, M. H. Koelink, et al., "Optical Properties of Human Dermis *In Vitro* and *In Vivo*," *Appl. Opt.*, vol. 32, 1993, pp. 435–447.

229. S. L. Jacques and L. Wang, "Monte Carlo Modeling of Light Transport in Tissues," in *Optical-Thermal Response of Laser-Irradiated Tissue*, A. J. Welch and M. J. C. van Gemert (eds.), Plenum Press, New York, 1995, pp. 73–100.

230. L.-H. Wang, S. L. Jacques, L.-Q. Zheng, "MCML–Monte Carlo Modeling of Light Transport in Multi-Layered Tissues," *Comput. Meth. Progr. Biomed.*, vol. 47, 1995, pp. 131–146.

231. S. R. Arridge, M. Hiraoka, and M. Schweiger, "Statistical Basis for the Determination of Optical Pathlength in Tissue," *Phys. Med. Biol.*, vol. 40, 1995, pp. 1539–1558.

232. T. L. Troy, D. L. Page, and E. M. Sevick-Muraca, "Optical Properties of Normal and Diseased Breast Tissues: Prognosis for Optical Mammography," *J. Biomed. Opt.*, vol. 1, no. 3, 1996, pp. 342–355.

233. E. Okada, M. Firbank, M. Schweiger, et al., "Theoretical and Experimental Investigation of Near-Infrared Light Propagation in a Model of the Adult Head," *Appl. Opt.*, vol. 36, no. 1, 1997, pp. 21–31.

234. V. G. Kolinko, F. F. M. de Mul, J. Greve, and A. V. Priezzhev, "On Refraction in Monte-Carlo Simulations of Light Transport through Biological Tissues," *Med. Biol. Eng. Comp.* vol. 35, 1997, pp. 287–288.

235. L.-H. Wang, "Rapid Modeling of Diffuse Reflectance of Light in Turbid Slabs," *J. Opt. Soc. Am. A.*, vol. 15, no. 4, 1998, pp. 936–944.

236. C. R. Simpson, M. Kohl, M. Essenpreis, and M. Cope, "Near-Infrared Optical Properties of *Ex Vivo* Human Skin and Subcutaneos Tissues Measured Using the Monte Carlo Inversion Technique," *Phys. Med. Biol.*, vol. 43, 1998, pp. 2465–2478.

237. J. Laufer, C. R. Simpson, M. Kohl, M. Essenpreis, and M. Cope, "Effect of Temperature on the Optical Properties of *Ex Vivo* Human Dermis and Subdermis," *Phys. Med. Biol.* vol. 43, 1998, pp. 2479–2489.

238. P. M. Ripley, J. G. Laufer, A. D. Gordon, R. J. Connell, and S. G. Bown, "Near-Infrared Optical Properties of *Ex Vivo* Human Uterus Determined by the Monte Carlo Inversion Technique," *Phys. Med. Biol.*, vol. 44, 1999, pp. 2451–2462.

239. M. L. de Jode, "Monte Carlo Simulations of Light Distributions in an Embedded Tumour Model: Studies of Selectivity in Photodynamic Therapy," *Lasers Med. Sci.*, vol. 15, 2000, pp. 49–56.

240. R. Jeraj and P. Keall, "The Effect of Statistical Uncertainty on Inverse Treatment Planning Based on Monte Carlo Dose Calculation," *Phys. Med. Biol.*, vol. 45, 2000, pp. 3601–3613.

241. I. V. Meglinskii, "Monte Carlo Simulation of Reflection Spectra of Random Multilayer Media Strongly Scattering and Absorbing Light," *Quant. Electron.*, vol. 31, no. 12, 2001, pp. 1101–1107.

242. C. K. Hayakawa, J. Spanier, F. Bevilacqua, A. K. Dunn, J. S. You, B. J. Tromberg, and V. Venugopalan, "Perturbation Monte Carlo Methods to Solve Inverse Photon Migration Problems in Heterogeneous Tissues," *Opt. Lett.*, vol. 26, no. 17, 2001, pp. 1335–1337.

243. I. V. Meglinski and S. J. Matcher, "Quantitative Assessment of Skin Layers Absorption and Skin Reflectance Spectra Simulation in Visible and Near-Infrared Spectral Region," *Physiol. Meas.*, vol. 23, 2002, pp. 741–753.

244. S. J. Preece and E. Claridge, "Monte Carlo Modelling of the Spectral Reflectance of the Human Eye," *Phys. Med. Biol.*, vol. 47, 2002, pp. 2863–2877.

245. D. A. Boas, J. P. Culver, J. J. Stott, and A. K. Dunn, "Three Dimensional Monte Carlo Code for Photon Migration through Complex Heterogeneous Media Including the Adult Human Head," *Optics Express*, vol. 10, no. 3, 2002, pp. 159–170.

246. I. V. Meglinski and S. J. Matcher, "Computer Simulation of the Skin Reflectance Spectra, *Comput. Meth. Progr. Biomed.*, vol. 70, 2003, pp. 179–186.

247. F. F. M. de Mul, "Monte-Carlo Simulations of Light Scattering in Turbid Media," Chapter 12 in *Coherent-Domain Optical Methods: Biomedical Diagnostics, Environmental and Material Science*, V. V. Tuchin (ed.), Kluwer Academic Publishers, Boston, vol. 1, 2004, pp. 465–532.

248. A. V. Voronov, E. V. Tret'akov, and V. V. Shuvalov, "Fast Path-Integration Technique Simulation of Light Propagation through Highly Scattering Objects," *Quant. Electr.*, vol. 34, no. 6, 2004, pp. 547–553.

249. M. J. Wilson and R. K. Wang, "A Path-Integral Model of Light Scattered by Turbid Media," *J. Phys. B: At. Mol. Opt. Phys.*, vol. 34, 2001, pp. 1453–1472.

250. L. T. Perelman, J. Wu, I. Itzkan, and M. S. Feld, "Photon Migration in Turbid Media Using Path Integrals," *Phys. Rev. Lett.*, vol. 72, 1994, pp. 1341–1344.

251. A. Y. Polishchuk and R. R. Alfano, "Fermat Photons in Turbid Media: An Exact Analytic Solution for Most Favorable Paths—A Step Toward Optical Tomography," *Opt. Lett.*, vol. 20, 1995, pp. 1937–1939.

252. S. L. Jacques, "Path Integral Description of Light Transport in Tissue," *Ann. NY Acad. Sci.*, vol. 838, 1998, pp. 1–13.

253. V. V. Lyubimov, A. G. Kalintsev, A. B. Konovalov, O. V. Lyamtsev, O. V. Kravtsenyuk, A. G. Murzin, O. V. Golubkina, G. B. Mordvinov, L. N. Soms, and L. M. Yavorskaya, "Application of the Photon Average Trajectories Method to Real-Time Reconstruction of Tissue Inhomogeneities

in Diffuse Optical Tomography of Strongly Scattering Media," *Phys. Med. Biol.*, vol. 47, 2002, pp. 2109–2128.

254. A. H. Gandjbakhche, V. Chernomordik, J. C. Hebden, and R. Nossal, "Time-Dependent Contrast Functions for Quantitative Imaging in Time-Resolved Transillumination Experiments," *Appl. Opt.*, vol. 37, 1998, pp. 1973–1981.

255. A. N. Yaroslavsky, S. R. Utz, S. N. Tatarintsev, and V. V. Tuchin, "Angular Scattering Properties of Human Epidermal Layers," *Proc. SPIE* 2100, 1994, pp. 38–41.

256. M. A. Everett, E. Yeargers, R. M. Sayre, and R. L. Olson, "Penetration of Epidermis by Ultraviolet Rays," *Photochem. Photobiol.*, vol. 5, 1966, pp. 533–542.

257. M. Keijzer, J. M. Pickering, and M. J. C. van Gemert, "Laser Beam Diameter for Port Wine Stain Treatment," *Lasers Surg. Med.*, vol. 11, 1991, pp. 601–605.

258. M. J. C. van Gemert, D. J. Smithies, W. Verkruysse, et al., "Wavelengths for Port Wine Stain Laser Treatment: Influence of Vessel Radius and Skin Anatomy," *Phys. Med. Biol.*, vol. 42, 1997, pp. 41–50.

259. A. Kienle and R. Hibst, "A New Optimal Wavelength for Treatment of Port Wine Stains?" *Phys. Med. Biol.*, vol. 40, 1995, pp. 1559–1576.

260. A. Kienle, L. Lilge, and M. S. Patterson, "Investigation of Multi-Layered Tissue with *In Vivo* Reflectance Measurements," *Proc. SPIE* 2326, 1994, pp. 212–214.

261. V. V. Tuchin, Yu. N. Scherbakov, A. N. Yakunin, and I. V. Yaroslavsky, "Numerical Technique for Modeling of Laser-Induced Hyperthermia," in *Laser-Induced Interstitial Thermotherapy*, G. Müller and A. Roggan (eds.), SPIE Press, Bellingham, WA, 1995, pp. 100–113.

262. Yu. N. Scherbakov, A. N. Yakunin, I. V. Yaroslavsky, and V. V. Tuchin, "Thermal Processes Modeling During Uncoagulating Laser Radiation Interaction with Multi-Layer Biotissue. 1. Theory and Calculating Models. 2. Numerical Results," *Opt. Spectrosc.*, vol. 76, no. 5, 1994, pp. 754–765.

263. S. Willmann, A. Terenji, I. V. Yaroslavsky, T. Kahn, P. Hering, and H.-J. Schwarzmaier, "Determination of the Optical Properties of a Human Brain Tumor Using a New Microspectrophotometric Technique," *Proc. SPIE* 3598, 1999, pp. 233–239.

264. A. N. Yaroslavsky, P. C. Schulze, I. V. Yaroslavsky, R. Schober, F. Ulrich, and H.-J. Schwarzmaier, "Optical Properties of Selected Native and Coagulated Human Brain Tissues *In Vitro* in the Visible and Near Infrared Spectral Range," *Phys. Med. Biol.*, vol. 47, 2002, pp. 2059–2073.

265. W. M. Star, B. C. Wilson, and M. C. Patterson, "Light Delivery and Dosimetry in Photodynamic Therapy of Solid Tumors," in *Photodynamic Therapy, Basic Principles and Clinical Applications*, B. W. Henderson and T. J. Dougherty (eds.), Marcel-Dekker, New York, 1992, pp. 335–368.

266. B. Nemati, H. G. Rylander III, and A. J. Welch, "Optical Properties of Conjuctiva, Sclera, and the Ciliary Body and Their Consequences for Transs-

cleral Cyclophotocoagulation," *Appl. Opt.* , vol. 35, no. 19, 1966, pp. 3321–3327.

267. B. Nemati, A. Dunn, A. J. Welch, and H. G. Rylander III, "Optical Model for Light Distribution during Transscleral Cyclophotocoagulation," *Appl. Opt.*, vol. 37, no. 4, 1998, pp. 764–771.

268. M. J. C. van Gemert, A. J. Welch, J. W. Pickering, et al., "Wavelengths for Laser Treatment of Port Wine Stains and Telangiectasia," *Lasers Surg. Med.*, vol. 16, 1995, pp. 147–155.

269. I. N. Minin, *Theory of Radiative Transfer in Atmosphere of Planets*, Nauka, Moscow, 1988.

270. W. Cai, B. B. Das, F. Liu, et al., "Time-Resolved Optical Diffusion Tomographic Image Reconstruction in Highly Scattering Media," *Proc. Math. Acad. Sci. USA*, vol. 93, 1996, pp. 13561–13564.

271. S. R. Arridge and J. C. Hebden, "Optical Imaging in Medicine II: Modelling and Reconstruction," *Phys. Med. Biol.*, vol. 42, 1997, pp. 841–854.

272. M. S. Patterson, B. Chance, and B. C. Wilson, "Time Resolved Reflectance and Transmittance for the Non-Invasive Measurement of Tissue Optical Properties," *Appl. Opt.*, vol. 28, 1989, pp. 2331–2336.

273. S. L. Jacques, "Time-Resolved Reflectance Spectroscopy in Turbid Tissues," *IEEE Trans. Biomed. Eng.*, vol. 36, 1989, pp. 1155–1161.

274. S. J. Matcher, M. Cope, and D. T. Delpy, "*In Vivo* Measurements of the Wavelength Dependence of Tissue-Scattering Coefficients Between 760 and 900 nm Measured with Time-Resolved Spectroscopy," *Appl. Opt.*, vol. 36, no. 1, 1997, pp. 386–396.

275. W. Cui, N. Wang, and B. Chance, "Study of Photon Migration Depths with Time–Resolved Spectroscopy," *Opt. Lett.*, vol. 16, 1991, pp. 1632–1634.

276. M. Ferrari, Q. Wei, L. Carraresi, et al., "Time-Resolved Spectroscopy of the Human Forearm," *J. Photochem. Photobiol. B: Biol.*, vol. 16, 1992, pp. 141–153.

277. A. H. Hielscher, H. Liu, B. Chance, et al., "Time-Resolved Photon Emission from Layered Turbid Media," *Appl. Opt.*, vol. 35, 1996, pp. 719–728.

278. E. B. de Haller and C. Depeursinge, "Simulation of the Time-Resolved Breast Transillumination," *Med. Biol. Eng. Comp.*, vol. 31, 1993, pp. 165–170.

279. S. Andersson-Engels, R. Berg, S. Svanberg, and O. Jarlman, "Time-Resolved Transillumination for Medical Diagnostics," *Opt. Lett.*, vol. 15, 1990, pp. 1179–1181.

280. L. Wang, P. P. Ho, C. Liu, et al., "Ballistic 2–D Imaging through Scattering Walls using an Ultrafast Optical Kerr gate, *Science.*, vol. 253, 1991, pp. 769–771.

281. B. B. Das, K. M. Yoo, and R. R. Alfano, "Ultrafast Time-Gated Imaging in Thick Tissues: a Step toward Optical Mammography," *Opt. Lett.*, vol. 18, 1993, pp. 1092–1094.

282. V. V. Lubimov, "Image Transfer in a Slab of Scattering Medium and Estimation of the Resolution of Optical Tomography uses the First Arrived Photons of Ultrashort Pulses," *Opt. Spectrosc.*, vol. 76, 1994, pp. 814–815.

283. Y. Q. Yao, Y. Wang, Y. L. Pei, et al., "Frequency-Domain Optical Imaging of Absorption and Scattering Distributions by Born Iterative Method," *J. Opt. Soc. Am. A.*, vol. 14, no. 1, 1997, pp. 325–342.

284. R. Cubeddu, A. Pifferi, P. Taroni, et al., "Time-Resolved Imaging on a Realistic Tissue Phantom: μ_s' and μ_a Images Versus Time-Integrated Images," *Appl. Opt.*, vol. 35, 1996, pp. 4533–4540.

285. A. Yodh and B. Chance, "Spectroscopy and Imaging with Diffusing Light," *Physics Today*, March, 1995, pp. 34–40.

286. S. R. Arridge, M. Cope, and D. T. Delpy, "Theoretical Basis for the Determination of Optical Pathlengths in Tissue: Temporal and Frequency Analysis," *Phys. Med. Biol.*, vol. 37, 1992, pp. 1531–1560.

287. E. B. de Haller, "Time-Resolved Transillumination and Optical Tomography," *J. Biomed. Opt.*, vol. 1, no. 1, 1996, pp. 7–17.

288. H. Heusmann, J. Kolzer, and G. Mitic, "Characterization of Female Breast *In Vivo* by Time Resolved and Spectroscopic Measurements in Near Infrared Spectroscopy," *J. Biomed. Opt.*, vol. 1, no. 4, 1996, pp. 425–434.

289. K. Suzuki, Y. Yamashita, K. Ohta, et al., "Quantitative Measurement of Optical Parameters in Normal Breast using Time-Resolved Spectroscopy: *In Vivo* Results of 30 Japanese Women," *J. Biomed. Opt.*, vol. 1, no. 3, 1996, pp. 330–334.

290. A. Taddeucci, F. Martelli, M. Barilli, et al., "Optical Properties of Brain Tissue," *J. Bimed. Opt.*, vol. 1, no. 1, 1996, pp. 117–130.

291. R. K. Wang and M. J. Wilson, "Vertex/Propagator Model for Least-Scattered Photons Traversing a Turbid Medium," *J. Opt. Soc. Am. A*, vol. 18, no. 1, 2001, pp. 224–231.

292. J.-M. Tualle, E. Tinet, J. Prat, and S. Avrillier, "Light Propagation Near-Turbid–Turbid Planar Interfaces, *Opt. Communs*, vol. 183, 2000, pp. 337–346.

293. Y. Tsuchiya, "Photon Path Distribution and Optical Responses of Turbid Media: Theoretical Analysis Based on the Microscopic Beer-Lambert Law," *Phys. Med. Biol.*, vol. 46, 2001, pp. 2067–2084.

294. G. Zacharakis, A. Zolindaki, V. Sakkalis, G. Filippidis, T. G. Papazoglou, D. D. Tsiftsis, and E. Koumantakis, "*In Vitro* Optical Characterization and Discrimination of Female Breast Tissue During Near Infrared Femtosecond Laser Pulses Propagation," *J. Biomed. Opt.*, vol. 6, no. 4, 2001, pp. 446–449.

295. R. Elaloufi, R. Carminati, and J.-J. Greffet, "Time-Dependent Transport through Scattering Media: from Radiative Transfer to Diffusion," *J. Opt. A: Pure Appl. Opt.*, vol. 4, 2002, pp. S103–S108.

296. D. Arifler, M. Guillaud, A. Carraro, A. Malpica, M. Follen, and R. Richards-Kortum, "Light Scattering from Normal and Dysplastic Cervical Cells at Different Epithelial Depths: Finite-Difference Time-Domain Modeling with

a Perfectly Matched Layer Boundary Condition," *J. Biomed. Opt.*, vol. 8, no. 3, 2003, pp. 484–494.

297. V. Chernomordik, A. H. Gandjbakhche, J. C. Hebden, and G. Zaccanti, "Effect of Lateral Boundaries on Contrast Functions in Time-Resolved Transillumination Measurements," *Med. Phys.*, vol. 26, no. 9, 1999, pp. 1822–1831.

298. V. Chernomordik, A. Gandjbakhche, M. Lepore, R. Esposito, and I. Delfino, "Depth Dependence of the Analytical Expression for the Width of the Point Spread Function (Spatial Resolution) in Time-Resolved Transillumination," *J. Biomed. Opt.*, vol. 6, no. 4, 2001, pp. 441–445.

299. V. Chernomordik, D. W. Hattery, I. Gannot, G. Zaccanti, and A. Gandjbakhche, "Analytical Calculation of the Mean Time Spent by Photons Inside an Absorptive Inclusion Embedded in a Highly Scattering Medium," *J. Biomed. Opt.*, vol. 7, no. 3, 2002, pp. 486–492.

300. V. Chernomordik, D. W. Hattery, D. Grosenick, H. Wabnitz, H. Rinneberg, K. T. Moesta, P. M. Schlag, and A. Gandjbakhche, "Quantification of Optical Properties of a Breast Tumor Using Random Walk Theory, *J. Biomed. Opt.*, vol. 7, no. 1, 2002, pp. 80–87.

301. I. V. Yaroslavsky, A. N. Yaroslavsky, J. Rodriguez, and H. Battarbee, "Propagation of Pulses and Photon-Density Waves in Turbid Media," Chapter 3, in *Handbook of Optical Biomedical Diagnostics*, vol. PM107, V. V. Tuchin (ed.), SPIE Press, Bellingham, WA, 2002, pp. 217–263.

302. J. Rodriguez, I. V. Yaroslavsky, H. Battarbee, and V. V. Tuchin, "Time-Resolved Imaging in Diffusive Media," Chapter 6, in *Handbook of Optical Biomedical Diagnostics*, vol. PM107, V. V. Tuchin (ed.), SPIE Press, Bellingham, WA, 2002, pp. 357–404.

303. S. Fantini and M. A. Franceschini, "Frequency-Domain Techniques for Tissue Spectroscopy and Imaging," Chapter 7, in *Handbook of Optical Biomedical Diagnostics*, vol. PM107, V. V. Tuchin (ed.), SPIE Press, Bellingham, WA, 2002, pp. 405–453.

304. D. J. Papaioannou, G. W. Hooft, S. B. Colak, and J. T. Oostveen, "Detection Limit in Localizing Objects Hidden in a Turbid Medium Using an Optically Scanned Phased Array," *J. Biomed. Opt.*, vol. 1, no. 3, 1996, pp. 305–310.

305. B. W. Pogue and M. S. Patterson, "Error Assessment of a Wavelength Tunable Frequency Domain System for Noninvasive Tissue Spectroscopy," *J. Biomed. Opt.*, vol. 1, no. 3, 1996, pp. 311–323.

306. J. B. Fishkin, O. Coquoz, E. R. Anderson, et al., "Frequency-Domain Photon Migration Measurements of Normal and Malignant Tissue Optical Properties in a Human Subject," *Appl. Opt.*, vol. 36, no. 1, 1997, pp. 10–20.

307. B. J. Tromberg, L. O. Svaasand, T.-T. Tsay, and R. C. Haskell, "Properties of Photon Density Waves in Multiple-Scattering Media," *Appl. Opt.*, vol. 32, 1993, pp. 607–616.

308. B. J. Tromberg, O. Coquoz, J. B. Fishkin, et al., "Non-Invasive Measurements of Breast Tissue Optical Properties using Frequency-Domain Photon Migration," *Phil. Trans. R. Soc. Lond. B.*, vol. 352, 1997, pp. 661–668.

309. D. A. Boas, M. A. O'Leary, B. Chance, and A. G. Yodh, "Scattering and Wavelength Transduction of Diffuse Photon Density Waves," *Phys. Rev. E.*, vol. 47, 1993, pp. R2999–R3002.

310. D. A. Boas, M. A. O'Leary, B. Chance, and A. G. Yodh, "Scattering of Diffuse Photon Density Waves by Spherical Inhomogeneities within Turbid Media: Analytic Solution and Applications," *Proc. Natl. Acad. Sci. USA.*, 91, 1994, pp. 4887–4891.

311. J. R. Lakowicz and K. Berndt, "Frequency-Domain Measurements of Photon Migration in Tissues," *Chem. Phys. Lett.*, vol. 166, 1990, pp. 246–252.

312. M. S. Patterson, J. D. Moulton, B. C. Wilson, et al., "Frequency-Domain Reflectance for the Determination of the Scattering and Absorption Properties of Tissue," *Appl. Opt.*, vol. 30, 1991, pp. 4474–4476.

313. J. M. Schmitt, A. Knüttel, and J. R. Knutson, "Interference of Diffusive Light Waves," *J. Opt. Soc. Am. A.*, vol. 9, 1992, pp. 1832–1843.

314. J. B. Fishkin and E. Gratton, "Propagation of Photon-Density Waves in Strongly Scattering Media Containing an Absorbing Semi-Infinite Plane Bounded by a Strait Edge," *J. Opt. Soc. Am. A.*, vol. 10, 1993, pp. 127–140.

315. L. O. Svaasand, B. J. Tromberg, R. C. Haskell, et al., "Tissue Characterization and Imaging using Photon Density Waves," *Opt. Eng.*, vol. 32, 1993, pp. 258–266.

316. Yu. T. Masurenko, "Spectral-Correlation Method for Imaging of Strongly Scattering Objects," *Opt. Spectrosc.*, vol. 76, 1994, pp. 816–821.

317. H. B. Jiang, K. D. Paulsen, U. L. Osterberg, and M. S. Patterson, "Frequency-Domain Optical-Image Reconstruction in Turbid Media—an Experimental Study of Single-Target Tetectability," *Appl. Opt.*, vol. 36, 1997, pp. 52–63.

318. S. J. Madsen, P. Wyst, L. O. Svaasand, et al., "Determination of the Optical Properties of the Human Uterus using Frequency-Domain Photon Migration and Steady-State Techniques," *Phys. Med. Biol.*, vol. 39, no. 8, 1994, pp. 1191–1202.

319. X. D. Li, T. Durduran, A. G. Yodh, et al., "Diffraction Tomography for Biochemical Imaging with Diffuse Photon-Density Waves," *Opt. Lett.*, vol. 22, 1997, pp. 573–575.

320. C. L. Matson, N. Clark, L. McMackin, and J. S. Fender, "Three-Dimensional Tumor Localization in Thick Tissue with the Use of Diffuse Photon-Density Waves," *Appl. Opt.*, vol. 36, 1997, pp. 214–220.

321. B. Chance, "Optical Method," *Annual Rev. Biophys. Biophys. Chem.*, vol. 20, 1991, pp. 1–28.

322. B. W. Pogue and M. S. Patterson, "Frequency-Domain Optical Absorption Spectroscopy of Finite Tissue Volumes Using Diffusion Theory," *Phys. Med. Biol.*, vol. 39, 1994, pp. 1157–1180.

323. B. W. Pogue, M. S. Patterson, H. Jiang, and K. D. Paulsen, "Initial Assessment of a Simple System for Frequency Domain Diffuse Optical Tomography," *Phys. Med. Biol.*, vol. 40, 1995, pp. 1709–1729.

324. X. Wu, L. Stinger, and G. W. Faris, "Determination of Tissue Properties by Immersion in a Matched Scattering Fluid," *Proc. SPIE* 2979, 1997, pp. 300–306.

325. S. Fantini, M. A. Franceschini, J. B. Fishkin, et al., "Quantitative Determination of the Absorption and Spectra of Chromophores in Strongly Scattering Media: a Light-Emitting-Diode Based Technique," *Appl. Opt.*, vol. 32, 1994, pp. 5204–5212.

326. M. A. Franceschini, K. T. Moesta, and S. Fantini, "Frequency-Domain Techniques Enhance Optical Mammography: Initial Clinical Results," *Proc. Natl. Acad. Sci. USA*, vol. 94, 1997, pp. 6468–6473.

327. I. V. Yaroslavsky, A. N. Yaroslavskaya, V. V. Tuchin, and H.-J. Schwarzmaier, "Effect of the Scattering Delay on Time-Dependent Photon Migration in Turbid Media," *Appl. Opt.*, vol. 36, no. 22, 1997, pp. 6529–6538.

328. W. W. Mantulin, S. Fantini, M. A. Franceschini, S. A. Walker, J. S. Maier, and E. Gratton, "Tissue Optical Parameter Map Generated with Frequency-Domain Spectroscopy," *Proc. SPIE* 2396, 1995, pp. 323–330.

329. H. Wabnitz and H. Rinneberg, "Imaging in Turbid Media by Photon Density Waves: Spatial Resolution and Scaling Relations," *Appl. Opt.*, vol. 36, no. 1, 1997, pp. 67–73.

330. D. A. Boas, M. A. O'Leary, B. Chance, and A. Yodh, "Detection and Characterization of Optical Inhomogeneities with Diffuse Photon Density Waves: a Signal-to-Noise Analysis," *Appl. Opt.*, vol. 36, 1997, pp. 75–92.

331. A. Knüttel, J. M. Schmitt, and J. R. Knutson, "Spatial Localization of Absorbing Bodies by Interfering Diffusive Photon-Density Waves," *Appl. Opt.*, vol. 32, 1933, pp. 381–389.

332. S. Fantini, M. A. Franceschini, and E. Gratton,"Semi-Infinite-Geometry Boundary Problem for Light Migration in Highly Scattering Media: a Frequency-Domain Study in the Diffusion Approximation," *J. Opt. Soc. Am. B.*, vol. 11, 1994, pp. 2128–2138.

333. J. A. Moon and J. Reintjes, "Image Resolution by Use of Multiply Scattered Light," *Opt. Lett.*, vol. 19, 1994, pp. 521–523.

334. A. Weersink, J. E. Hayward, K. R. Diamond, and M. S. Patterson, "Accuracy of Noninvasive *In Vivo* Measurements of Photosensitizer Uptake Based on a Diffusion Model of Reflectance Spectroscopy," *Photochem. Photobiology*, vol. 66, no. 3, 1997, pp. 326–335.

335. V. V. Tuchin, "Fundamentals of Low-Intensity Laser Radiation Interaction with Biotissues: Dosimetry and Diagnostical Aspects," *Bullet. Russian Acad. Sci., Phys. ser.*, vol. 59, no. 6, 1995, pp. 120–143.

336. Y. Chen, C. Mu, X. Intes, and B. Chance, "Signal-to-Noise Analysis for Detection Sensitivity of Small Absorbing Heterogeneity in Turbid Media with Single-Source and Dual-Interfering-Source," *Optics Express*, vol. 9, no. 4, 2001, pp. 212–224.

337. T. Durduran, J. P. Culver, M. J. Holboke, X. D. Li, L. Zubkov, B. Chance, D. N. Pattanayak, and A. G. Yodh, "Algorithms for 3D Localization and

Imaging Using Near-Field Diffraction Tomography with Diffuse Light," *Optics Express*, vol. 4, no. 8, 1999, pp. 247–262.

338. S. J. Matcher, "Signal Quantification and Localization in Tissue Near-Infrared Spectroscopy," Chapter 9, in *Handbook of Optical Biomedical Diagnostics*, vol. PM107, V. V. Tuchin (ed.), SPIE Press, Bellingham, WA, 2002, pp. 487–584.

339. J. S. Maier, S. A. Walker, S. Fantini, M. A. Franceschini, and E. Gratton, "Possible Correlation between Blood Glucose Concentration and the Reduced Scattering Coefficient of Tissues in the Near Infrared," *Opt. Lett.*, vol. 19, 1994, pp. 2062–2064.

340. M. Kohl, M. Cope, M. Essenpreis, and D. Böcker, "Influence of Glucose Concentration on Light Scattering in Tissue-Simulating Phantoms," *Opt. Lett.*, vol. 19, 1994, pp. 2170–2172.

341. J. T. Bruulsema, J. E. Hayward, T. J. Farrell, M. S. Patterson, L. Heinemann, M. Berger, T. Koschinsky, J. Sandahal-Christiansen, H. Orskov, M. Essenpreis, G. Schmelzeisen-Redeker, and D. Böcker, "Correlation between Blood Glucose Concentration in Diabetics and Noninvasively Measured Tissue Optical Scattering Coefficient," *Opt. Lett.*, vol. 22, no. 3, 1997, pp. 190–192.

342. M. G. Erickson, J. S. Reynolds, and K. J. Webb, "Comparison of Sensitivity for Single-Source and Dual-Interfering-Source Configurations in Optical Diffusion Imaging," *J. Opt. Soc. Am. A*, vol. 14, no. 11, 1997, pp. 3083–3092.

343. V. V. Tuchin, "Coherent Optical Techniques for the Analysis of Tissue Structure and Dynamics," *J. Biomed. Opt.*, vol. 4, no. 1, 1999, pp. 106–124.

344. P. Bruscaglioni, G. Zaccanti, and Q. Wei, "Transmission of a Pulsed Polarized Light Beam through Thick Turbid Media: Numerical Results,"*Appl. Opt.*, vol. 32, 1993, pp. 6142–6150.

345. D. Bicout, C. Brosseau, A. S. Martinez, and J. M. Schmitt, "Depolarization of Multiply Scattering Waves by Spherical Diffusers: Influence of the Size Parameter," *Phys. Rev. E.*, vol. 49, 1994, pp. 1767–1770.

346. M. Dogariu and T. Asakura, "Photon Pathlength Distribution from Polarized Backscattering in Random Media," *Opt. Eng.*, vol. 35, 1996, pp. 2234–2239.

347. A. Dogariu, C. Kutsche, P. Likamwa, G. Boreman, and B. Moudgil, "Time-Domain Depolarization of Waves Retroreflected from Dense Colloidal Media," *Opt. Lett.*, vol. 22, 1997, pp. 585–587.

348. A. H. Hielsher, J. R. Mourant, and I. J. Bigio, "Influence of Particle Size and Concentration on the Diffuse Backscattering of Polarized Light from Tissue Phantoms and Biological Cell Suspensions," *Appl. Opt.*, vol. 36, 1997, pp. 125–135.

349. A. Ambirajan and D. C. Look, "A Backward Monte Carlo Study of the Multiple Scattering of a Polarized Laser Beam," *J. Quant. Spectrosc. Radiat. Transfer*, vol. 58, 1997, pp. 171–192.

350. M. J. Rakovic and G. W. Kattawar, "Theoretical Analysis of Polarization Patterns from Incoherent Backscattering of Light," *Appl. Opt.*, vol. 37, no. 15, 1998, pp. 3333–3338.

351. M. J. Racovic, G. W. Kattavar, M. Mehrubeoglu, B. D. Cameron, L. V. Wang, S. Rasteger, and G. L. Cote, "Light Backscattering Polarization Patterns from Turbid Media: Theory and Experiment," *Appl. Opt.*, vol. 38, 1999, pp. 3399–3408.

352. G. Yao and L. V. Wang, "Propagation of Polarized Light in Turbid Media: Simulated Animation Sequences," *Optics Express*, vol. 7, no. 5, 2000, pp. 198–203.

353. D. A. Zimnyakov, Yu. P. Sinichkin, P. V. Zakharov, and D. N. Agafonov, "Residual Polarization of Non-Coherently Backscattered Linearly Polarized Light: the Influence of the Anisotropy Parameter of the Scattering Medium," *Waves in Random Media*, vol. 11, 2001, pp. 395–412.

354. D. A. Zimnyakov and Yu. P. Sinichkin, "Ultimate Degree of Residual Polarization of Incoherently Backscattered Light of Multiple Scattering of Linearly Polarized Light," *Opt. Spectrosc.*, vol. 91, 2001, pp. 103–108.

355. D. A. Zimnyakov, Yu. P. Sinichkin, I. V. Kiseleva and D. N. Agafonov, "Effect of Absorption of Multiply Scattering Media on the Degree of Residual Polarization of Backscattered Light," *Opt. Spectrosc.*, vol. 92, 2002, pp. 765–771.

356. I. A. Vitkin and R. C. N. Studinski, "Polarization Preservation in Diffusive Scattering from *In Vivo* Turbid Biological Media: Effects of Tissue Optical Absorption in the Exact Backscattering Direction," *Optics Communs*, vol. 190, 2001, pp. 37–43.

357. G. Bal and M. Moscoso, "Theoretical and Numerical Analysis of Polarization for Time-Dependent Radiative Transfer Equations," *J. Quant. Spectrosc. & Radiat. Transf.*, vol. 70, 2001, pp. 75–98.

358. A. A. Kokhanovsky, "Photon Transport in Asymmetric Random Media," *J. Opt. A: Pure Appl. Opt.*, vol. 4, 2002, pp. 521–526.

359. A. A. Kokhanovsky, "Reflection and Polarization of Light by Semi-Infinite Turbid Media: Simple Approximations," *J. Colloid & Interface Sci.*, vol. 251, 2002, pp. 429–433.

360. L. Dagdug, G. H. Weiss, and A. H. Gandjbakhche, "Effects of Anisotropic Optical Properties on Photon Migration in Structured Tissues," *Phys. Med. Biol.*, vol. 48, 2003, pp. 1361–1370.

361. H. H. Tynes, G. W. Kattawar, E. P. Zege, I. L. Katsev, A. S. Prikhach, and L. I. Chaikovskaya, "Monte Carlo and Multicomponent Approximation Methods for Vector Radiative Transfer by Use of Effective Mueller Matrix Calculations," *Appl. Opt.*, vol. 40, no. 3, 2001, pp. 400–412.

362. I. M. Stockford, S. P. Morgan, P. C. Y. Chang, and J. G. Walker, "Analysis of the Spatial Distribution of Polarized Light Backscattering," *J. Biomed. Opt.*, vol. 7, no. 3, 2002, pp. 313–320.

363. S. Bartel and A. H. Hielscher, "Monte Carlo Simulations of the Diffuse Backscattering Mueller Matrix for Highly Scattering Media," *Appl. Opt.*, vol. 39, 2000, pp. 1580–1588.

364. X. Wang and L. V. Wang, "Propagation of Polarized Light in Birefringent Turbid Media: Time-Resolved Simulations," *Optics Express*, vol. 9, no. 5, 2001, pp. 254–259.

365. K. Y. Yong, S. P. Morgan, I. M. Stockford, and M. C. Pitter, "Characterization of Layered Scattering Media Using Polarized Light Measurements and Neural Networks," *J. Biomed. Opt.*, vol. 8, no. 3, 2003, pp. 504–511.

366. I. L. Maksimova, S. V. Romanov, and V. F. Izotova, "The Effect of Multiple Scattering in Disperse Media on Polarization Characteristics of Scattered Light," *Opt. Spectrosc.*, vol. 92, no. 6, 2002, pp. 915–923.

367. X. Wang and L. V. Wang, "Propagation of Polarized Light in Birefringent Turbid Media: A Monte Carlo Study," *J. Biomed. Opt.*, vol. 7, no. 3, 2002, pp. 279–290.

368. S. V. Gangnus, S. J. Matcher, and I. V. Meglinski, "Monte Carlo Modeling of Polarized Light Propagation in Biological Tissues," *Laser Phys.*, vol. 14, 2004, pp. 886–891.

369. C. J. Hourdakis and A. Perris, "A Monte Carlo Estimation of Tissue Optical Properties for the Use in Laser Dosimetry," *Phys. Med. Biol.*, vol. 40, 1995, pp. 351–364.

370. A. Yodh, B. Tromberg, E. Sevick-Muraca, and D. Pine (eds.), "Special Section on Diffusing Photons in Turbid Media," *J. Opt. Soc. Am. A.*, vol. 14, 1997, pp. 136–342.

371. L. O. Svaasand and Ch. J. Gomer, "Optics of Tissue," in *Dosimetry of Laser Radiation in Medicine and Biology*, SPIE Press, Bellingham, vol. IS5, 1989, pp. 114–132.

372. H. Horinaka, K. Hashimoto, K. Wada, and Y. Cho, "Extraction of Quasi-Straightforward–Propagating Photons from Diffused Light Transmitting through a Scattering Medium by Polarization Modulation," *Opt. Lett.*, vol. 20, 1995, pp. 1501–1503.

373. S. P. Morgan, M. P. Khong, and M. G. Somekh, "Effects of Polarization State and Scatterer Concentration Optical Imaging through Scattering Media," *Appl. Opt.*, vol. 36, 1997, pp. 1560–1565.

374. A. B. Pravdin, S. P. Chernova, and V. V. Tuchin, "Polarized Collimated Tomography for Biomedical Diagnostics," *Proc. SPIE* 2981, 1997, pp. 230–234.

375. M. R. Ostermeyer, D. V. Stephens, L. Wang, and S. L. Jacques, "Nearfield Polarization Effects on Light Propagation in Random Media," OSA TOPS 3, Optical Society of America, Washington, DC, 1996, pp. 20–25.

376. R. R. Anderson, "Polarized Light Examination and Photography of the Skin," *Arch. Dermatol.*, vol. 127, 1991, pp. 1000–1005.

377. A. W. Dreher and K. Reiter, "Polarization Technique Measures Retinal Nerve Fibers," *Clin. Vis. Sci.*, vol. 7, 1992, pp. 481–485.

378. N. Kollias, "Polarized Light Photography of Human Skin," in *Bioengineering of the Skin: Skin Surface Imaging and Analysis*, K.-P. Wilhelm, P. Elsner, E. Berardesca, and H. I. Maibach (eds.), CRC Press, Boca Raton, 1997, pp. 95–106.

379. S. G. Demos and R. R. Alfano, "Optical Polarization Imaging," *Appl. Opt.*, vol. 36, 1997, pp. 150–155.

380. G. Yao and L. V. Wang, "Two-Dimensional Depth-Resolved Mueller Matrix Characterization of Biological Tissue by Optical Coherence Tomography," *Opt. Lett.*, vol. 24, 1999, pp. 537–539.

381. B. D. Cameron, M. J. Racovic, M. Mehrubeoglu, G. Kattavar, S. Rasteger, L. V. Wang, and G. Cote, "Measurement and Calculation of the Two-Dimensional Backscattering Mueller Matrix of a Turbid Medium," *Opt. Lett.*, vol. 23, 1998, pp. 485–487; Errata, *Opt. Lett.*, vol. 23, 1998, p. 1630.

382. S. L. Jacques, R. J. Roman, and K. Lee, "Imaging Superficial Tissues with Polarized Light," *Lasers Surg. Med.*, vol. 26, 2000, pp. 119–129.

383. S. L. Jacques, J. C. Ramella-Roman, and K. Lee, "Imaging Skin Pathology with Polarized Light," *J. Biomed. Opt.*, vol. 7, no. 3, 2002, pp. 329–340.

384. M. Moscoso, J. B. Keller, and G. Papanicolaou, "Depolarization and Blurring of Optical Images by Biological Tissue," *J. Opt. Soc. Am.*, vol. 18, no. 4, 2001, pp. 948–960.

385. M. H. Smith, "Optimizing a Dual-Rotating-Retarder Mueller Matrix Polarimeter," *Proc. SPIE* 4481, 2001.

386. M. H. Smith, "Interpreting Mueller Matrix Images of Tissues," *Proc. SPIE* 4257, 2001, pp. 82–89.

387. L. L. Deibler and M. H. Smith, "Measurement of the Complex Refractive Index of Isotropic Materials with Mueller Matrix Polarimetry," *Appl. Opt.*, vol. 40, no. 22, 2001, pp. 3659–3667.

388. X.-R. Huang and R. W. Knighton, "Linear Birefringence of the Retinal Nerve Fiber Layer Measured *In Vitro* with a Multispectral Imaging Micropolarimeter, *J. Biomed. Opt.*, vol. 7, no. 2, 2002, pp. 199–204.

389. A. P. Sviridov, D. A. Zimnyakov, Yu. P. Sinichkin, L. N. Butvina, A. I. Omel'chenko, G. Sh. Makhmutova, and V. N. Bagratashvili, "IR Fourier Spectroscopy of *In-Vivo* Human Skin and Polarization of Backscattered Light in the Case of Skin Ablation by AIG: Nd laser radiation," *J. Appl. Spectrosc.*, vol. 69, 2002, pp. 484–488.

390. A. N. Yaroslavsky, V. Neel, and R. R. Anderson, "Demarcation of Non-melanoma Skin Cancer Margins in Thick Excisions Using Multispectral Polarized Light Imaging," *J. Invest. Dermatol.*, vol. 121, 2003, pp. 259–266.

391. E. E. Gorodnichev, A. I. Kuzovlev, and D. B. Rogozkin, "Depolarization of Light in Small-Angle Multiple Scattering in Random Media," *Laser Physics*, vol. 9, 1999, pp. 1210–1227.

392. I. Freund, M. Kaveh, R. Berkovits, and M. Rosenbluh, "Universal Polarization Correlations and Microstatistics of Optical Waves in Random Media," *Phys. Rev. B.*, vol. 42, no. 4, 1990, pp. 2613–2616.

393. D. Eliyahu, M. Rosenbluh, and I. Freund, "Angular Intensity and Polarization Dependence of Diffuse Transmission through Random Media," *J. Opt. Soc. Am. A.*, vol. 10, no. 3, 1993, pp. 477–491.

394. G. Jarry, E. Steiner, V. Damaschini, M. Epifanie, M. Jurczak, and R. Kaizer, "Coherence and Polarization of Light Propagating through Scattering Media and Biological Tissues," *Appl. Opt.*, vol. 37, 1998, pp. 7357–7367.

395. D. A. Zimnyakov and V. V. Tuchin, "About Interrelations of Distinctive Scales of Depolarization and Decorrelation of Optical Fields in Multiple Scattering," *JETP Lett.*, vol. 67, 1998, pp. 455–460.

396. D. A. Zimnyakov, V. V. Tuchin, and A. G. Yodh, "Characteristic Scales of Optical Field Depolarization and Decorrelation for Multiple Scattering Media and Tissues," *J. Biomed. Opt.*, vol. 4, 1999, pp. 157–163.

397. F. Bettelheim, "On the Optical Anisotropy of Lens Fibre Cells," *Exp. Eye Res.*, vol. 21, 1975, pp. 231–234.

398. J. Y. T. Wang and F. A. Bettelheim, "Comparative Birefringence of Cornea," *Comp. Bichem. Physiol., Part A: Mol. Integr. Physiol.*, vol. 51, 1975, pp. 89–94.

399. R. P. Hemenger, "Birefringence of a Medium of Tenuous Parallel Cylinders," *Appl. Opt.*, vol. 28, no. 18, 1989, pp. 4030–4034.

400. D. J. Maitland and J. T. Walsh, "Quantitative Measurements of Linear Birefringence during Heating of Native Collagen," *Laser Surg. Med.*, vol. 20, 1997, pp. 310–318.

401. H. B. Klein Brink, "Birefringence of the Human Crystalline Lens *in vivo*," *J. Opt. Soc. Am. A*, vol. 8, 1991, pp. 1788–1793.

402. R. P. Hemenger, "Refractive Index Changes in the Ocular Lens Result from Increased Light Scatter," *J. Biomed. Opt.*, vol. 1, 1996, pp. 268–272.

403. V. F. Izotova, I. L. Maksimova, I. S. Nefedov, and S. V. Romanov, "Investigation of Mueller matrices of Anisotropic Nonhomogeneous Layers in Application to Optical Model of Cornea," *Appl. Opt.*, vol. 36, no. 1, 1997, pp. 164–169.

404. J. S. Baba, B. D. Cameron, S. Theru, and G. L. Coté, "Effect of Temperature, *p*H, and Corneal Birefringence on Polarimetric Glucose Monitoring in the Eye," *J. Biomed. Opt.*, vol. 7, no. 3, 2002, pp. 321–328.

405. G. J. van Blokland, "Ellipsometry of the Human Retina *In Vivo*: Preservation of Polarization," *J. Opt. Soc. Am. A*, vol. 2, 1985, pp. 72–75.

406. H. B. Klein Brink and G. J. van Blokland, "Birefringence of the Human Foveal Area Assessed *In Vivo* with Mueller-Matrix Ellipsometry," *J. Opt. Soc. Am. A*, vol. 5, 1988, pp. 49–57.

407. R. C. Haskell, F. D. Carlson, and P. S. Blank, "Form Birefringence of Muscle," *Biophys. J.*, vol. 56, 1989, pp. 401–413.

408. S. Bosman, "Heat-Induced Structural Alterations in Myocardium in Relation to Changing Optical Properties," *Appl. Opt.*, vol. 32, no. 4, 1993, pp. 461–463.

409. G. V. Simonenko, T. P. Denisova, N. A. Lakodina, and V. V. Tuchin, "Measurement of an Optical Anisotropy of Biotissues" *Proc. SPIE* 3915, 2000, pp. 152–157.

410. G. V. Simonenko, V. V. Tuchin, and N. A. Lakodina, "Measurement of the Optical Anisotropy of Biological Tissues with the Use of a Nematic Liquid Crystal Cell," *J. Opt. Technol.*, vol. 67, no. 6, 2000, pp. 559–562.

411. O. V. Angel'skii, A. G. Ushenko, A. D. Arkhelyuk, S. B. Ermolenko, and D. N. Burkovets, "Scattering of Laser Radiation by Multifractal Biological Structures," *Opt. Spectrosc.*, vol. 88, no. 3, 2000, pp. 444–447.

412. M. R. Hee, D. Huang, E. A. Swanson, and J. G. Fujimoto, "Polarization-Sensitive Low-Coherence Reflectometer for Birefringence Characterization and Ranging," *J. Opt. Soc. Am. B*, vol. 9, 1992, pp. 903–908.

413. J. F. de Boer, T. E. Milner, M. J. C. van Gemert, and J. S. Nelson, "Two-Dimensional Birefringence Imaging in Biological Tissue by Polarization-Sensitive Optical Coherence Tomography," *Opt. Lett.*, vol. 22, no. 12, 1997, pp. 934–936.

414. M. J. Everett, K. Schoenerberger, B. W. Colston, Jr., and L. B. Da Silva, "Birefringence Characterization of Biological Tissue by Use of Optical Coherence Tomography," *Opt. Lett.*, vol. 23, no. 3, 1998, pp. 228–230.

415. J. F. de Boer, T. E. Milner, and J. S. Nelson, "Determination of the Depth Resolved Stokes Parameters of Light Backscattered from Turbid Media Using Polarization Sensitive Optical Coherence Tomography," *Opt. Lett.*, vol. 24, 1999, pp. 300–302.

416. J. F. de Boer and T. E. Milner, "Review of Polarization Sensitive Optical Coherence Tomography and Stokes Vector Determination," *J. Biomed. Opt.*, vol. 7, no. 3, 2002, pp. 359–371.

417. C. K. Hitzenberger, E. Gotzinger, M. Sticker, M. Pircher, and A. F. Fercher, "Measurement and Imaging of Birefringence and Optic Axis Orientation by Phase Resolved Polarization Sensitive Optical Coherence Tomography," *Opt. Express*, vol. 9, 2001, pp. 780–790.

418. S. Jiao and L. V. Wang, "Jones-Matrix Imaging of Biological Tissues with Quadruple-Channel Optical Coherence Tomograthy," *J. Biomed. Opt.*, vol. 7, no. 3, 2002, pp. 350–358.

419. M. G. Ducros, J. F. de Boer, H. Huang, L. Chao, Z. Chen, J. S. Nelson, T. E. Milner, and H. G. Rylander, "Polarization Sensitive Optical Coherence Tomography of the Rabbit Eye," *IEEE J. Sel. Top. Quantum Electron.*, vol. 5, 1999, pp. 1159–1167.

420. M. G. Ducros, J. D. Marsack, H. G. Rylander III, S. L. Thomsen, and T. E. Milner, "Primate Retina Imaging with Polarization-Sensitive Optical Coherence Tomography," *J. Opt. Soc. Am. A*, vol. 18, 2001, pp. 2945–2956.

421. C. E. Saxer, J. F. de Boer, B. H. Park, Y. Zhao, C. Chen, and J. S. Nelson, "High Speed Fiber Based Polarization Sensitive Optical Coherence Tomography of *In Vivo* Human Skin," *Opt. Lett.*, vol. 26, 2001, pp. 1069–1071.

422. B. H. Park C. E. Saxer, S. M. Srinivas, J. S. Nelson, and J. F. de Boer, "*In Vivo* Burn Depth Determination by High-Speed Fiber-Based Polarization Sensitive Optical Coherence Tomography," *J. Biomed. Opt.*, vol. 6, 2001, pp. 474–479.

423. X. J. Wang, T. E. Milner, J. F. de Boer, Y. Zhang, D. H. Pashley, and J. S. Nelson, "Characterization of Dentin and Enamel by Use of Optical Coherence Tomography," *Appl. Opt.*, vol. 38, 1999, pp. 2092–2096.

424. A. Baumgartner, S. Dichtl, C. K. Hitzenberger, H. Sattmann, B. Robl, A. Moritz, A. F. Fercher, and W. Sperr, "Polarization-Sensitive Optical Coherence Tomography of Dental Structures," *Caries Res.*, vol. 34, no. 1, 2000, pp. 59–69.

425. A. Kienle, F. K. Forster, R. Diebolder, and R. Hibst, "Light Propagation in Dentin: Influence of Microstructure on Anisotropy," *Phys. Med. Biol.*, vol. 48, 2003, N7–N14.

426. D. Fried, J. D. B. Featherstone, R. E. Glena, and W. Seka, "The Nature of Light Scattering in Dental Enamel and Dentin at Visible and Near-Infrared Wavelengths," *Appl. Opt.*, vol. 34, no. 7, 1995, pp. 1278–1285.

427. G. B. Altshuler, "Optical Model of the Tissues of the Human Tooth," *J. Opt. Technol.*, vol. 62, 1995, pp. 516–520.

428. R. C. N. Studinski and I. A. Vitkin, "Methodology for Examining Polarized Light Interactions with Tissues and Tissuelike Media in the Exact Backscattering Direction," *J. Biomed. Opt.*, vol. 5, no. 3, 2000, pp. 330–337.

429. K. C. Hadley and I. A. Vitkin, "Optical Rotation and Linear and Circular Depolarization Rates in Diffusively Scattered Light from Chiral, Racemic, and Achiral Turbid Media," *J. Biomed. Opt.*, vol. 7, no. 3, 2002, pp. 291–299.

430. J. Applequist, "Optical Activity: Biot's Bequest," *Am. Sci.*, vol. 75, 1987, pp. 59–67.

431. J. D. Bancroft and A. Stevens (eds.), *Theory and Practice of Histological Techniques*, Churchill Livingstone, Edinburgh, New York, 1990.

432. D. M. Maurice, *The Cornea and Sclera. The Eye*, H. Davson (ed.), Academic Press, Orlando, 1984, pp. 1–158.

433. R. W. Hart and R. A. Farrell, "Light Scattering in the Cornea," *J. Opt. Soc. Am.*, vol. 59, no. 6, 1969, pp. 766–774.

434. L. D. Barron, *Molecular Light Scattering and Optical Activity*, Cambridge Univ., London, 1982.

435. R. L. McCally and R. A. Farrell, "Light Scattering from Cornea and Corneal Transparency," in *Noninvasive Diagnostic Techniques in Ophthalmology*, B. R. Master (ed.), Springer-Verlag, New York, 1990, pp. 189–210.

436. I. L. Maksimova and L. P. Shubochkin, "Light-Scattering Matrices for a Close-Packed Binary System of Hard Spheres," *Opt. Spectrosc.*, vol. 70, no. 6, 1991, pp. 745–748.

437. V. Shankaran, M. J. Everett, D. J. Maitland, and J. T. Walsh, Jr., "Polarized Light Propagation through Tissue Phantoms Containing Densely Packed Scatterers," *Opt. Lett.*, vol. 25, no. 4, 2000, pp. 239–241.

438. V. Shankaran, J. T. Walsh, Jr., and D. J. Maitland, "Comparative Study of Polarized Light Propagation in Biological Tissues," *J. Biomed. Opt.*, vol. 7, no. 3, 2002, pp. 300–306.

439. V. V. Tuchin, "Optics of the Human Sclera: Photon Migration, Imaging and Spectroscopy," OSA TOPS 21, Optical Society of America, Washington, DC, 1998, pp. 99–104.

440. I. L. Maksimova, "Scattering of Radiation by Regular and Random Systems Comprised of Parallel Long Cylindrical Rods," *Opt. Spectrosc.*, vol. 93, no. 4, 2002, pp. 610–619.

441. A. G. Ushenko and V. P. Pishak, "Laser Polarimetry of Biological Tissues: Principles and Applications" in *Coherent-Domain Optical Methods: Biomedical Diagnostics, Environmental and Material Science*, vol. 1, V. V. Tuchin (ed.), Kluwer Academic Publishers, Boston, 2004, pp. 94–138.

442. N. G. Khlebtsov, I. L. Maksimova, V. V. Tuchin, and L. Wang, "Introduction to Light Scattering by Biological Objects," Chapter 1 in *Handbook of Optical Biomedical Diagnostics*, vol. PM107, V. V. Tuchin (ed.), SPIE Press, Bellingham, WA, 2002, pp. 31–167.

443. W. A. Shurcliff, *Polarized Light. Production and Use*, Harvard Univ., Cambridge, Mass., 1962.

444. W. A. Shurcliff and S. S. Ballard, *Polarized Light*, Van Nostrand, Princeton, 1964.

445. E. L. O'Neill, *Introduction to Statistical Optics*, Addison-Wesley, Reading, Mass., 1963.

446. D. S. Kliger, J. W. Lewis, and C. E. Randall, *Polarized Light in Optics and Spectroscopy*, Academic, Boston, 1990.

447. E. Collet, *Polarized Light. Fundamentals and Applications*, Dekker, New York, 1993.

448. R. M. A. Azzam and N. M. Bashara, *Ellipsometry and Polarized Light*, Elsevier Science, Amsterdam, 1994.

449. C. Brosseau, *Fundamentals of Polarized Light: A Statistical Optics Approach*, Wiley, New York, 1998.

450. I. L. Maksimova, S. N. Tatarintsev, and L. P. Shubochkin, "Multiple Scattering Effects in Laser Diagnostics of Bioobjects," *Opt. Spectrosc.*, vol. 72, 1992, pp. 1171–1177.

451. V. F. Izotova, I. L. Maksimova, and S. V. Romanov, "Utilization of Relations Between Elements of the Mueller Matrices for Estimating Properties of Objects and the Reliability of Experiments," *Opt. Spectrosc.*, vol. 80, no. 5, 1996, pp. 753–759.

452. V. V. Tuchin, "Biomedical Spectroscopy," in *Encyclopedia of Optical Engineering*, R. G. Driggers (ed.), Marcel-Dekker, New York, 2003, pp. 166–182; www.dekker.com/servlet/product/DOI/101081EEOE120009763

453. V. V. Tuchin, "Light-Tissue Interactions" in *Biomedical Photonics Handbook*, Tuan Vo-Dinh (ed.), CRC Press, Boca Raton, 2003, pp. 3-1–3-26.

454. D. Fried, "Optical Methods for Caries Detection, Diagnosis, and Therapeutic Intervention," in *Biomedical Photonics Handbook*, Tuan Vo-Dinh (ed.), CRC Press, Boca Raton, 2003, pp. 50-1–50-27.

455. S. E. Braslavsky and K. Heihoff, "Photothermal methods" in *Handbook of Organic Photochemistry*, J. C. Scaiano (ed.), CRC Press, Boca Raton, 1989.

456. V. E. Gusev and A. A. Karabutov, *Laser Optoacoustics*, AIP Press, New York, 1993.

457. A. Mandelis and K. H. Michaelian (eds.), "Special Section on Photoacoustic and Photothermal Science and Engineering," *Opt. Eng.*, vol. 36, no. 2, 1997, pp. 301–534.

458. A. A. Karabutov and A. A. Oraevsky, "Time-Resolved Detection of Optoacoustic Profiles for Measurement of Optical Energy Distribution in Tissues," in *Handbook of Optical Biomedical Diagnostics*, Chapter 10, vol. PM107, V. V. Tuchin (ed.), SPIE Press, Bellingham, WA, 2002, pp. 585–674.

459. A. A. Oraevsky and A. A. Karabutov, "Optoacoustic Tomography," in *Biomedical Photonics Handbook*, Chapter 34, Tuan Vo-Dinh (ed.), CRC Press, Boca Raton, 2003, pp. 34-1–34.

460. S. Nagai and M. Izuchi, "Quantitative Photoacoustic Imaging of Biological Tissues," *Jap. J. Appl. Phys.*, vol. 27, no. 3, 1988, pp. L423–L425.

461. A. M. Ashurov, U. Madvaliev, V. V. Proklov, et al., "Photoacoustic Scanning Microscope," *Sci. Res. Instr.*, no. 2, 1988, pp. 154–157.

462. M. G. Sowa and H. H. Mantsch, "FT-IR Step-Scan Photoacoustic Phase Analysis and Depth Profiling of Calcified Tissue," *Appl. Spectr.*, vol. 48, no. 3, 1994, pp. 316–319.

463. R. A. Kruger and P. Liu, "Photoacoustic Ultrasound: Pulse Production and Detection in 0.5% Liposyn," *Med. Phys.*, vol. 21, no. 7, 1994, pp. 1179–1184.

464. R. A. Kruger, L. Pingyu, Y. Fang, and C. R. Appledorn, "Photoacoustic Ultrasoud-Reconstruction Tomography," *Med. Phys.*, vol. 22, no. 10, 1995, pp. 1605–1609.

465. A. A. Oraevsky, "Laser Optoacoustic Imaging for Diagnosis of Cancer," *IEEE/LEOS Newsletter*, vol. 10, no. 12, 1996, pp. 17–20.

466. A. A. Karabutov, N. B. Podymova, and V. S. Letokhov, "Time-Resolved Laser Optoacoustic Tomography of Inhomogeneous Media," *Appl. Phys. B.*, vol. 63, 1996, pp. 545–563.

467. A. A. Oraevsky, S. J. Jacques, and F. K. Tittel, "Measurement of Tissue Optical Properties by Time-Resolved Detection of Laser-Induced Transient Stress," *Appl. Opt.*, vol. 36, no. 1, 1997, pp. 402–415.

468. R. O. Esenaliev, K. V. Larin, I. V. Larina, M. Motamedi, and A. A. Oraevsky, "Optical Properties of Normal and Coagulated Tissues: Measurements Using Combination of Optoacoustic and Diffuse Reflectance Techniques," *Proc. SPIE* 3726, 1999, pp. 560–566.

469. A. A. Karabutov, E. V. Savateeva, N. B. Podymova, and A. A. Oraevsky, "Backward Mode Detection of Laser-Induced Wide-Band Ultrasonic Tran-

sients with Optoacoustic Transducer," *J. Appl. Phys.*, vol. 87, no. 4, 2000, pp. 2003–2014.

470. V. G. Andreev, A. A. Karabutov, and A. A. Oraevsky, "Detection of Ultrawide-Band Ultrasound Pulses in Optoacoustic Tomography," *IEEE Trans. Ultrason. Ferroelectr. Freq. Control*, vol. 50, no. 10, 2003, pp. 1383–1390.

471. G. Paltauf and H. Schmidt-Kloiber, "Pulsed Optoacoustic Characterization of Layered Media," *J. Appl. Phys.*, vol. 88, no. 3, 2000, pp. 1624–1631.

472. K. P. Köstli, M. Frenz, H. P. Weber, G. Paltauf, and H. Schmidt-Kloiber, "Optoacoustic Infrared Spectroscopy of Soft Tissue," *J. Appl. Phys.*, vol. 88, no. 3, 2000, pp. 1632–1637.

473. K. P. Köstli, M. Frenz, H. P. Weber, G. Paltauf, and H. Schmidt-Kloiber, "Optoacoustic Tomography: Time-Gated Measurement of Pressure Distributions and Image Reconstruction," *Appl. Opt.*, vol. 40, no. 22, 2001, pp. 3800–3809.

474. J. A. Viator, G. Au, G. Paltauf, S. L. Jacques, S. A. Prahl, H. Ren, Z. Chen, and J. S. Nelson, "Clinical Testing of a Photoacoustic Probe for Port Wine Stain Depth Determination," *Lasers Surg. Med.*, vol. 30, 2002, pp. 141–148.

475. R. A. Kruger, W. L. Kiser, D. R. Reinecke, and G. A. Kruger, "Thermoacoustic Computed Tomography Using a Conventional Linear Transducer Array," *Med. Phys.*, vol. 30, 2003, pp. 856–860.

476. R. A. Kruger, W. L. Kiser, D. R. Reinecke, G. A. Kruger, and K. D. Miller, "Thermoacoustic Optical Molecular Imaging of Small Animals," *Molecular Imag.*, vol. 2, 2003, pp. 113–123.

477. C. G. A. Hoelen, F. F. M. de Mul, R. Pongers, and A. Dekker, "Three-Dimensional Photoacoustic Imaging of Blood Vessels in Tissue," *Opt. Lett.*, vol. 23, 1998, pp. 648–650.

478. C. G. A. Hoelen and F. F. M. de Mul, "Image Reconstruction for Photoacoustic Scanning of Tissue Structures," *Appl. Opt.*, vol. 39, no. 31, 2000, pp. 5872–5883.

479. C. G. A. Hoelen, A. Dekker, and F. F. M. de Mul, "Detection of Photoacoustic Transients Originating from Microstructures in Optically Diffuse Media such as Biological Tissue," *IEEE Trans. Ultrason. Ferroelectr. Freq. Control*, vol. 48, no. 1, 2001, pp. 37–47.

480. M. C. Pilatou, N. J. Voogd, F. F. M. de Mul, L. N. A. van Adrichem, and W. Steenbergen, "Analysis of Three-Dimensional Photoacoustic Imaging of a Vascular Tree *In Vitro*," *Rev. Sci. Instrum.*, vol. 74, no. 10, 2003, pp. 4495–4499.

481. R. G. M. Kolkman, E. Hondebrink, W. Steenbergen, T. G. van Leeuwen, and F. F. M. de Mul, "Photoacoustic Imaging of Blood Vessels with a Double-Ring Sensor Featuring a Narrow Angular Aperture," *J. Biomed. Opt.*, vol. 9, no. 6, 2004, pp. 1327–1335.

482. G. Ku and L.-H. V. Wang, "Scanning Electromagnetic-Induced Thermoacoustic Tomography: Signal, Resolution, and Contrast," *Med. Phys.*, vol. 28, 2001, pp. 4–10.

483. X. D. Wang, Y. J. Pang, G. Ku, X. Y. Xie, G. Stoica, and L. V. Wang, "Non-invasive Laser-Induced Photoacoustic Tomography for Structural and Functional *In Vivo* Imaging of the Brain," *Nat. Biotechnol.*, vol. 21, 2003, pp. 803–806.

484. X. D. Wang, Y. J. Pang, G. Ku, G. Stoica, and L. V. Wang, "Three-Dimensional Laser-Induced Photoacoustic Tomography of Mouse Brain with the Skin and Skull Intact," *Opt. Lett.*, vol. 28, no. 19, 2003, pp. 1739–1741.

485. J. J. Niederhauser, D. Frauchiger, H. P. Weber, and M. Frenz, "Real-Time Optoacoustic Imaging Using a Schlieren Transducer," *Appl. Phys. Lett.*, vol. 81, 2002, pp. 571–573.

486. B. P. Payne, V. Venugopalan, B. B. Mikić, and N. S. Nishioka, "Optoacoustic Determination of Optical Attenuation Depth Using Intereferometric Detection," *J. Biomed. Opt.*, vol. 8, no. 2, 2003, pp. 264–272.

487. B. P. Payne, V. Venugopalan, B. B. Mikić, and N. S. Nishioka, "Optoacoustic Tomography Using Time-Resolved Intereferometric Detection of Surface Displacement," *J. Biomed. Opt.*, vol. 8, no. 2, 2003, pp. 273–280.

488. U. Oberheide, I. Bruder, H. Welling, W. Ertmer, and H. Lubatschowski, "Optoacoustic Imaging for Optimization of Laser Cyclophotocoagulation," *J. Biomed. Opt.*, vol. 8, no. 2, 2003, pp. 281–287.

489. G. Schüle, G. Hüttman, C Framme, J. Roider, and R. Brinkmann, "Noninvasive Optoacoustic Temperature Determination at the Fundus of the Eye during Laser Irradiation," *J. Biomed. Opt.*, vol. 9, no. 1, 2004, pp. 173–179.

490. R. E. Imhof, C. J. Whitters, and D. J. S. Birch, "Opto-Thermal *In Vivo* Monitoring of Sunscreens on Skin," *Phys. Med. Biol.*, vol. 35, no. 1, 1990, pp. 95–102.

491. S. A. Prahl, I. A. Vitkin, U. Bruggemann, B. C. Wilson, and R. R. Anderson, "Determination of Optical Properties of Turbid Media Using Pulsed Photothermal Radiometry," *Phys. Med. Biol.*, vol. 37, 1992, pp. 1203–1217.

492. S. L. Jacques, J. S. Nelson, W. H. Wright, and T. E. Milner, "Pulsed Photothermal Radiometry of Port–Wine–Stain Lesions," *Appl. Opt.*, vol. 32, 1993, pp. 2439–2446.

493. I. A. Vitkin, B. C. Wilson, and R. R. Anderson, "Analysis of Layered Scattering Materials by Pulsed Photothermal Radiometry: Application to Photon Propagation in Tissue," *Appl. Opt.*, vol. 34, 1995, pp. 2973–2982.

494. T. E. Milner, D. M. Goodman, B. S. Tanenbaum, and J. S. Nelson, "Depth Profiling of Laser-Heated Chromophores in Biological Tissues by Pulsed Photothermal Radiometry," *J. Opt. Soc. Am. A*, vol. 12, 1995, pp. 1479–1488.

495. T. E. Milner, D. M. Goodman, B. S. Tanenbaum, B. Anvari, and J. S. Nelson, "Noncontact Determination of Thermal Diffusivity in Biomaterials Using Infrared Imaging Radiometry," *J. Biomed. Opt.*, vol. 1, 1996, pp. 92–97.

496. D. Fried, S. R. Visuri, J. D. B. Featherstone, J. T. Walsh, W. Seka, R. E. Glena, S. M. McCormack, and H. A. Wigdor, "Infrared Radiome-

try of Dental Enamel During Er: YAG and Er: YSGG Laser Irradiation," *J. Biomed. Opt.*, vol. 1, no. 4, 1996, pp. 455–465.

497. U. S. Sathyam, and S. A. Prahl, "Limitations in Measurement of Subsurface Temperatures Using Pulsed Photothermal Radiometry," *J. Biomed. Opt.*, vol. 2, no. 3, 1997, pp. 251–261.

498. B. Li, B. Majaron, J. A. Viator, T. E. Milner, Z. Chen, Y. Zhao, H. Ren, and J. S. Nelson, "Accurate Measurement of Blood Vessel Depth in Port Wine Stained Human Skin *In Vivo* Using Pulsed Photothermal Radiometry," *J. Biomed. Opt.*, vol. 9, no. 2, 2004, pp. 299–307.

499. B. Choi, B. Majaron, and J. S. Nelson, "Computational Model to Evaluate Port Wine Stain Depth Profiling Using Pulsed Photothermal Radiometry," *J. Biomed. Opt.*, vol. 9, no. 5, 2004, pp. 961–966.

500. L. Nicolaides, A. Mandelis, and S. H. Abrams, "Novel Dental Dynamic Depth Profilometric Imaging Using Simultaneous Frequency-Domain Infrared Photothermal Radiometry and Laser Luminescence," *J. Biomed. Opt.*, vol. 5, 2000, pp. 31–39.

501. R. J. Jeon, A. Mandelis, V. Sanchez, and S. H. Abrams, "Nonintrusive, Noncontacting Frequency-Domain Photothermal Radiometry and Luminescence Depth Profilometry of Carious and Artificial Subsurface Lesions in Human Teeth," *J. Biomed. Opt.*, vol. 9, no. 4, 2004, pp. 804–819.

502. D. H. Douglas-Hamilton and J. Conia, "Thermal Effects in Laser-Assisted Pre-Embryo Zona Drilling," *J. Biomed. Opt.*, vol. 6, no. 2, 2001, pp. 205–213.

503. D. Lapotko, T. Romanovskaya, and V. Zharov, "Photothermal Images of Live Cells in Presence of Drug," *J. Biomed. Opt.*, vol. 7, no. 3, 2002, pp. 425–434.

504. V. Zharov, "Far-Field Photothermal Microscopy beyond the Diffraction Limit," *Opt. Lett.*, vol. 28, 2003, pp. 1314–1316.

505. V. Zharov, V. Galitovsky, and M. Viegas, "Photothermal Detection of Local Thermal Effects during Selective Nanophotothermolysis," *Appl. Phys. Lett.*, vol. 83, no. 24, 2003, pp. 4897–4899.

506. V. Galitovskiy, P. Chowdhury, and V. P. Zharov, "Photothermal Detection of Nicotine-Induced Apoptotic Effects in a Pancreatic Cancer Cells," *Life Sciences*, vol. 75, 2004, pp. 2677–2687.

507. V. P. Zharov, E. I. Galanzha, and V. V. Tuchin, "Integrated Photothermal Flow Cytometry *In Vivo*," *J. Biomed. Opt.*, vol. 10, 2005, pp. 647–655.

508. V. P. Zharov, E. I. Galanzha, and V. V. Tuchin, "Photothermal Image Flow Cytometry *in Vivo*," *Opt. Lett.*, vol. 30, no. 6, 2005, pp. 107–110.

509. A. D. Yablon, N. S. Nishioka, B. B. Mikić, and V. Venugopalan, "Measurement of Tissue Absorption Coefficients by Use of Interferometric Photothermal Spectroscopy," *Appl. Opt.*, vol. 38, 1999, pp. 1259–1272.

510. M. L. Dark, L. T. Perelman, I. Itzkan, J. L. Schaffer, and M. S. Feld, "Physical Properties of Hydrated Tissue Determined by Surface Interferometry of Laser-Induced Thermoelastic Deformation," *Phys. Med. Biol.*, vol. 45, 2000, pp. 529–539.

511. A. A. Oraevsky, S. L. Jacques, R. O. Esenaliev, and F. K. Tittel, "Pulsed Laser Ablation of Soft Tissues, Gels and Aqueous Solutions at Temperatures Below 100°C," *Lasers Surg. Med.*, vol. 18, no. 3, 1995, pp. 231–240.

512. K. V. Larin, I. V. Larina, and R. O. Esenaliev, "Monitoring of Tissue Coagulation During Thermotherapy Using Optoacoustic Technique," *J. Phys. D: Appl. Phys.*, vol. 38, 2005, pp. 2645–2653.

513. C. H. Schmitz, U. Oberheide, S. Lohmann, H. Lubatschowski, and W. Ertmer, "Pulsed Photothermal Radiometry as a Method for Investigating Blood Vessel-Like Structures," *J. Biomed. Opt.*, vol. 6, no. 2, 2001, pp. 214–223.

514. Z. Zhao, S. Nissilä, O. Ahola, and R. Myllylä, "Production and Detection Theory of Pulsed Photoacoustic Wave with Maximum Amplitude and Minimum Distortion in Absorbing Liquid," *IEEE Trans. Instrum. Measur.*, vol. 47, no. 2, 1998, pp. 578–583.

515. I. V. Larina, K. V. Larin, and R. O. Esenaliev, "Real-Time Optoacoustic Monitoring of Temperature in Tissues," *J. Phys. D: Appl. Phys.*, vol. 38, 2005, pp. 2633–2639.

516. R. O. Esenaliev, I. V. Larina, K. V. Larin, D. J. Deyo, M. Motamedi, and D. S. Prough, "Optoacoustic Technique for Noninvasive Monitoring of Blood Oxygenation: A Feasibility Study," *Appl. Opt.*, vol. 41, no. 22, 2002, pp. 4722–4731.

517. L. V. Wang, "Ultrasound-Mediated Biophotonics Imaging: A Review of Acousto-Optical Tomography and Photo-Acoustic Tomography," *Disease Markers*, vol. 19, 2003, 2004, pp. 123–138.

518. K. Maslov, G. Stoica, and L. V. Wang, "*In vivo* Dark-Field Reflection-Mode Photoacoustic Microscopy," *Opt. Lett.*, vol. 30, no. 6, 2005, pp. 625–627.

519. L. Wang and X. Zhao, "Ultrasound-Modulated Optical Tomography of Absorbing Objects Buried in Dense Tissue-Simulating Turbid Media," *Appl. Opt.*, vol. 36, no. 28, 1997, pp. 7277–7282.

520. L. Wang, "Ultrasonic Modulation of Scattered Light in Turbid Media and a Potential Novel Tomography in Biomedicine," *Photochem. Photobiol.*, vol. 67, no. 1, 1998, pp. 41–49.

521. L. V. Wang and G. Ku, "Frequency-Swept Ultrasound-Modulated Optical Tomography of Scattering Media," *Opt. Let.*, vol. 23, no. 12, 1998, pp. 975–977.

522. J. Selb, S. Lévêque-Fort, A. Dubois, B. C. Forget, L. Pottier, F. Ramaz, and C. Boccara, "Ultrasonically Modulated Optical Imaging," in *Biomedical Photonics Handbook*, Chapter 35, Tuan Vo-Dinh (ed.), CRC Press, Boca Raton, 2003, pp. 35-1–12.

523. M. Kempe, M. Larionov, D. Zaslavsky, and A. Z. Genack, "Acousto-Optic Tomography with Multiply Scattered Light," *J. Opt. Soc. Am. A*, vol. 14, no. 5, 1997, pp. 1151–1158.

524. L.-H. V. Wang, "Mechanisms of Ultrasonic Modulation of Multiply Scattered Coherent Light: An Analytical Mode," *Phys. Rev. Lett.*, vol. 8704, 2001, pp. 3903-(1–4).

525. J. P. Gore and L. X. Xu, "Thermal Imaging for Biological and Medical Diagnostics," in *Biomedical Photonics Handbook*, Chapter 17, Tuan Vo-Dinh (ed.), CRC Press, Boca Raton, 2003, pp. 17-1–12.

526. A. L. McKenzie, "Physics of Thermal Processes in Laser-Tissue Interaction," *Phys. Med. Biol.*, vol. 35, 1990, pp. 1175–1209.

527. A. J. Welch and M. J. C. van Gemert (eds.), *Optical-Thermal Response of Laser Irradiated Tissue*, Plenum Press, 1995.

528. C. H. G. Wright, S. F. Barrett, and A. J. Welch, "Laser-Tissue Interaction," in *Lasers in Medicine*, D. R. Vij and K. Mahesh (eds.), Kluwer, Boston, 2002.

529. S. Weinbaum and L. M. Jiji, "A New Simplified Bioheat Equation for the Effect of Blood Flow on Local Average Tissue Temperature," *J. Biomech. Eng.*, vol. 107, 1985, pp. 131–139.

530. Z. F. Cui and J. C. Barbenel, "The Influence of Model Parameter Values on the Prediction of Skin Surface Temperature: I. Resting and Surface Insulation," *Phys. Med. Biol.*, vol. 35, 1990, pp. 1683–1697.

531. M. Nitzan and B. Khanokh, "Infrared Radiometry of Thermally Insulated Skin for the Assessment of Skin Blood Flow," *Opt. Eng.*, vol. 33, 1994, pp. 2953–2957.

532. L. V. Wang and Q. Shen, "Sonoluminescent Tomography of Strongly Scattering Media," *Opt. Lett.*, vol. 23, no. 7, 1998, pp. 561–563.

533. Q. Shen and L. V. Wang, "Two-Dimensional Imaging of Dense Tissue-Simulating Turbid Media by Use of Sonoluminescence," *Appl. Opt.*, vol. 38, no. 1, 1999, pp. 246–252.

534. O. Khalil, "Non-Invasive Glucose Measurement Technologies: An Update from 1999 to the Dawn of the New Millenium," *Diabetes Technol. Ther.*, vol. 6, no. 5, 2004, pp. 660–697.

535. K. M. Quan, G. B. Christison, H. A. Mackenzie, and P. Hodgson, "Glucose Determination by a Pulsed Photoacoustic Technique: An Experimental Study Using a Gelatin–Based Tissue Phantom," *Phys. Med. Biol.*, vol. 38, 1993, pp. 1911–1922.

536. H. A. MacKenzie, H. S. Ashton, S. Spiers, Y. Shen, S. S. Freeborn, J. Hannigan, J. Lindberg, and P. Rae, "Advances in Photoacoustic Noninvasive Glucose Testing," *Clin. Chem.*, vol. 45, 1999, pp. 1587–1595.

537. A. A. Bednov, A. A. Karabutov, E. V. Savateeva, W. F. March, and A. A. Oraevsky, "Monitoring Glucose *In Vivo* by Measuring Laser-Induced Acoustic Profiles," *Proc. SPIE* vol. 3916, 2000, pp. 9–18.

538. A. A. Bednov, E. V. Savateeva, and A. A. Oraevsky, "Glucose Monitoring in Whole Blood by Measuring Laser-Induced Acoustic Profiles," *Proc. SPIE* vol. 4960, 2003, pp. 21–29.

539. Z. Zhao and R. Myllylä, "Photoacoustic Blood Glucose and Skin Measurement Based on Optical Scattering Effect," *Proc. SPIE*, vol. 4707, 2002, pp. 153–157.

540. M. Kinnunen and R. Myllylä, "Effect of Glucose on Photoacoustic Signals at the Wavelemgth of 1064 and 532 nm in Pig Blood and Intralipid," *J. Phys. D: Appl. Phys.*, vol. 38, 2005, pp. 2654–2661.

541. Z. Zhao, *Pulsed Photoacoustic Techniques and Glucose Determination in Human Blood and Tissue*, Ph.D. Dissertation, Oulu, Finland, University of Oulu, 2002. Available online at: http://herkules.oulu.fi/isbn9514266900/index.html.

542. Glucon, Inc.: http://www.glucon.com/

543. Y. Shen, Z. Lu, S. Spiers, H. A. MacKenzie, H. S. Ashton, J. Hannigan, S. S. Freeborn, and J. Lindberg, "Measurement of the Optical Absorption Coefficient of a Liquid by Use of a Time-Resolved Photoacoustic Technique," *Appl. Opt.*, vol. 39, 2000, pp. 4007–4012.

544. D. C. Klonoff, J. R. Braig, B. B. Sterling, C. Kramer, D. S. Goldberger, and Y. Trebino, "Mid-Infrared Spectroscopy for Non-invasive Blood Glucose Monitoring," *IEEE Laser Electro-Opt. Soc. Newslett.*, vol. 12, 1998, pp. 13–14.

545. P. Zheng, C. E. Kramer, C. W. Barnes, J. R. Braig, and B. B. Sterling, "Noninvasive Glucose Determination by Oscillating Thermal Gradient Spectrometry," *Diabetes. Technol. Ther.*, vol. 2, 2000, pp. 17–25.

546. C. D. Malchoff, K. Shoukri, J. I. Landau, and J. M. Buchert, "A Novel Noninvasive Blood Glucose Monitor," *Diabetes Care*, vol. 25, 2002, pp. 2268–2275.

547. R. O. Esenaliev, Y. Y. Petrov, O. Hartrumpf, D. J. Deyo, and D. S. Prough, "Contineous, Noninvasive Monitoring of Total Hemoglobin Concentration by an Optoacoustic Technique," *Appl. Opt.*, vol. 43, 2004, pp. 3401–3407.

548. I. Petrova, R. O. Esenaliev, Y. Y. Petrov, H.-P. F. Brecht, C. H. Svensen, J. Olsson, D. J. Deyo, and D. S. Prough, "Optoacoustic Monitoring of Blood Hemoglobin Concentration: A Pilot Clinical Study," *Opt. Lett.*, vol. 30, 2005, pp. 1677–1679.

549. R. O. Esenaliev, K. V. Larin, I. V. Larina, and M. Motamedi, "Noninvasive Monitoring of Glucose Concentration with Optical Coherent Tomography," *Opt. Lett.*, vol. 26, no. 13, 2001, pp. 992–994.

550. K. V. Larin, M. S. Eledrisi, M. Motamedi, R. O. Esenaliev, "Noninvasive Blood Glucose Monitoring with Optical Coherence Tomography: a Pilot Study in Human Subjects," *Diabetes Care*, vol. 25, no. 12, 2002, pp. 2263–2267.

551. K. V. Larin, M. Motamedi, T. V. Ashitkov, and R. O. Esenaliev, "Specificity of Noninvasive Blood Glucose Sensing Using Optical Coherence Tomography Technique: a Pilot Study," *Phys Med Biol*, vol. 48, 2003, pp. 1371–1390.

552. D. M. Zhestkov, A. N. Bashkatov, E. A. Genina, and V. V. Tuchin, "Influence of Clearing Solutions Osmolarity on the Optical Properties of RBC," *Proc. SPIE* 5474, 2004, pp. 321–330.

553. B. Yin, D. Xing, Y. Wang, Y. Zeng, Y. Tan, and Q. Chen, "Fast Photoacoustic Imaging System Based on 320-element Linear Transducer Array," *Phys. Med. Biol.*, vol. 49, 2004, pp. 1339–1346.

554. Y. Su, F. Zhang, K. Xu, J. Yao, and R. K. Wang, "A Photoacoustic Tomography System for Imaging of Biological Tissues," *J. Phys. D: Appl. Phys.*, vol. 38, 2005, pp. 2640–2644.

555. P. A. Fomitchov, A. K. Kromine, and S. Krishnaswamy, "Photoacoustic Probes for Nondestructive Testing and Biomedical Applications," *Appl. Opt.*, vol. 41, no. 22, 2002, pp. 4451–4459.

556. M. Jaeger, J. J. Niederhauser, M. Hejazi, and M. Frenz, "Diffraction-Free Acoustic Detection for Optoacoustic Depth Profiling of Tissue Using an Optically Transparent Polyvinylidene Fluoride Pressure Transducer Operated in Backward and Forward Mode," *J. Biomed. Opt.*, vol. 10, no. 2, 2005, pp. 024035-1-7.

557. G. Ku, X. Wang, G. Stoica, and L. V. Wang, "Multiple-Bandwidth Photoacoustic Tomography," *Phys. Med. Biol.*, vol. 49, 2004, pp. 1329–1338.

558. B. Beauvoit, T. Kitai, B. Chance, "Contribution of the Mitochondrial Compartment to the Optical Properties of the Rat Liver: a Theoretical and Practical Approach," *Biophys. J.*, vol. 67, 1994, pp. 2501–2510.

559. V. G. Vereshchagin and A. N. Ponyavina, "Statistical Characteristic and Transparency of Thin Closely Packed Disperse Layer," *J. Appl. Spectr. (USSR)*, vol. 22, no. 3, 1975, pp. 518–524.

560. V. Twersky "Interface Effects in Multiple Scattering by Large, Low Refracting, Absorbing Particles," *J. Opt. Soc. Am.*, vol. 60, no. 7, 1970, 908–914.

561. M. Lax, "Multiple Scattering of Waves II. The Effective Field in Dense System," *Phys. Rev.*, vol. 85, no. 4, 1952, pp. 621–629.

562. L. Tsang, J. A. Kong, and R. T. Shin, *Theory of Microwave Remote Sensing*, Wiley, New York, 1985.

563. K. M. Hong, "Multiple Scattering of Electromagnetic Waves by a Crowded Monolayer of Spheres: Application to Migration Imaging Films," *J. Opt. Soc. Am.*, vol. 70, no. 7, 1980, pp. 821–826.

564. A. Ishimaru and Y. Kuga "Attenuation Constant of a Coherent Field in a Dense Distribution of Particles," *J. Opt. Soc. Am.*, vol. 72, no. 10, 1982, pp. 1317–1320.

565. A. N. Ponyavina, "Selection of Optical Radiation in Scattering by Partially Ordered Disperse Media," *J. Appl. Spectrosc.*, vol. 65, no. 5, 1998, pp. 721–733.

566. T. R. Smith, "Multiple Scattering in the Cornea," *J. Mod. Opt.*, vol. 35, no. 1, 1988, pp. 93–101.

567. V. Twersky, "Absorption and Multiple Scattering by Biological Suspensions," *J. Opt. Soc. Am.*, vol. 60, 1970, pp. 1084–1093.

568. J. M. Steinke and A. P. Shephard, "Diffusion Model of the Optical Absorbance of Whole Blood," *J. Opt. Soc. Am. A*, vol. 5, 1988, pp. 813–822.

569. I. F. Cilesiz and A. J. Welch, "Light Dosimetry: Effects of Dehydration and Thermal Damage on the Optical Properties of the Human Aorta," *Appl. Opt.*, vol. 32, 1993, pp. 477–487.

570. W.-C. Lin, M. Motamedi, and A. J. Welch, "Dynamics of Tissue Optics During Laser Heating of Turbid Media," *Appl. Opt.*, vol. 35, no. 19, 1996, pp. 3413–3420.

571. G. Vargas, E. K. Chan, J. K. Barton, H. G. Rylander III, and A. J. Welch, "Use of an Agent to Reduce Scattering in Skin," *Laser. Surg. Med.*, vol. 24, 1999, pp. 133–141.

572. R. M. P. Doornbos, R. Lang, M. C. Aalders, F. W. Cross, and H. J. C. M. Sterenborg, "The Determination of *In Vivo* Human Tissue Optical Properties and Absolute Chromophore Concentrations Using Spatially Resolved Steady-State Diffuse Reflectance Spectroscopy," *Phys. Med. Biol.*, vol. 44, 1999, pp. 967–981.

573. J. R. Lakowicz, *Principles of Fluorescence Spectroscopy*, 2nd ed., Kluwer Academic/Plenum Publ., New York, 1999.

574. H. Schneckenburger, R. Steiner, W. Strauss, K. Stock, and R. Sailer, "Fluorescence Technologies in Biomedical Diagnostics," Chapter 15 in *Optical Biomedical Diagnostics*, V. V. Tuchin (ed.), SPIE Press, Bellingham, WA, 2002, pp. 825–874.

575. Yu. P. Sinichkin, N. Kollias, G. Zonios, S. R. Utz, and V. V. Tuchin, "Reflectance and Fluorescence Spectroscopy of Human Skin *In Vivo*," Chapter 13 in *Optical Biomedical Diagnostics*, V. V. Tuchin (ed.), SPIE Press, Bellingham, WA, 2002, pp. 725–785.

576. S. Svanberg, "New Developments in Laser Medicine," *Phys. Scripta*, vol. T72, 1997, pp. 69–75.

577. R. R. Richards-Kortum, R. P. Rava, R. E. Petras, M. Fitzmaurice, M. Sivak, and M. S. Feld, "Spectroscopic Diagnosis of Colonic Dysplasia," *Photochem. Photobiol.*, vol. 53, 1991, pp. 777–786.

578. H. J. C. M. Sterenborg, M. Motamedi, R. F. Wagner, J. R. M. Duvic, S. Thomsen, and S. L. Jacques, "*In Vivo* Fluorescence Spectroscopy and Imaging of Human Skin Tumors," *Lasers Med. Sci.*, vol. 9, 1994, pp. 344–348.

579. H. Zeng, C. MacAulay, D. I. McLean, and B. Palcic, "Spectroscopic and Microscopic Characteristics of Human Skin Autofluorescence Emission," *Photochem. Photobiol.*, vol. 61, 1995, pp. 639–645.

580. Yu. P. Sinichkin, S. R. Utz, A. H. Mavlutov, and H. A. Pilipenko, "*In Vivo* Fluorescence Spectroscopy of the Human Skin: Experiments and Models," *J. Biomed. Opt.*, vol. 3, 1998, pp. 201–211.

581. R. Drezek, K. Sokolov, U. Utzinger, I. Boiko, A. Malpica, M. Follen, and R. Richards-Kortum, "Understanding the Contributions of NADH and Collagen to Cervical Tissue Fluorescence Spectra: Modeling, Measurements, and Implications," *J. Biomed. Opt.*, vol. 6, no. 4, 2001, pp. 385–396.

582. L. C. Lucchina, N. Kollias, R. Gillies, S. B. Phillips, J. A. Muccini, M. J. Stiller, R. J. Trancik, and L. A. Drake, "Fluorescence Photography in the Evaluation of Acne," *J. Am. Acad. Dermatol.*, vol. 35, 1996, pp. 58–63.

583. N. S. Soukos, S. Som, A. D. Abernethy, K. Ruggiero, J. Dunham, C. Lee, A. G. Doukas, and J. M. Goodson, "Phototargeting Oral Black-Pigmented Bacteria," *Antimicrob. Agents Chemother.*, vol. 49, 2005, pp. 1391–1396.

584. P. Kask, K. Palo, N. Fay, L. Brand, U. Mets, D. Ullmann, J. Jungmann, J. Pschorr, and K. Gall, "Two-Dimensional Fluorescence Intensity Distribution Analysis: Theory and Applications," *Biophys. J.*, vol. 78, 2000, pp. 1703–1713.

585. D. E. Hyde, T. J. Farrell, M. S. Patterson, and B. C. Wilson, "A Diffusion Theory Model of Spatially Resolved Fluorescence from Depth-Dependent Fluorophore Concentrations," *Phys. Med. Biol.*, vol. 46, 2001, pp. 369–383.

586. K. Sokolov, J. Galvan, A. Myakov, A. Lacy, R. Lotan, and R. Richards-Kortum, "Realistic Three-Dimensional Epithelial Tissue Phantoms for Biomedical Optics," *J. Biomed. Opt.*, vol. 7, no. 1, 2002, pp. 148–156.

587. M. J. Eppstein, D. J. Hawrysz, A. Godavarty, and E. M. Sevick-Muraca, "Three-Dimensional, Bayesian Image Reconstruction from Sparse and Noisy Data Sets: Near-Infrared Fluorescence Tomography," *Proc. Natl. Acad. Sci. USA*, vol. 99, no. 15, 2002, pp. 9619–9624.

588. A. Eidsath, V. Chernomordik, A. Gandjbakhche, P. Smith, and A. Russo, "Three-Dimensional Localization of Fluorescent Masses Deeply Embedded in Tissue," *Phys. Med. Biol.*, vol. 47, 2002, pp. 4079–4092.

589. N. C. Biswal, S. Gupta, N. Ghosh, and A. Pradhan, "Recovery of Turbidity Free Fluorescence from Measured Fluorescence: an Experimental Approach," *Optics Express*, vol. 11, no. 24, 2003, pp. 3320–3331.

590. Q. Liu, C. Zhu, and N. Ramanujam, "Experimental Validation of Monte Carlo Modeling of Fluorescence in Tissues in the UV-Visible Spectrum," *J. Biomed. Opt.*, vol. 8, no. 2, 2003, pp. 223–236.

591. C. Zhu, Q. Liu, and N. Ramanujam, "Effect of Fiber Optic Probe Geometry on Depth-Resolved Fluorescence Measurements from Epithelial Tissues: a Monte Carlo Simulation," *J. Biomed. Opt.*, vol. 8, no. 2, 2003, pp. 237–247.

592. D. Y. Churmakov, I. V. Meglinski, S. A. Piletsky, and D. A. Greenhalgh, "Analysis of Skin Tissues Spatial Fluorescence Distribution by the Monte Carlo Simulation," *Appl. Phys. D*, vol. 36, 2003, pp. 1722–1728.

593. H. Schneckenburger, M. H. Gschwend, R. Sailer, H.-P. Mock, and W. S. L. Strauss, "Time-gated fluorescence microscopy in molecular and cellular biology," *Cell. Mol. Biol.*, vol. 44, 1998, pp. 795–805.

594. O. O. Abugo, P. Herman, and J. R. Lakowicz, "Fluorescence Properties of Albumin Blue 633 and 670 in Plasma and Whole Blood, *J. Biomed. Opt.*, vol. 6, no. 3, 2001, pp. 359–365.

595. J. E. Budaj, S. Achilefu, R. B. Dorshow, and R. Rajagopalan, "Novel Fluorescent Contrast Agents for Optical Imaging of *In Vivo* Tumors Based on a Receptor-Targeted Dye-Peptide Conjugate Platform," *J. Biomed. Opt.*, vol. 6, no. 2, 2001, pp. 122–133.

596. K. Suhling, J. Siegel, D. Phillips, P. M. W. French, S. Leveque-Fort, S. E. D. Webb, and D. M. Davis, "Imaging the Environment of Green Fluorescent Protein," *Biophys. J.*, vol. 83, 2002, pp. 3589–3595.

597. I. Gannot, A. Garashi, G. Gannot, V. Chernomordik, and A. Gandjbakhche, "*In Vivo* Quantitative Three-Dimensional Localization of Tumor Labeled with Exogenous Specific Fluorescence Markers," *Appl. Opt.*, vol. 42, no. 16, 2003, pp. 3073–3080.

598. D. Hattery, V. Chernomordik, M. Loew, I. Gannot, and A. Gandjbakhche, "Analytical Solutions for Time-Resolved Fluorescence Lifetime Imaging in a Turbid Medium Such as Tissue," *J. Opt. Soc. Am. A*, vol. 18, no. 7, 2001, pp. 1523–1530.

599. M. Sadoqi, P. Riseborough, and S. Kumar, "Analytical Models for Time Resolved Fluorescence Spectroscopy in Tissues," *Phys. Med. Biol.*, vol. 46, 2001, pp. 2725–2743.

600. E. Kuwana and E. M. Sevick-Muraca, "Fluorescence Lifetime Spectroscopy in Multiply Scattering Media with Dyes Exhibiting Multiexponential Decay Kinetics," *Biophys. J.*, vol. 83, 2002, pp. 1165–1176.

601. K. Vishwanath, B. Pogue, and M.-A. Mycek, "Quantitative Fluorescence Lifetime Spectroscopy in Turbid Media: Comparison of Theoretical, Experimental and Computational Methods," *Phys. Med. Biol.*, vol. 47, 2002, pp. 3387–3405.

602. C. S. Betz, M. Mehlmann, K. Rick, H. Stepp, G. Grevers, R. Baumgartner, and A. Leunig, "Autofluorescence Imaging and Spectroscopy of Normal and Malignant Mucosa in Patients with Head and Neck Cancer," *Lasers Surg. Med.*, vol. 25, 1999, pp. 323–334.

603. N. Anastassopoulou, B. Arapoglou, P. Demakakos, M. I. Makropoulou, A. Paphiti, and A. A. Serafetinides, "Spectroscopic Characterisation of Carotid Atherosclerotic Plaque by Laser Induced Fluorescence," *Lasers Surg. Med.*, vol. 28, 2001, pp. 67–73.

604. S. K. Chang, M. Y. Dawood, G. Staerkel, U. Utzinger, E. N. Atkinson, R. R. Richards-Kortum, and M. Follen, "Fluorescence Spectroscopy for Cervical Precancer Detection: is there Variance Across the Menstrual Cycle?" *J. Biomed. Opt.*, vol. 7, no. 4, 2002, pp. 595–602.

605. R. Hage, P. R. Galhanone, R. A. Zangaro, K. C. Rodrigues, M. T. T. Pacheco, A. A. Martin, M. M. Netto, F. A. Soares, and I. W. da Cunha, "Using the Laser-Induced Fluorescence Spectroscopy in the Differentiation Between Normal and Neoplastic Human Breast Tissue," *Lasers Med. Sci.*, vol. 18, 2003, pp. 171–176.

606. S. Andersson-Engels, G. Canti, R. Cubeddu, C. Eker, C. Klinteberg, A. Pifferi, K. Svanberg, S. Svanberg, P. Taroni, G. Valentini, and I. Wang, "Preliminary Evaluation of Two Fluorescence Imaging Methods for the Detection and the Delineation of Basal Cell Carcinomas of the Skin," *Lasers Surg. Med.*, vol. 26, 2000, pp. 76–82.

607. T. Wu, J. Y. Qu, T.-H. Cheung, K. W.-K. Lo, and M.-Y. Yu, "Preliminary Study of Detecting Neoplastic Growths *In Vivo* with Real Time Calibrated Autofluorescence Imaging," *Optics Express*, vol. 11, no. 4, 2003, pp. 291–298.

608. L. Rovati and F. Docchio, "Autofluorescence Methods in Ophthalmology," *J. Biomed. Opt.*, vol. 9, no. 1, 2004, pp. 9–21.

609. W. Denk, "Two-Photon Excitation in Functional Biological Imaging," *J. Biomed. Opt.*, vol. 1, no. 3, 1996, pp. 296–304.

610. D. W. Piston, B. R. Masters, and W. W. Webb, "Three-Dimensionally Resolved NAD(P)H Cellular Metabolic Redox Imaging of the *In Situ* Cornea with Two-Photon Excitation Laser Scanning Microscopy," *J. Microscopy*, vol. 178, 1995, pp. 20–27.

611. K. M. Berland, P. T. So, and E. Gratton, "Two-Photon Fluorescence Correlation Spectroscopy: Method and Application to the Intracellular Environment," *Biophys. J.*, vol. 68, 1995, pp. 649–701.

612. S. W. Hell, K. Bahlmann, M. Schrader, et al., "Three-Photon Excitation in Fluorescence Microscopy," *J. Biomed. Opt.*, vol. 1, no. 1, 1996, pp. 71–74.

613. B. R. Masters, P. T. C. So, and E. Gratton, "Multi-Photon Excitation Fluorescence Microscopy and Spectroscopy of *In Vivo* Human Skin," *Biophys. J.*, vol. 72, 1997, pp. 2405–2412.

614. B. R. Masters, "Confocal Laser Scanning Microscopy," in *Coherent-Domain Optical Methods: Biomedical Diagnostics, Environmental and Material Science*, Chapter 21, V. V. Tuchin (ed.), Kluwer Academic Publishers, Boston, vol. 2, 2004, pp. 364–415.

615. P. T. C. So, C. Y. Dong, and B. R. Masters, "Two-Photon Excitation Fluorescence Microscopy," in *Biomedical Photonics Handbook*, Tuan Vo-Dinh (ed.), CRC Press, Boca Rotan, Florida, 2003, pp. 11-1-17.

616. Z. X. Zhang, G. J. Sonek, X. B. Wei, C. Sun, M. W. Berns, and B. J. Tromberg, "Cell Viability and DNA Denaturation Measurements by Two-Photon Fluorescence Excitation in CW Al:GaAs Diode Laser Optical Traps," *J. Biomed. Opt.*, vol. 4, no. 2, 1999, pp. 256–259.

617. Ch. J. Bardeen, V. V. Yakovlev, J. A. Squier, K. R. Wilson, S. D. Carpenter, and P. M. Weber, "Effect of Pulse Shape on the Efficiency of Multiphoton Processes: Implications for Biological Microscopy," *J. Biomed. Opt.*, vol. 4, no. 3, 1999, pp. 362–367.

618. A. K. Dunn, V. P. Wallace, M. Coleno, M. W. Berns, and B. J. Tromberg, "Influence of Optical Properties on Two-Photon Fluorescence Imaging in Turbid Samples," *Appl. Opt.*, vol. 39, no. 7, 2000, pp. 1194–1201.

619. Y. Ozaki, "Medical Application of Raman Spectroscopy," *Appl. Spectrosc. Rev.*, vol. 24, no. 3, 1988, pp. 259–312.

620. L. T. Perelman, M. D. Modell, E. Vitkin, and E. B. Hanlon, "Light Scattering Spectroscopy: from Elastic to Inelastic," Chapter 9, in *Coherent-Domain Optical Methods: Biomedical Diagnostics, Environmental and Material Science*, vol. 1, V. V. Tuchin (ed.), Kluwer Academic Publishers, Boston, 2004, pp. 355–395.

621. A. T. Tu, *Raman Spectroscopy in Biology*, John Wiley & Sons Ltd., New York, 1982.

622. G. W. Lucassen, G. N. A. van Veen, and J. A. J. Jansen, "Band Analysis of Hydrated Human Skin Stratum Corneum Attenuated Total Reflectance Fourier Transform Infrared Spectra *In Vivo*", *J. Biomedical Optics*, vol. 3, 1998, pp. 267–280.

623. G. J. Puppels, "Confocal Raman Microspectroscopy," in *Fluorescent and Luminescent Probes for Biological Activity*, W. Mason (ed.), Academic Press, London, 1999, pp. 377–406.

624. G. W. Lucassen, P. J. Caspers, and G. J. Puppels, "Infrared and Raman Spectroscopy of Human Skin *In Vivo*," Chapter 14, in *Optical Biomedical Diagnostics*, V. V. Tuchin (ed.), SPIE Press, Bellingham, 2002, pp. 787–823.

625. R. Petry, M. Schmitt, and J. Popp, "Raman Spectroscopy—a Prospective Tool in the Life Sciences," *Chemphyschem.*, vol. 4, 2003, pp. 14–30.

626. U. Utzinger, D. L. Heintselman, A. Mahadevan-Jansen, A. Malpica, M. Follen, and R. Richards-Kortum, "Near-Infrared Raman Spectroscopy for *In Vivo* Detection of Cervical Precancers," *Appl. Spectrosc.*, vol. 55, 2001, pp. 955–959.

627. C. S. Schatz and R. P. Van Duyne, "Electromagnetic Mechanism of Surface-Enhanced Spectroscopy," in *Handbook of Vibrational Spectroscopy*, J. M. Chalmers and P. R. Grifiths (eds.), John Wiley & Sons Ltd., Chichester, 2002, pp.

628. N. Skrebova Eikje, Y. Ozaki, K. Aizawa, and S. Arase, "Fiber Optic Near-Infrared Raman Spectroscopy for Clinical Noninvasive Determination of Water Content in Diseased Skin and Assessment of Cutaneous Edima," *J. Biomed. Opt.*, vol. 10, 2005, pp. 014013-1-13.

629. J. R. Mourant, R. R. Gibson, T. M. Johnson, S. Carpenter, K. W. Short, Y. R. Yamada, and J. P. Freyer, "Methods for Measuring the Infrared Spectra of Biological Cells," *Phys. Med. Biol.*, vol. 48, 2003, pp. 243–257.

630. M. Kohl, M. Essenpreis, and M. Cope, "The Influence of Glucose Concentration upon the Transport of Light in Tissue-Simulating Phantoms," *Phys. Med. Biol.*, vol. 40, 1995, pp. 1267–1287.

631. J. Qu and B. C. Wilson, "Monte Carlo Modeling Studies of the Effect of Physiological Factors and other Analytes on the Determination of Glucose Concentration *In Vivo* by Near Infrared Optical Absorption and Scattering Measurements," *J. Biomed. Opt.*, vol. 2, no. 3, 1997, pp. 319–325.

632. G. C. Beck, N. Akgun, A. Ruck, and R. Steiner, "Developing Optimized Tissue Phantom Systems for Optical Biopsies," *Proc. SPIE* 3197, 1997, pp. 76–85.

633. G. C. Beck, N. Akgun, A. Ruck, and R. Steiner, "Design and Characterization of a Tissue Phantom System for Optical Diagnostics," *Lasers Med. Sci.*, vol. 13, 1998, pp. 160–171.

634. D. D. Royston, R. S. Poston, and S. A. Prahl, Optical properties of scatering and absorbing materials used in the development of optical phantoms at 1064 nm," *J. Biomed. Opt.*, vol. 1, no. 1, 1996, pp. 110–123.

635. J. J. Burmeister, H. Chung, and M. A. Arnold, "Phantoms for noninvasive blood glucose sensing with near infrared transmission spectroscopy," *Photochem. Photobiol.*, vol. 67, no. 1, 1998, pp. 50–55.

636. A. B. Pravdin, S. P. Chernova, T. G. Papazoglu, and V. V. Tuchin, "Tissue Phantoms," Chapter 5 in *Handbook of Optical Biomedical Diagnostics*, vol. PM107, V. V. Tuchin (ed.), SPIE Press, Bellingham, WA, 2002, pp. 311–352.

637. R. A. J. Groenhuis, H. A. Ferwerda, and J. J. Ten Bosch, "Scattering and Absorption of Turbid Materials Determined from Reflection Measurements. 1. Theory," *Appl. Opt.*, vol. 22, 1983, pp. 2456–2462.

638. R. A. J. Groenhuis, J. J. Ten Bosch, and H. A. Ferwerda, "Scattering and Absorption of Turbid Materials Determined from Reflection Measurements. 2. Measuring Method and Calibration," *Appl. Opt.*, vol. 22, 1983, pp. 2463–2467.

639. D. Chursin, V. Shuvalov, and I. Shutov, "Optical Tomograph with Photon Counting and Projective Reconstruction of the Parameters of Absorbing 'Phantoms' in Extended Scattering Media," *Quantum Electronics*, vol. 29, no. 10, 1999, pp. 921–926.

640. G. Wagnieres, S. Cheng, M. Zellweger, N. Utke, D. Braichotte, J. Ballini, and H. Bergh, "An Optical Phantom with Tissue-Like Properties in the Visible for use in PDT and Fluorescence Spectroscopy," *Phys. Med. Biol.*, vol. 42, 1997, pp. 1415–1426.

641. S. Chernova, A. Pravdin, Y. Sinichkin, V. Kochubey, V. Tuchin, and S. Vari, "Correlation of Fluorescence and Reflectance Spectra of Tissue Phantoms with their Structure and Composition," *Proc. SPIE* 3598, 1999, pp. 294–300.

642. S. Chernova, O. Kasimov, L. Kuznetsova, T. Moskalenko, and A. B. Pravdin, "*Ex Vivo* and Phantom Fluorescence Spectra of Human Cervical tissue," *Proc. SPIE* 4001, 2000, pp. 290–298.

643. S. Chernova, A. Pravdin, Y. Sinichkin, V. Tuchin, and S. Vari, "Layered Gel-Based Phantoms Mimicking Fluorescence of Cervical Tissue," *Proc. OWLS V*, Springer, Berlin, 2000, pp. 301–306.

644. Y. Mendelson and J. Kent, "An *In Vitro* Tissue Model for Evaluating the Effect of Carboxyhemoglobin Concentration on Pulse Oximetry," *IEEE Trans. Biomed. Eng.*, vol. 36, no. 6, 1989, pp. 625–627.

645. V. Sankaran, J. Walsh, and D. Maitland, "Polarized Light Propagation in Biological Tissue and Tissue Phantoms," *Proc. SPIE* 4001, 2000, pp. 54–62.

646. B. Pogue, L. Lilge, M. Patterson, B. Wilson, and T. Hasan, "Absorbed Photodynamic Dose from Pulsed Versus Continuous Wave Light Examined with Tissue-Simulating Dosimeters," *Appl. Opt.*, vol. 36, no. 28, 1997, pp. 7257–7269.

647. M. Wolf, M. Keel, V. Dietz, K. Siebenthal, H. Bucher, and O. Baenziger, "The Influence of a Clear Layer on Near-Infrared Spectrophotometry Mea-

surements Using a Liquid Neonatal Head Phantom," *Phys. Med. Biol.*, vol. 44, 1999, pp. 1743–1753.

648. S. J. Matcher, "Signal Quantification and Localization in Tissue Near-Infrared Spectroscopy", Chapter 9 in *Handbook of Optical Biomedical Diagnostics*, vol. PM107, V. V. Tuchin (ed.), SPIE Press, Bellingham, WA, 2002, pp. 487–586.

649. R. R. Anderson and J. A. Parrish, "The Optics of Human Skin," *J. Invest. Dermatology*, vol. 77, 1981, pp. 13–19.

650. M. Hammer, A. Roggan, D. Schweitzer, and G. Müller, "Optical Properties of Ocular Fundus Tissues—an *In Vitro* Study Using the Double-Integrating-Sphere Technique and Inverse Monte Carlo Simulation," *Phys. Med. Biol.*, vol. 40, 1995, pp. 963–978.

651. W. M. Star, "The Relationship between Integrating Sphere and Diffusion Theory Calculations of Fluence Rate at the Wall of a Spherical Cavity," *Phys. Med. Biol.*, vol. 40, 1995, pp. 1–8.

652. S. A. Prahl, M. J. C. van Gemert, and A. J. Welch, "Determining the Optical Properties of Turbid Media by Using the Adding-Doubling Method," *Appl. Opt.*, vol. 32, 1993, pp. 559–568.

653. P. Marquet, F. Bevilacqua, C. Depeursinge, and E. B. de Haller, "Determination of Reduced Scattering and Absorption Coefficients by a Single Charge-Coupled-Device Array Measurement. 1. Comparison between Experiments and Simulations," *Opt. Eng.*, vol. 34, 1995, pp. 2055–2063.

654. F. Bevilacqua, P. Marquet, C. Depeursinge, and E. B. de Haller, "Determination of Reduced Scattering and Absorption Coefficients by a Single Charge-Coupled-Device Array Measurement. 2. Measurements on Biological Tissue," *Opt. Eng.*, vol. 34, 1995, pp. 2064–2069.

655. F. Bevilacqua, D. Piguet, P. Marquet, J. D. Gross, B. J. Tromberg, and C. Depeursinge, "*In Vivo* Local Determination of Tissue Optical Properties," *Proc. SPIE* 3194, 1997, pp. 262–268.

656. I. V. Yaroslavsky and V. V. Tuchin, "An Inverse Monte Carlo Method for Spectrophotometric Data Processing," *Proc. SPIE*, vol. 2100, 1994, pp. 57–68.

657. R. Marchesini, A. Bertoni, S. Andreola, et al., "Extinction and Absorption Coefficients and Scattering Phase Functions of Human Tissues *In Vitro*," *Appl. Opt.*, vol. 28, 1989, pp. 2318–2324.

658. S. L. Jacques, C. A. Alter, and S. A. Prahl, "Angular Dependence of the He-Ne Laser Light Scattering by Human Dermis," *Lasers Life Sci.*, vol. 1, 1987, pp. 309–333.

659. I. Driver, C. P. Lowdell, and D. V. Ash, "*In Vivo* Measurement of the Optical Interaction Coefficients of Human Tumours at 630 nm," *Phys. Med. Biol.*, vol. 36, 1991, pp. 805–813.

660. V. G. Peters, D. R. Wyman, M. S. Patterson, and G. L. Frank, "Optical Properties of Normal and Diseased Human Tissues in the Visible and Near Infrared," *Phys. Med. Biol.*, vol. 35, 1990, pp. 1317–1334.

661. M. Seyfried, "Optical Radiation Interaction with Living Tissue," in *Radiation Measurement in Photobiology*, Academic, New York, 1989, pp. 191–223.

662. A. Roggan, O. Minet, C. Schröder, and G. Müller, "The Determination of Optical Tissue Properties with Double Integrating Sphere Technique and Monte Carlo Simulations," *Proc. SPIE* 2100, 1994, pp. 42–56.

663. J. W. Pickering, C. J. M. Moes, H. J. C. M. Sterenborg, et al., "Two Integrating Spheres with an Intervening Scattering Sample," *J. Opt. Soc. Am. A.*, vol. 9, 1992, pp. 621–631.

664. J. W. Pickering, S. A. Prahl, N. van Wieringen, et al., "Double–Integrating Sphere System for Measuring the Optical Properties of Tissue," *Appl. Opt.*, vol. 32, 1993, pp. 399–410.

665. A. N. Yaroslavsky, I. V. Yaroslavsky, T. Goldbach, and H.-J. Schwarzmaier, "Inverse Hybrid Technique for Determining the Optical Properties of Turbid Media from Integrating-Sphere Measurements," *Appl. Opt.*, vol. 35, no. 34, 1996, pp. 6797–6809.

666. H.-J. Schwarzmaier, A. N. Yaroslavsky, I. V. Yaroslavsky, et al., "Optical Properties of Native and Coagulated Human Brain Structures," *Proc. SPIE* 2970, 1997, pp. 492–499.

667. E. K. Chan, B. Sorg, D. Protsenko, M. O'Neil, M. Motamedi, and A. J. Welch, "Effects of Compression on Soft Tissue Optical Properties," *IEEE J. Select. Tops Quant. Electr.*, vol. 2, no. 4, 1996, pp. 943–950.

668. E. Chan, T. Menovsky, and A. J. Welch, "Effect of Cryogenic Grinding of Soft-Tissue Optical Properties," *Appl. Opt.*, vol. 35, no. 22, 1996, pp. 4526–4532.

669. I. Fine, E. Loewinger, A. Weinreb, and D. Weinberger, "Optical Properties of the Sclera," *Phys. Med. Biol.*, vol. 30, 1985, pp. 565–571.

670. L.-H. Wang and S. L. Jacques, "Use of Laser Beam with an Oblique Angle of Incidence to Measure the Reduced Scattering Coefficient of a Turbid Medium," *Appl. Opt.*, vol. 34, 1995, pp. 2362–2366.

671. S.-P. Liu, L.-H. Wang, S. L. Jacques, and F. K. Tittel, "Measurement of Tissue Optical Properties by the Use of Oblique-Incidence Optical Fiber Reflectometry," *Appl. Opt.*, vol. 36, 1997, pp. 136–143.

672. A. N. Yaroslavsky, A. Vervoorts, A. V. Priezzhev, I. V. Yaroslavsky, J. G. Moser, and H.-J. Schwarzmaier, "Can Tumor Cell Suspension Serve as an Optical Model of Tumor Tissue in situ?" *Proc. SPIE* 3565, 1999, pp. 165–173.

673. A. N. Yaroslavsky, I. V. Yaroslavsky, and H.-J. Schwarzmaier, "Small-Angle Approximation to Determine Radiance Distribution of a Finite Beam Propagating through Turbid Medium," *Proc. SPIE*, vol. 3195, 1998, pp. 110–120.

674. D. W. Ebert, C. Roberts, S. K. Farrar, W. M. Johnston, A. S. Litsky, and A. L. Bertone, "Articular Cartilage Optical Properties in the Spectral Range 300–850 nm," *J. Biomed. Opt.*, vol. 3, 1998, pp. 326–333.

675. D. K. Sardar, M. L. Mayo, and R. D. Glickman, "Optical Characterization of Melanin," *J. Biomed. Opt.*, vol. 6, 2001, pp. 404–411.

676. M. Hammer and D. Schweitzer, "Quantitative Reflection Spectroscopy at the Human Ocular Fundus," *Phys. Med. Biol.*, vol. 47, 2002, pp. 179–191.

677. T. L. Troy and S. N. Thennadil, "Optical Properties of Human Skin in the Near Infrared Wavelength Range of 1000 to 2200 nm," *J. Biomed. Opt.*, vol. 6, 2001, pp. 167–176.

678. Y. Du, X. H. Hu, M. Cariveau, X. Ma, G. W. Kalmus, and J. Q. Lu, "Optical Properties of Porcine Skin Dermis between 900 nm and 1500 nm," *Phys. Med. Biol.*, vol. 46, 2001, pp. 167–181.

679. S. A. Prahl, "Light Transport in Tissues," PhD Thesis, Univ. of Texas, Austin, 1988.

680. S. A. Prahl, "The Inverse Adding-Doubling Program," htttp://omlc.ogi.edu/software/iad/index.html.

681. O. Khalil, S.-j. Yeh, M. G. Lowery, X. Wu, C. F. Hanna, S. Kantor, T.-W. Jeng, J. S. Kanger, R. A. Bolt, and F. F. de Mul, "Temperature Modulation of the Visible and Near Infrared Absorption and Scattering Coefficients of Human Skin," *J. Biomed. Opt.*, vol. 8(2), 2003, pp. 191–205.

682. T. J. Pfefer, L. S. Matchette, C. L. Bennett, J. A. Gall, J. N. Wilke, A. J. Durkin, and M. N. Ediger, "Reflectance-Based Determination of Optical Properties in Highly Attenuating Tissue," *J. Biomed. Opt.*, vol. 8(2), 2003, pp. 206–215.

683. F. Thueler, I. Charvet, F. Bevilacqua, M. St. Ghislain, G. Ory, P. Marquet, P. Meda, B. Vermeulen, and C. Depeursinge, "*In Vivo* Endoscopic Tissue Diagnostics Based on Spectroscopic Absorption, Scattering, and Phase Function Properties," *J. Biomed. Opt.*, vol. 8, no. 3, 2003, pp. 495–503.

684. S. L. Jacques, "Simple Monte Carlo Code," http://omlc.ogi.edu/software/index.html.

685. G. Kumar and J. M. Schmitt, "Optimal Probe Geometry for Near-Infrared Spectroscopy of Biological Tissue," *Appl. Opt.*, vol. 36, no. 10, 1997, pp. 2286–2293.

686. A. Kienle, F. K. Forster, and R. Hibst, "Influence of the Phase Function on Determination of the Optical Properties of Biological Tissue by Spatially Resolved Reflectance," *Opt. Lett.*, vol. 26, 2001, pp. 1571–1573.

687. C. K. Hayakawa, J. Spanier, F. Bevilacqua, A. K. Dunn, J. S. You, B. J. Tromberg, and V. Venugopalan, "Perturbation Monte Carlo Methods to Solve Inverse Photon Migration Problems in Heterogeneous Tissues," *Opt. Lett.*, vol. 26, 2001, pp. 1335–1337.

688. F. Bevilacqua and C. Depeursinge, "Monte Carlo Study of Diffuse Reflectance at Source-Detector Separations Close to One Transport Mean Free Path," *J. Opt. Soc. Am. A*, vol. 16(2), 1999, pp. 2935–2945.

689. W. Steenbergen, R. Kolkman, and F. de Mul, "Light-Scattering Properties of Undiluted Human Blood Subjected to Simple Shear," *J. Opt. Soc. Am. A*, vol. 16, no. 12, 1999, pp. 2959–2967.

690. S. T. Flock, B. C. Wilson, and M. S. Patterson, "Total Attenuation Coefficient and Scattering Phase Function of Tissues and Phantom Materials at 633 nm," *Med. Phys.*, vol. 14, 1987, pp. 835–841.

691. A. Roggan, K. Dörschel, O. Minet, D. Wolff, and G. Müller, "The Optical Properties of Biological Tissue in the Near Infrared Wavelength Range—Review and Measurements," in *Laser-Induced Interstitial Thermotherapy*, vol. PM25, G. Müller and A. Roggan (eds.), SPIE Press, Bellingham, WA, 1995, pp. 10–44.

692. S. Nickell, M. Hermann, M. Essenpreis, T. J. Farrell, U. Krämer, and M. S. Patterson, "Anisotropy of Light Propagation in Human Skin," *Phys. Med. Biol.*, vol. 45, 2000, pp. 2873–2886.

693. S. Stolik, J. A. Delgado, A. Pérez, and L. Anasagasti, "Measrement of the Penetration Depths of Red and Near Infrared Light in Human "*Ex Vivo*" Tissues," *J. Photochem. Photobiol. B: Biol.*, vol. 57, 2000, pp. 90–93.

694. S. Tauber, R. Baumgartner, K. Schorn, and W. Beyer, "Lightdosimetric Quantitative Analysis of the Human Petrous Bone: Experimental Study for Laser Irradiation of the Cochlea," *Laser Sur. Med.*, vol. 28, 2001, pp. 18–26.

695. F. Bevilacqua, D. Piguet, P. Marquet, J. D. Gross, B. J. Tromberg, and C. Depeursinge, "*In Vivo* Local Determination of Tissue Optical Properties: Applications to Human Brain," *Appl. Opt.*, vol. 38, 1999, pp. 4939–4950.

696. J. Mobley and Tuan Vo-Dinh, "Optical Properties of Tissues," in *Biomedical Photonics Handbook*, Tuan Vo-Dinh (ed.), CRC Press, Boca Raton, 2003, pp. 2-1–2-75.

697. A. Roggan, D. Schäder, U. Netz, J.-P. Ritz, C.-T. Germer, and G. Müller, "The Effect of Preparation Technique on the Optical Parameters of Biological Tissue," *Appl. Phys.* B, vol. 69, 1999, pp. 445–453.

698. W. Gottschalk, Ein Messverfahren zur Bestimmung der Optischen Parameter Biologischer Gevebe *In Vitro*, Dissertation 93 HA8984 Universitaet Fridriciana, Karlsruhe, 1992.

699. N. Ghosh, S. K. Mohanty, S. K. Majumder, and P. K. Gupta, "Measurement of Optical Transport Properties of Normal and Malignant Human Breast Tissue," *Appl. Opt.*, vol. 40, 2001, pp. 176–184.

700. R. Hornung, T. H. Pham, K. A. Keefe, M. W. Berns, Y. Tadir, and B. J. Tromberg, "Quantitative Near-Infrared Spectroscopy of Cervical Dysplasia In Vivo," *Hum. Reprod.*, vol. 14, 1999, pp. 2908–2916.

701. E. Gratton, S. Fantini, M. A. Franceschini, G. Gratton, and M. Fabiani, "Measurements of Scattering and Absorption Changes in Muscle and Brain," *Phil. Trans. R. Soc. Lond.* B, vol. 352, 1997, pp. 727–735.

702. T. J. Farrell, M. S. Patterson, and M. Essenpreis, "Influence of Layered Tissue Architecture on Estimates of Tissue Optical Properties Obtained from Spatially-Resolved Diffuse Reflectometry," *Appl. Opt.*, vol. 37, 1998, pp. 1958–1972.

703. A. N. Bashkatov, "Controlling of Optical Properties of Tissues at Action by Osmotically Active Immersion Liquids," Cand. Science Thesis, Saratov State Univ., Saratov, 2002.

704. E. A. Genina, A. N. Bashkatov, V. I. Kochubey, and V. V. Tuchin, "Optical Clearing of Human *Dura Mater*," *Opt. Spectrosc.*, vol. 98, no. 3, 2005, pp. 470–476.

705. A. N. Bashkatov, E. A. Genina, V. I. Kochubey, and V. V. Tuchin, "Optical Properties of the Subcutaneous Adipose Tissue in the Spectral Range 400–2500 nm," *Opt. Spectrosc.*, vol. 99, no. 5, 2005, pp. 836–842.

706. A. N. Bashkatov, E. A. Genina, V. I. Kochubey, V. V. Tuchin, E. E. Chikina, A. B. Knyazev, and O. V. Mareev, "Optical Properties of Mucous Membrane in the Spectral Range 350 to 2000 nm," *Opt. Spectrosc.*, vol. 97, no. 6, 2004, pp. 978–983.

707. S. Fantini, S. A. Walker, M. A. Franceschini, M. Kaschke, P. M. Schlag, and K. T. Moesta, "Assessment of the Size, Position, and Optical Properties of Breast Tumors *In Vivo* by Noninvasive Optical Methods," *Appl. Opt.*, vol. 37, 1998, pp. 1982–1989.

708. S. Fantini, D. Hueber, M. A. Franceschini, E. Gratton, W. Rosenfeld, P. G. Stubblefield, D. Maulik, and M. R. Stankovic, "Non-Invasive Optical Monitoring of the Newborn Piglet Brain using Continuous-Wave and Frequency-Domain Spectroscopy," *Phys. Med. Biol.*, vol. 44, 1999, 1543–1563.

709. J. F. Black, J. K. Barton, G. Frangineas, and H. Pummer, "Cooperative Phenomena in Two-Pulse Two-Color Laser Photocoagulation of Cutaneous Blood Vessels," *Proc. SPIE* 4244, 2001, pp. 13–24.

710. S. L. Jacques, "Origins of Tissue Optical Properties in the UVA, Visible and NIR Regions," in *Advances in Optical Imaging and Photon Migration*, R. R. Alfano and J. G. Fujimoto (eds.), OSA TOPS 2, Optical Society of America, Washington, DC, 1996, pp. 364–371.

711. D. Levitz, L. Thrane, M. H. Frosz, P. E. Andersen, C. B. Andersen, S. Andersson-Engels, J. Valanciunaite, J. Swartling, and P. R. Hansen, "Determination of Optical Scattering Properties of Highly-Scattering Media in Optical Coherence Tomography Images," *Optics Express* vol. 12, 2004, pp. 249–259, http://www.opticsexpress.org/abstract.cfm?URI=OPEX-12-2-249.

712. A. Knüttel and M. Boehlau-Godau, "Spatially Confined and Temporally Resolved Refractive Index and Scattering Evaluation in Human Skin Performed with Optical Coherence Tomography," *J. Biomed. Opt.*, vol. 5, 2000, pp. 83–92.

713. A. Knüttel, S. Bonev, and W. Knaak, "New Method for Evaluation of *In Vivo* Scattering and Refractive Index Properties Obtained with Optical Coherence Tomography," *J. Biomed. Opt.*, vol. 9, 2004, pp. 265–273.

714. M. J. Holboke, B. J. Tromberg, X. Li, N. Shah, J. Fishkin, D. Kidney, J. Butler, B. Chance, and A. G. Yodh, "Three-Dimensional Diffuse Optical Mammography with Ultrasound Localization in a Human Subject," *J. Biomed. Opt.*, vol. 5, 2000, pp. 237–247.

715. B. J. Tromberg, N. Shah, R. Lanning, A. Cerussi, J. Espinoza, T. Pham, L. Svaasand, and J. Butler, "Non-invasive *In Vivo* Characterization of Breast Tumors using Photon Migration Spectroscopy," *Neoplasia*, vol. 2, 2000, pp. 26–40.

716. I. V. Turchin, E. A. Sergeeva, L. S. Dolin, and V. A. Kamensky, "Estimation of Biotissue Scattering Properties from OCT Images Using a Small-Angle Approximation of Transport Theory," *Laser Physics*, vol. 13, 2003, pp. 1524–1529.

717. L. S. Dolin, F. I. Feldchtein, G. V. Gelikonov, V. M. Gelikonov, N. D. Gladkova, R. R. Iksanov, V. A. Kamensky, R. V. Kuranov, A. M. Sergeev, N. M. Shakhova, and I. V. Turchin, "Fundamentals of OCT and Clinical Applications of Endoscopic OCT," Chapter 17, in *Coherent-Domain Optical Methods: Biomedical Diagnostics, Environmental and Material Science*, vol. 2, V. V. Tuchin (ed.), Kluwer Academic Publishers, Boston, 2004, pp. 211–270.

718. I. V. Turchin, V. A. Kamensky, E. A. Sergeeva, and N. M. Shakhova, "OCT Image Processing Algorithm for Differentiation Biological Tissue Pathologies," *13 International Laser Physics Workshop*, Book of Abstracts, Trieste, Italy, 2004, p. 189.

719. M. Firbank, M. Hiraoka, M. Essenpreis, and D. T. Delpy, "Measurement of the Optical Properties of the Skull in the Wavelength Range 650–950 nm," *Phys. Med. Biol.*, vol. 38, 1993, pp. 503–510.

720. N. Ugryumova, S. J. Matcher, and D. P. Attenburrow, "Measurement of Bone Mineral Density via Light Scattering," *Phys. Med. Biol.*, vol. 49, 2004, pp. 469–283.

721. A. I. Kholodnykh, I. Y. Petrova, K. V. Larin, M. Motamedi, and R. O. Esenaliev, "Precision of Measurement of Tissue Optical Properties with Optical Coherence Tomography," *Appl. Opt.*, vol. 42, 2003, pp. 3027–3037.

722. P. Rol, P. Niederer, U. Dürr, P.-D. Henchoz, and F. Fankhauser, "Experimental Investigation on the Light Scattering Properties of the Human Sclera," *Laser Light Ophthalmol.*, vol. 3, 1990, pp. 201–212.

723. P. O. Rol, Optics for Transscleral Laser Applications: Dissertation No. 9655 for Doctor of Natural Sciences, Swiss Federal Institute of Technology, Zurich, Switzerland, 1992.

724. A. N. Yaroslavsky, I. V. Yaroslavsky, T. Goldbach, and H.-J. Schwarzmaier, "Optical Properties of Blood in the Near-Infrared Spectral Range," *Proc. SPIE* 2678, 1996, pp. 314–324.

725. A. N. Yaroslavsky, A. V. Priezzhev, J. Rodriguez, I. V. Yaroslavsky, and H. Battarbee, "Optics of Blood," in *Handbook of Optical Biomedical Diagnostics*, Chapter 2, vol. PM107, V. V. Tuchin (ed.), SPIE Press, Bellingham, WA, 2002, pp. 169–216.

726. A. N. Bashkatov, E. A. Genina, V. I. Kochubey, and V. V. Tuchin, "Optical Properties of Human Skin, Subcutaneous and Mucous Tissues in the Wavelength Range from 400 to 2000 nm," *J. Phys. D: Appl. Phys.*, vol. 38, 2005, pp. 2543–2555.

727. L. O. Reynolds and N. J. McCormick, "Approximate Two-Parameter Phase Function for Light scattering," *J. Opt. Soc. Am.*, vol. 70, 1980, pp. 1206–1212.

728. P. W. Barber and S. C. Hill, *Light Scattering by Particles: Computational Methods*, World Scientific, Singapore, 1990.

729. A. N. Yaroslavsky, I. V. Yaroslavsky, T. Goldbach, and H.-J. Schwarzmaier, "Different Phase Function Approximations to Determine Optical Properties of Blood: A Comparison," *Proc. SPIE* 2982, 1997, pp. 324–330.

730. M. Hammer, D. Schweitzer, B. Michel, E. Thamm, and A. Kolb, "Single Scattering by Red Blood Cells," *Appl. Opt.*, vol. 37, no. 31, 1998, 7410–7418.

731. I. J. Bigio and J. R. Mourant, "Ultraviolet and Visible Spectroscopies for Tissue Diagnostics: Fluorescence Spectroscopy and Elastic-Scattering Spectroscopy," *Phys. Med. Biol.*, vol. 42, 1997, pp. 803–814.

732. L. T. Perelman and V. Backman, "Light Scattering Spectroscopy of Epithelial Tissues: Principles and Applications," Chapter 12, in *Optical Biomedical Diagnostics*, vol. PM107, V. V. Tuchin (ed.), SPIE Press, Bellingham, WA, 2002, pp. 675–724.

733. V. Backman, M. Wallace, L. T. Perelman, R. Gurjar, G. Zonios, M. G. Müller, Q. Zhang, T. Valdez, J. T. Arendt, H. S. Levin, T. McGillican, K. Badizadegan, M. Seiler, S. Kabani, I. Itzkan, M. Fitzmaurice, R. R. Dasari, J. M. Crawford, J. Van Dam, and M. S. Feld, "Detection of Preinvasive Cancer Cells. Early-Warning Changes in Precancerous Epithelial Cells Can be Spotted *In Situ*," *Nature*, vol. 406, no. 6791, 2000, pp. 35–36.

734. C. Yang, L. T. Perelman, A. Wax, R. R. Dasari, and M. S. Feld, "Feasibility of Field-Based Light Scattering Spectroscopy," *J. Biomed. Opt.*, vol. 5, 2000, pp. 138–143.

735. G. Zonios, L. T. Perelman, V. Backman, R. Manoharan, M. Fitzmaurice, and M. S. Feld. "Diffuse Reflectance Spectroscopy of Human Adenomatous Colon Polyps *In Vivo*," *Appl. Opt.*, vol. 38, 1999, pp. 6628–6637.

736. G. Marguez, L. V. Wang, S.-P. Lin, J. A. Swartz, and S. Thomsen, "Anisotropy in the Absorption and Scattering Spectra of the Chicken Breast Tissue," *Appl. Opt.*, vol. 37, no. 4, 1998, pp. 798–804.

737. V. V. Tuchin, X. Xu, and R. K. Wang, "Dynamic Optical Coherence Tomography in Optical Clearing, Sedimentation and Aggregation Study of Immersed Blood," *Appl. Opt.-OT*, vol. 41, no. 1, 2002, pp. 258–271.

738. A. N. Bashkatov, E. A. Genina, V. I. Kochubey, N. A. Lakodina, and V. V. Tuchin, "Optical Clearing of Human Cranial Bones by Administration of Immersion Agents," NATO Advanced Study Inst. on Biophotonics: From Fundamental Principles to Health, Environment, Security and Defense Applications, Ottawa, Ontario, Canada, September 29–October 9, 2004.

739. A. N. Bashkatov, E. A. Genina, V. I. Kochubey, N. A. Lakodina, and V. V. Tuchin, "Optical Properties of Human Cranial Bones in the Spectral Range from 800 to 2000 nm," *Proc. SPIE* 6163, 2006, pp. 616310-1–11.

740. L. Reynolds, C. Johnson, and A. Ishimaru, "Diffuse Reflectance from a Finite Blood Medium: Applications to the Modeling of Fiber Optic Catheters," *Appl. Opt.*, vol. 15, 1976, pp. 2059–2067.

741. J. M. Steinke and A. P. Shepherd, "Comparison of Mie Theory and the Light Scattering of Red Blood Cells," *Appl. Opt.*, vol. 27, 1988, pp. 4027–4033.

742. D. K. Sardar and L. B. Levy, "Optical Properties of Whole Blood," *Lasers Med. Sci.*, vol. 13, 1998, pp. 106–111.

743. R. N. Pittman, "*In Vivo* Photometric Analysis of Hemoglobin," *Annals Biomed. Eng.*, vol. 14, 1986, pp. 119–137.

744. M. A. Bartlett and H. Jiang, "Effect of Refractive Index on the Measurement of Optical Properties in Turbid Media," *Appl. Opt.*, vol. 40, 2001, pp. 1735–1741.

745. D. Arifler, M. Guillaud, A. Carraro, A. Malpica, M. Follen, and R. Richards-Kortum, "Light Scattering from Normal and Dysplastic Cervical Cells at Different Epithelial Depths: Finite-Difference Time-Domain Modeling with a Perfectly Matched Layer Boundary Condition," *J. Biomed. Opt.*, vol. 8, 2003, pp. 484–494.

746. S. Cheng, H. Y. Shen, G. Zhang, C. H. Huang, and X. J. Huang, "Measurement of the Refractive Index of Biotissue at Four Laser Wavelengths," *Proc. SPIE* 4916, 2002, pp. 172–176.

747. H. Liu and S. Xie, "Measurement Method of the Refractive Index of Biotissue by Total Internal Reflection," *Appl. Opt.*, vol. 35, 1996, pp. 1793–1795.

748. V. V. Tuchin, D. M. Zhestkov, A. N. Bashkatov, and E. A. Genina, "Theoretical Study of Immersion Optical Clearing of Blood in Vessels at Local Hemolysis," *Optics Express*, vol. 12, 2004, pp. 2966–2971.

749. R. Barer, K. F. A. Ross, and S. Tkaczyk, "Refractometry of Living Cells," *Nature*, vol. 171, no. 4356, 1953, pp. 720–724.

750. M. Haruna, K. Yoden, M. Ohmi, and A. Seiyama, "Detection of Phase Transition of a Biological Membrane by Precise Refractive-Index Measurement Based on the Low Coherence Interferometry," *Proc. SPIE* 3915, 2000, pp. 188–193.

751. D. J. Faber, M. C. G. Aalders, E. G. Mik, B. A. Hooper, M. J. C. van Gemert, and T. G. van Leeuwen, "Oxygen Saturation-Dependent Absorption and Scattering of Blood," *Phys. Rev. Lett.*, vol. 93, 2004, pp. 028102-1–4.

752. V. V. Tuchin, R. K. Wang, E. I. Galanzha, N. A. Lakodina, and A. V. Solovieva, "Monitoring of Glycated Hemoglobin in a Whole Blood by Refractive Index Measurement with OCT," Conference Program CLEO/QELS, Baltimore, June 1–6, Optical Society of America, Washington, DC, 2003, p. 120.

753. V. V. Tuchin, R. K. Wang, E. I. Galanzha, J. B. Elder, and D. M. Zhestkov, "Monitoring of Glycated Hemoglobin by OCT Measurement of Refractive Index," *Proc. SPIE* 5316, 2004, pp. 66–77.

754. G. Mazarevica, T. Freivalds, and A. Jurka, "Properties of Erythrocyte Light Refraction in Diabetic Patients," *J. Biomed. Opt.*, vol. 7, 2002, pp. 244–247.

755. S. F. Shumilina, "Dispersion of Real and Imaginary Part of the Complex Refractive Index of Hemoglobin in the Range 450 to 820 nm," *Bullet. Beloruss. SSR Acad. Sci., Phys.-Math. Ser.*, no. 1, 1984, pp. 79–84.

756. J. Hempe, R. Gomez, R. McCarter, and S. Chalew, "High and Low Hemoglobin Glycation Phenotypes in Type 1 Diabetes. A Challenge for Interpretation of Glycemic Control," *J. Diabetes Complications* vol. 16, 2002, pp. 313–320.

757. M. V. Volkenshtein, *Molecualar Optics*, Moscow, Fizmatlit, 1951.

758. G. V. Maksimov, O. G. Luneva, N. V. Maksimova, E. Matettuchi, E. A. Medvedev, V. Z. Pashchenko, and A. B. Rubin, "Role of Viscosity and Permeability of the Erythrocyte Plasma Membrane in Changes in Oxygen-Binding Properties of Hemoglobin During Diabetes Mellitus," *Bull. Exp. Biol. Med.*, vol. 140, no. 5, 2005, pp. 510–513.

759. A. N. Bashkatov, E. A. Genina, V. I. Kochubey, Yu. P. Sinichkin, A. A. Korobov, N. A. Lakodina, and V. V. Tuchin, "*In Vitro* Study of Control of Human Dura Mater Optical Properties by Acting of Osmotical Liquids," *Proc. SPIE* 4162, 2000, pp. 182–188.

760. A. N. Bashkatov, E. A. Genina, V. I. Kochubey, and V. V. Tuchin, "Estimation of Wavelength Dependence of Refractive Index of Collagen Fibers of Scleral Tissue," *Proc. SPIE* 4162, 2000, pp. 265–268.

761. A. N. Bashkatov, E. A. Genina, V. I. Kochubey, M. M. Stolnitz, T. A. Bashkatova, O. V. Novikova, A. Yu. Peshkova, and V. V. Tuchin, "Optical Properties of Melanin in the Skin and Skin-Like Phantoms," *Proc. SPIE* 4162, 2000, pp. 219–226.

762. Y. Kamai and T. Ushiki, "The Three-Dimensional Organization of Collagen Fibrils in the Human Cornea and Sclera," *Invest. Ophthalmol. & Visual Sci.*, vol. 32, 1991, pp. 2244–2258.

763. V. N. Grisimov, "Refractive Index of the Ground Material of Dentin," *Opt. Specrosc.*, vol. 77, 1994, pp. 272–273.

764. X. Wang, T. E. Milner, M. C. Chang, and J. S. Nelson, "Group Refractive Index Measurement of Dry and Hydrated Type I Collagen Films Using Optical Low-Coherence Reflectometry," *J. Biomed. Opt.*, vol. 1, no. 2, 1996, pp. 212–216.

765. W. V. Sorin and D. F. Gray, "Simalteneous Thickness and Group Index Measurements Using Optical Low-Coherence Refractometry," *IEEE Photon. Technol. Lett.*, vol. 4, 1992, pp. 105–107.

766. X. J. Wang, T. E. Milner, R. P. Dhond, W. V. Sorin, S. A. Newton, and J. S. Nelson, "Characterization of Human Scalp Hairs by Optical Low-Coherence Reflectometry," *Opt. Lett.*, vol. 20, 1995, pp. 524–526.

767. G. J. Tearney, M. E. Brezinski, J. F. Southern, B. E. Bouma, M. R. Hee, and J. G. Fujimoto, "Determination of the Refractive Index of Highly Scattering Human Tissue by Optical Coherence Tomography," *Opt. Lett.*, vol. 20, 1995, pp. 2258–2260.

768. M. Ohmi, Y. Ohnishi, K. Yoden, and M. Haruna, "*In Vitro* Simalteneous Measurement of Refractive Index and Thickness of Biological Tissue by the Low Coherence Interferometry," *IEEE Trans. Biomed. Eng.*, vol. 47, 2000, pp. 1266–1270.

769. X. Wang, C. Zhang, L. Zhang, L. Xue, and J. Tian, "Simalteneous Refractive Index and Thickness Measurement of Biotissue by Optical Coherence Tomography," *J. Biomed. Opt.*, vol. 7, 2002, pp. 628–632.

770. S. A. Alexandrov, A. V. Zvyagin, K. K. M. B. D. Silva, and D. D. Sampson, "Bifocal Optical Coherence Refractometry of Turbid Media," *Opt. Lett.*, vol. 28, 2003, pp. 117–119.

771. A. V. Zvyagin, K. K. M. B. D. Silva, S. A. Alexandrov, T. R. Hillman, J. J. Armstrong, T. Tsuzuki, and D. D. Sampson, "Refractive Index Tomography of Turbid Media by Bifocal Optical Coherence Refractometry," *Optics Express*, vol. 11, 2003, pp. 3503–3517.

772. Y. L. Kim, J. T. Walsh Jr., T. K. Goldstick, and M. R. Glucksberg, "Variation of Corneal Refractive Index with Hydration," *Phys. Med. Biol.*, vol. 49, 2004, pp. 859–868.

773. W. Drexler, C. K. Hitzenberger, A. Baumgartner, O. Findl, H. Sattmann, and A. F. Fercher, "Investigation of Dispersion Effects in Ocular Media by Multiple Wavelength Partial Coherence Interferometry," *Exp. Eye Res.*, vol. 66, 1998, pp. 25–33.

774. R. C. Lin, M. A. Shure, A. M. Rollins, J. A. Izatt, and D. Huang, "Group Index of the Human Cornea at 1.3-μm Wavelength Obtained In Vitro by Optical Coherence Domain Reflectometry," *Opt. Lett.*, vol. 29, 2004, pp. 83–85.

775. G. V. Gelikonov, V. M. Gelikonov, S. U. Ksenofontov, A. N. Morosov, A. V. Myakov, Yu. P. Potapov, V. V. Saposhnikova, E. A. Sergeeva, D. V. Shabanov, N. M. Shakhova, and E. V. Zagainova, "Compact Optical Coherence Microscope," Chapter 20, in *Coherent-Domain Optical Methods: Biomedical Diagnostics, Environmental and Material Science*, vol. 2, V. V. Tuchin (ed.), Kluwer Academic Publishers, Boston, 2004, pp. 345–362.

776. J. M. Schmitt, M. Yadlowsky, and R. F. Bonner, "Subsurface Imaging of Living Skin with Optical Coherence Microscopy," *Dermatology*, vol. 191, 1995, pp. 93–98.

777. G. Vargas, K. F. Chan, S. L. Thomsen, and A. J. Welch, "Use of Osmotically Active Agents to Alter Optical Properties of Tissue: Effects on the Detected Fluorescence Signal Measured through Skin," *Lasers Surg. Med.*, vol. 29, 2001, pp. 213–220.

778. D. W. Leonard and K. M. Meek, "Refractive Indices of the Collagen Fibrils and Extrafibrillar Material of the Corneal Stroma," *Biophys. J.*, vol. 72, 1997, pp. 1382–1387.

779. K. M. Meek, S. Dennis, and S. Khan, "Changes in the Refractive Index of the Stroma and its Extrafibrillar Matrix When the Cornea Swells," *Biophys. J.*, vol. 85, 2003, pp. 2205–2212.

780. R. A. Farrell and R. L. McCally, "Corneal Transparency" in *Principles and Practice of Ophthalmology*, D. A. Albert and F. A. Jakobiec (eds.), W. B. Saunders, Philadelphia, PA, 2000, pp. 629–643.

781. D. E. Freund, R. L. McCally, and R. A. Farrell, "Effects of Fibril Orientations on Light Scattering in the Cornea," *J. Opt. Soc. Am. A.*, vol. 3, 1986, pp. 1970–1982.

782. R. A. Farrell, D. E. Freund, and R. L. McCally, "Hierarchical Structure and Light Scattering in the Cornea," *Mat. Res. Soc. Symp. Proc.*, vol. 255, 1992, pp. 233–246.

783. R. A. Farrell, D. E. Freund, and R. L. McCally, "Research on Corneal Structure," *Johns Hopkins APL Techn. Digest.*, vol. 11, 1990, pp. 191–199.

784. M. S. Borcherding, L. J. Blasik, R. A. Sittig, J. W. Bizzel, M. Breen, and H. G. Weinstein, "Proteoglycans and Collagen Fiber Organization in Human Corneoscleral Tissue," *Exp Eye Res.*, vol. 21, 1975, pp. 59–70.

785. M. Spitznas, "The Fine Structure of Human Scleral Collagen," *Am. J. Ophthalmol.*, vol. 71, no. 1, 1971, pp. 68–75.

786. Y. Huang and K. M. Meek, "Swelling Studies on the Cornea and Sclera: the Effect of pH and Ionic Strength," *Biophys. J.*, vol. 77, 1999, pp. 1655–1665.

787. S. Vaezy and J. I. Clark, "Quantitative Analysis of the Microstructure of the Human Cornea and Sclera Using 2-D Fourier Methods," *J. Microsc.*, vol. 175, no. 2, 1994, pp. 93–99.

788. Z. S. Sacks, R. M. Kurtz, T. Juhasz, and G. A. Mourau, "High Precision Subsurface Photodisruption in Human Sclera," *J. Biomed. Opt.*, vol. 7, no. 3, 2002, pp. 442–450.

789. F. A. Bettelheim, "Physical Basis of Lens Transparency," in *The Ocular Lens: Structure, Function and Pathology*, H. Maisel (ed.), Marcel-Dekker, New York, 1985.

790. S. Zigman, G. Sutliff, and M. Rounds, "Relationships between Human Cataracts and Environmental Radiant Energy. Cataract Formation, Light scattering and Fluorescence," *Lens Eye Toxicity Res.*, vol. 8, 1991, pp. 259–280.

791. J. Xu, J. Pokorny, and V. C. Smith, "Optical Density of the Human Lens," *J. Opt. Soc. Am. A.*, vol. 14, no. 5, 1997, pp. 953–960.

792. B. K. Pierscionek and R. A. Weale, "Polarising Light Biomicroscopy and the Relation between Visual Acuity and Cataract," *Eye*, vol. 9, 1995, pp. 304–308.

793. B. K. Pierscionek, "Aging Changes in the Optical Elements of the Eye," *J. Biomed. Opt.*, vol. 1, no. 3, 1996, pp. 147–156.

794. J. A. van Best and E. V. M. J. Kuppens, "Summary of Studies on the Blue–Green Autofluorescence and Light Transmission of the Ocular Lens," *J. Biomed. Opt.*, vol. 1, no. 3, 1996, pp. 243–250.

795. F. A. Bettelheim, A. C. Churchill, W. G. Robinson, Jr., and J. S. Zigler, Jr., "Dimethyl Sulfoxide Cataract: a Model for Optical Anisotropy Fluctuations," *J. Biomed. Opt.*, vol. 1, no. 3, 1996, pp. 273–279.

796. N.-T. Yu, B. S. Krantz, J. A. Eppstein, K. D. Ignotz, M. A. Samuels, J. R. Long, and J. F. Price, "Development of Noninvasive Diabetes Screening Device Using the Ratio of Fluorescence to Rayleigh Scattered Light," *J. Biomed. Opt.*, vol. 1, no. 3, 1996, pp. 280–288.

797. M. J. Costello, T. N. Oliver, and L. M. Cobo, "Cellular Architecture in Aged-Related Human Nuclear Cataracts," *Invest. Ophthal. & Vis. Sci.*, vol. 3, no. 11, 1992, pp. 2244–2258.

798. I. L. Maksimova, D. A. Zimnyakov, and V. V. Tuchin, "Controlling of Tissue Optical Properties I. Spectral Characteristics of Eye Sclera," *Opt. Spectrosc.*, vol. 89, 2000, pp. 78–86.

799. D. A. Zimnyakov, I. L. Maksimova, and V. V. Tuchin, "Controlling of Tissue Optical Properties II. Coherent Methods of Tissue Structure Study," *Opt. Spectrosc.*, vol. 88, 2000, pp. 1026–1034.

800. J. Dillon, "The Photophysics and Photobiology of the Eye," *J. Photochem. Photobiol. B: Biol.*, vol. 10, 1991, pp. 23–40.

801. G. B. Benedek, "Theory of Transparency of the Eye," *Appl. Opt.*, vol. 10, no. 3, 1971, pp. 459–473.

802. A. Tardieu and M. Delaye, "Eye Lens Proteins and Transparency from Light Transmission Theory to Solution X-ray Structural Analysis," *Ann. Rev. Biophys. Chem.*, vol. 17, 1988, pp. 47–70.

803. A. V. Krivandin, "On the Supramolecular Structure of Eye Lens Crystallins. The Study by Small-Angle X-ray Scattering," *Biophysica*, vol. 46, no. 6, 1997, pp. 1274–1278.

804. S. Vaezy and J. I. Clark, "Characterization of the Cellular Microstructures of Ocular Lens Using 2-D Power Law Analysis," *Ann. Biomed. Eng.*, vol. 23, 1995, pp. 482–490.

805. J. M. Ziman, *Models of Disorder: The Theoretical Physics of Homogeneously Disordered Systems*, Cambridge Univer. Press, London, New York, Melbourne, 1979.

806. M. S. Wertheim, "Exact Solution of the Percus-Yevick Integral Equation for Hard Spheres," *Phys. Rev. Lett.*, vol. 10, no. 8, 1963, pp. 321–323.

807. J. L. Lebovitz, "Exact Solution of Generalized Percus-Yevick Equation for a Mixture of Hard Spheres," *Phys. Rev.*, vol. 133, no. 4A, 1964, pp. 895–899.

808. R. J. Baxter, "Ornstein-Zernike Relation and Percus-Yevick Approximation for Fluid Mixtures," *J. Chem. Phys.*, vol. 52, no. 9, 1970, pp. 4559–4562.

809. A. P. Ivanov, V. A. Loiko, and V. P. Dik, *Light Propagation in Densely Packed Disperse Media*, Nauka i Tekhnika, Minsk, 1988.

810. V. G. Vereshchagin and A. N. Ponyavina, "Statistical Characteristic and Transparency of Thin Closely Packed Disperse Layer," *Zh. Prikl. Spektr. (J. Appl. Spectrosc.)*, vol. 22, no. 3, 1975, pp. 518–524.

811. N. L. Larionova, I. L. Maksimova, and V. V. Tuchin, "The Scattering Spectra and Color of Disperse Systems of Weakly Absorbing Particles," *Opt. Spectrosc.*, vol. 93, no. 2, 2002, pp. 273–281.

812. Z. S. Sacks, D. L. Craig, R. M. Kurtz, T. Juhasz, and G. Mourou, "Spatially Resolved Transmission of Highly Focused Beams Through Cornea and Sclera Between 1400 and 1800 nm," *Proc. SPIE* 3726, 1999, pp. 522–527.

813. T. J. T. P. van den Berg and K. E. W. P. Tan, "Light Transmittance of the Human Cornea from 320 to 700 nm for Different Ages," *Vision Res.*, vol. 33, 1994, pp. 1453–1456.

814. A. N. Korolevich, A. Ya. Khairulina, and L. P. Shubochkin, "Influence of Large Biological Cells Aggregation on Elements of the Light Scattering Matrix," *Opt. Spectrosc.*, vol. 77, 1994, pp. 278–282.

815. V. F. Izotova, I. L. Maksimova, and S. V. Romanov, "Analysis of accuracy of laser polarization nephelometer," *Opt. Spectrosc.*, Vol. 80, pp. 1001–1007 (1996).

816. P. S. Hauge, "Recent Developments in Instruments in Ellipsometry," *Surface Science*, vol. 96, 1980, pp. 108–140.

817. W. P. van de Merwe, D. R. Huffman, and B. V. Bronk, "Reproducibility and Sensitivity of Polarized Light Scattering for Identifying Bacterial Suspension," *Appl. Opt.*, vol. 28, 1989, pp. 5052–5057.

818. B. V. Bronk, S. D. Druger, J. Czege, and W. van de Merwe, "Measuring Diameters of Rod-Shaped Bacteria *In Vivo* with Polarized Light Scattering," *Biophys. J.*, vol. 69, 1995, pp. 1170–1177.

819. B. G. de Grooth, L. W. M. M. Terstappen, G. J. Puppels, and J. Greve, "Light-Scattering Polarization Measurements as a New Parameter in Flow Cytometry," *Cytometry*, vol. 8, 1987, pp. 539–544.

820. R. M. P. Doornbos, A. G. Hoekstra, K. E. I. Deurloo, B. G. de Grooth, P. M. A. Sloot, and J. Greve, "Lissajous-Like patterns in Scatter Plots of Calibration Beads," *Cytometry*, vol. 16, 1994, pp. 236–242.

821. O. J. Lokberg, "Speckles and Speckle Techniques for Biomedical Applications," *Proc. SPIE*, vol. 1524, 1991, pp. 35–47.

822. J. C. Dainty (ed.), *Laser Speckle and Related Phenomena*, 2nd ed., Springer-Verlag, New York, 1984.

823. S. A. Akhmanov, Yu. E. D'yakov, and A. S. Chirkin, *Introduction to Statistical Radiophysics and Optics*, Nauka, Moscow, 1981.

824. J. C. Dainty, "The Statistics of Speckle Patterns," in *Progress in Optics XIV*, E. Wolf (ed.), vol. 14, North Holland, 1976, pp. 3–48.

825. D. A. Zimnyakov, "Coherence Phenomena and Statistical Properties of Multiply Scattered Light," Chapter 4, in *Handbook of Optical Biomedical Diagnostics*, vol. PM107, V. V. Tuchin (ed.), SPIE Press, Bellingham, WA, 2002, pp. 265–319.

826. E. I. Galanzha, G. E. Brill, Y. Aisu, S. S. Ulyanov, and V. V. Tuchin, "Speckle and Doppler Methods of Blood and Lymph Flow Monitoring," Chapter 16, in *Handbook of Optical Biomedical Diagnostics*, vol. PM107, V. V. Tuchin (ed.), SPIE Press, Bellingham, WA, 2002, pp. 881–937.

827. D. A. Zimnyakov, J. D. Briers, and V. V. Tuchin, "Speckle Technologies for Monitoring and Imaging of Tissuelike Phantoms," Chapter 18, in *Handbook of Optical Biomedical Diagnostics*, vol. PM107, V. V. Tuchin (ed.), SPIE Press, Bellingham, WA, 2002, pp. 987–1036.

828. S. J. Kirkpatrick and D. D. Duncan, "Optical Assessment of Tissue Mechanics," Chapter 19, in *Handbook of Optical Biomedical Diagnostics*, vol. PM107, V. V. Tuchin (ed.), SPIE Press, Bellingham, WA, 2002, pp. 1037–1084.

829. D. A. Zimnyakov and V. V. Tuchin, "Speckle Correlometry" in *Biomedical Photonics Handbook*, Tuan Vo-Dinh (ed.), CRC Press, Boca Raton, 2003, pp. 14-1–14-23.

830. D. A. Zimnyakov, "Light Correlation and Polarization in Multiply Scattering Media: Industrial and Biomedical Applications," Chapter 1 in *Coherent-Domain Optical Methods: Biomedical Diagnostics, Environmental and Material Science*, V. V. Tuchin (ed.), Kluwer Academic Publishers, Boston, vol. 1, 2004, pp. 3–41.

831. Q. Luo, H. Cheng, Z. Wang, and V. V. Tuchin, "Laser Speckle Imaging of Cerebral Blood Flow," Chapter 5 in *Coherent-Domain Optical Methods: Biomedical Diagnostics, Environmental and Material Science*, V. V. Tuchin (ed.), Kluwer Academic Publishers, Boston, vol. 1, 2004, pp. 165–195.

832. V. P. Ryabukho, "Diffraction of Interference Fields on Random Phase Objects," Chapter 7 in *Coherent-Domain Optical Methods: Biomedical Diagnostics, Environmental and Material Science*, V. V. Tuchin (ed.), Kluwer Academic Publishers, Boston, vol. 1, 2004, pp. 235–318.

833. I. V. Fedosov, S. S. Ulyanov, E. I. Galanzha, V. A. Galanzha, and V. V. Tuchin, Laser Doppler and Speckle Techniques for Bioflow Measuremenys," Chapter 10 in *Coherent-Domain Optical Methods: Biomedical Diagnostics, Environmental and Material Science*, V. V. Tuchin (ed.), Kluwer Academic Publishers, Boston, vol. 1, 2004, pp. 397–435.

834. D. A. Zimnyakov, V. V. Tuchin, S. R. Utz, "Investigation of Statistical Properties of Partly Developed Speckle-Fields in Application to Skin Structure Diagnostics," *Opt. Spectrosc.*, vol. 76, 1994, pp. 838–844.

835. S. S. Ul'yanov, D. A. Zimnyakov, and V. V. Tuchin, "Fundamentals and Applications of Dynamic Speckles Induced by Focused Laser Beam Scattering," *Opt. Eng.*, vol. 33, no. 10, 1994, pp. 3189–3201.

836. S. S. Ul'yanov, V. P. Ryabukho, and V. V. Tuchin, "Speckle Interferometry for Biotissue Vibration measurement," *Opt. Eng.*, vol. 33, no. 3, 1994, pp. 908–914.

837. V. P. Ryabukho, V. L. Khomutov, V. V. Tuchin, D. V. Lyakin, and K. V. Konstantinov, "Laser Interferometer with an Object Sharply Focused Beam as a Tool for Optical Tomography," *Proc. SPIE* 3251, 1998, pp. 247–252.

838. A. P. Shepherd and P. Å. Öberg (eds.), *Laser Doppler Blood Flowmetry*, Kluwer, Boston, 1990.

839. M. E. Fein, A. H. Gluskin, W. W. Y. Goon, B. D. Chew, W. A. Crone, and H. W. Jones, "Evaluation of Optical Methods of Detecting Dental Pulp Vitality," *J. Biomed. Opt.*, vol. 2, no. 1, 1997, pp. 58–73.

840. F. F. M. de Mul, M. H. Koelink, A. L. Weijers, et al., "Self-Mixing Laser-Doppler velocimetry of Liquid Flow and Blood Perfusion of Tissue," *Appl. Opt.*, vol. 31, 1992, pp. 5844–5851.

841. M. H. Koelink, *Direct-Contact and Self-Mixing Laser Doppler Blood Flow Velocimetry*, PhD Thesis, Twente University, Enschede, The Netherlands, 2000.

842. J. Serup and B. E. Jemee (eds.), *Handbook of Non–Invasive Methods and the Skin*, CRC Press, Boca Raton et al., 1995.

843. V. P. Ryabukho, Yu. A. Avetisyan, A. E. Grinevich, D. A. Zimnyakov, and L. I. Golubentseva, "Effects of Speckle-Fields Correlation at Diffraction of Spatially-Modulated Laser Beam on a Random Phase screen," *Pis'ma Zh. Tekh. Fiz.*, vol. 20, no. 11, 1994, pp. 74–78.

844. V. P. Ryabukho, A. A. Chaussky, and V. V. Tuchin, "Interferometric Testing of the Random Phase Objects by Focused Spatially-Modulated Laser Beam," *Photon. Optoelectron.*, vol. 3, 1995, pp. 77–85.

845. V. P. Ryabukho, A. A. Chausskii, and O. A. Perepelitsyna, "Interference-Pattern Image Formation in an Optical System with a Random Phase Screen in the Space–Frequency Plane," *Opt. Spectrosc.*, vol. 92, 2002, pp. 191–198.

846. V. P. Ryabukho, A. A. Chaussky, V. L. Khomutov, and V. V. Tuchin, "Interferometric Testing of the Random Phase Objects (Biological Tissue Models) by a Spatially-Modulated Laser Beam," *Proc. SPIE* 2732, 1996, pp. 100–117.

847. E. Yu. Radchenko, G. G. Akchurin, V. V. Bakutkin, V. V. Tuchin, and A. G. Akchurin, "Measurement of Retinal Visual Acuity in Human Eyes," *Proc. SPIE* 4001, 1999, pp. 228–237.

848. S. Jutamulia and T. Asakura (eds.), Special Section on Optical Engineering in Ophthalmology, *Opt. Eng.*, vol. 34, no. 3, 1995, pp. 640–789.

849. R. R. Ansari, "Quasi-Elastic Light Scattering in Ophthalmology," Chapter 11 in *Coherent-Domain Optical Methods: Biomedical Diagnostics, Environmental and Material Science*, V. V. Tuchin (ed.), Kluwer Academic Publishers, Boston, vol. 1, 2004, pp. 437–464.

850. H. S. Dhadwal, R. R. Ansari, and M. A. DellaVecchia, "Coherent Fiber Optic Sensor for Early Detection of Cataractogenesis in the Human Eye Lens," *Opt. Eng.*, vol. 32, 1993, pp. 233–238.

851. M. Dieckman and K. Dierks, "Diagnostics Methods and Tissue Parameter Investigations Together with Measurement Results (*In Vivo*)," *Proc. SPIE* 2126, 1995, pp. 331–345.

852. S. S. Ul'yanov, "New Type of Manifestation of the Doppler Effect: an Application to Blood and Lymph Flow Measurements," *Opt. Eng.*, vol. 34, 1995, pp. 2850–2855.

853. S. S. Ul'yanov, V. V. Tuchin, A. A. Bednov, G. E. Brill, and E. I. Zakharova "The Application of Speckle-Interferometry Method for the Monitoring of Blood and Lymph Flow in Microvessels," *Lasers Med. Sci.*, vol. 11, 1996, pp. 97–107.

854. A. A. Bednov, S. S. Ulyanov, V. V. Tuchin, G. E. Brill, E. I. Zakharova, "Investigation of Lymph Flow Dynamics by Speckle-Interferometry Method," *Izvestiya VUZ, Appl. Nonlinear Dynamics*, vol. 4, no. 3, 1996, pp. 42–51; English translation: *Proc. SPIE* 3177, 1997, pp. 89–96.

855. S. S. Ulyanov, "Speckled Speckle Statistics with a Small Number of Scatterers: Implication for Blood Flow Measurement," *J. Biomed. Opt.*, vol. 3, 1998, pp. 237–245.

856. P. Starukhin, S. Ulyanov, E. Galanzha, and V. Tuchin, "Blood-Flow Measurements with a Small Number of Scattering Events," *Appl. Opt.*, vol. 39, no. 10, 2000, pp. 2823–2829.

857. S. S. Ulyanov and V. V. Tuchin, "Use of Low-Coherence Speckled Speckles for Bioflow Measurements," *Appl. Opt.*, vol. 39, no. 34, 2000, pp. 6385–6389.

858. I. V. Fedosov, V. V. Tuchin, E. I. Galanzha, A. V. Solov'eva, and T. V. Stepanova, "Recording of Lymph Flow Dynamics in Microvessels Using Correlation Properties of Scattered Coherent Radiation," *Quant. Electron.*, vol. 32, no. 11, 2002, pp. 970–974.

859. N. Konishi and H. Fujii, "Real-Time Visualization of Retinal Microcirculation by Laser Flowgraphy," *Opt. Eng.*, vol. 34, 1995, pp. 753–757.

860. Y. Tamaki, M. Araie, E. Kawamoto, S. Eguchi, and H. Fujii, "Noncontact, Two-Dimensional Measurement of Retinal Microcirculation Using Laser Speckle Phenomenon," *Inv. Ophthalmol. & Visual Sci.*, vol. 35, 1994, pp. 3825–3834.

861. T. J. H. Essex and P. O. Byrne, "A Laser Doppler Scanner for Imaging Blood Flow in Skin," *J. Biomed. Eng.*, vol. 13, 1991, pp. 189–194.

862. K. Wårdell, I. M. Braverman, D. G. Silverman, and G. E. Nilsson, "Spatial Heterogeneity in Normal Skin Perfusion Recorded with Laser Doppler Imaging and Flowmetry," *Microvascular Res.*, vol. 48, 1994, pp. 26–38.

863. J. D. Briers, G. Richards, and X. W. He, "Capillary Blood Flow Monitoring Using Speckle Contrast Analysis (LASCA)," *J. Biomed. Opt.*, vol. 4, no. 1, 1999, pp. 164–175.

864. G. Dacosta, "Optical remote sensing of heartbeats," *Opt. Communs*, Vol. 117, pp. 395–398 (1995).

865. D. A. Zimnyakov and V. V. Tuchin, "Laser Tomography" in *Lasers in Medicine*, D. R. Vij and K. Mahesh (eds.), Chapter 5, Kluwer Academic Publishers, Boston, Dordrecht, and London, 2002, pp. 147–194.

866. H. Cheng, Q. Luo, Q. Liu, Q. Lu, H. Gong, and S. Zeng, "Laser Speckle Imaging of Blood Flow in Microcirculation," *Phys. Med. Biol.*, vol. 49, 2004, pp. 1347–1357.

867. H. Cheng, Q. Luo, Z. Wang, H. Gong, S. Chen, W. Liang, and S. Zeng, "Efficient Characterization of Regional Mesentric Blood Flow by Use of Laser Speckle Imaging," *Appl. Opt.*, vol. 42, no. 28, 2004, pp. 5759–5764.

868. B. Choi, N. M. Kang, and J. S. Nelson, "Laser Speckle Imaging for Monitoring Blood Flow Dynamics in the *In Vivo* Redent Dorsal Skin Fold Model," *Microvasc. Res.*, vol. 68, 2004, pp. 143–146.

869. D. A. Weitz and D. J. Pine, "Diffusing-wave spectroscopy," Chapter 16 in *Dynamic Light Scattering. The Method and Some Applications*, W. Brown (ed.), Oxford University Press, New York, 1993, pp. 652–720.

870. I. V. Meglinskii, A. N. Korolevich, and V. V. Tuchin, "Investigation of Blood Flow Microcirculation by Diffusing Wave Spectroscopy," *Critical Reviews in Biomedical Engineering*, vol. 29, no. 3, 2001, pp. 535–548.

871. I. V. Meglinskii and V. V. Tuchin, "Diffusing Wave Spectroscopy: Application for Skin Blood Monitoring," Chapter 4 in *Coherent-Domain Optical Methods: Biomedical Diagnostics, Environmental and Material Science*, V. V. Tuchin (ed.), Kluwer Academic Publishers, Boston, vol. 1, 2004, pp. 139–164.

872. N. A. Fomin, *Speckle Photography for Fluid Mechanic Measurements: Experimental Fluid Mechanics*, Springer-Verlag, Berlin, 1998.

873. G. Maret and E. Wolf, "Multiple Light Scattering from Disordered Media. The Effect of Brownian Motion of Scatterers," *Z. Physik B–Condens. Matter*, vol. 65, 1987, pp. 409–413.

874. A. G. Yodh, P. D. Kaplan, and D. J. Pine, "Pulsed Diffusing-Wave Spectroscopy: High Resolution through Nonlinear Optical Gaiting," *Phys. Rev. B*, vol. 42, 1990, pp. 4744–4747.

875. D. A. Boas, L. E. Campbell, and A. G. Yodh, "Scattering and Imaging with Diffusing Temporal Field Correlations," *Phys. Rev. Lett.*, vol. 75, 1995, pp. 1855–1858.

876. D. J. Pine, D. A. Weitz, J. X. Zhu, and E. Hebolzheimer, "Diffusing-Wave Spectroscopy: Dynamic Light Scattering in the Multiple Scattering Limit," *J. Phys. France*, vol. 51, 1990, pp. 2101–2127.

877. X.-L. Wu, D. J. Pine, P. M. Chaikin, J. S. Huang, and D. A. Weitz, "Diffusing-Wave Spectroscopy in Shear Flow," *J. Opt. Soc. Am. B*, vol. 7, no. 1, 1990, pp. 15–20.

878. J. B. Pawley (ed.), *Handbook of Biological Confocal Microscopy*, Plenum Press, New York, 1990.

879. B. R. Masters (ed.), *Confocal Microscopy*, SPIE Milestone Ser. MS131, Bellingham, WA, 1996.

880. T. Wilson (ed.), *Confocal Microscopy*, Academic Press, London, 1990.

881. T. Wilson, "Confocal Microscopy," in *Biomedical Photonics Handbook*, Tuan Vo-Dinh (ed.), CRC Press, Boca Rotan, Florida, 2003, pp. 10-1–18.

882. T. F. Watson, "Application of High-Speed Confocal Imaging Techniques in Operative Dentistry," *Scanning*, vol. 16, 1994, pp. 168–173.

883. B. R. Masters and A. A. Thaer, "Real Time Scanning Slit Confocal Microscopy of the *In Vivo* Human Cornea," *Appl. Opt.*, vol. 33, 1994, pp. 695–701.

884. B. R. Masters and A. A. Thaer, "*In Vivo* Real-Time Confocal Microscopy of Wing Cells in the Human Cornea: a New Benchmark for *In Vivo* Corneal Microscopy," *Bioimages*, vol. 3, no. 1, 1995, pp. 7–11.

885. M. Rajadhyaksha, M. Grossman, D. Esterowitz, R. H. Webb, and R. R. Anderson, "*In Vivo* Confocal Scanning Laser Microscopy of Human Skin: Melanin Provides Strong Contrast," *J. Invest. Dermatol.*, vol. 104, 1995, pp. 946–952.

886. M. Rajadhyaksha and J. M. Zavislan, "Confocal Reflectance Microscopy of Unstained Tissue *In Vivo*," *Retinoids*, vol. 14, no. 1, 1998, pp. 26–30.

887. M. Rajadhyaksha, R. R. Anderson, and R. H. Webb, "Video-Rate Confocal Scanning Laser Microscope for Imaging Human Tissues *In Vivo*" *Appl. Opt.*, vol. 38, 1999, pp. 2105–2115.

888. B. Masters, "Confocal Microscopy of Biological Tissues," *Proc. SPIE* 2732, 1996, pp. 155–167.

889. M. Bohnke and B. R. Masters, "Confocal Microscopy of the Cornea," *Prog. Retinal Eye Res.*, vol. 18, no. 5, 1999, pp. 553–628.

890. D. C. Beebe and B. Masters, "Cell Lineage and the Differentiation of Corneal Epithelial Cells," *Invest. Ophthalmol. & Vis. Sci.*, vol. 37, no. 9, 1996, pp. 1815–1825.

891. B. R. Masters, "Three-Dimensional Confocal Microscopy of the Living *In Situ* Rabbit Cornea," *Optics Express*, vol. 3, no. 9, 1998, pp. 351–355; www.osa.org.

892. B. R. Masters, G. Gonnord, and P. Corcuff, "Three-Dimensional Microscopic Biopsy of *In Vivo* Human Skin: A New Technique Based on a Flexible Confocal Microscope," *J. Microsc.*, vol. 185, 1997, pp. 329–338.

893. Y. Kimura, P. Wilder-Smith, T. Krasieva, A. M. A. Arrastia-Jitosho, L.-H. L. Liaw, and K. Matsumoto, "Visualization and Quantification of Dentin Structure Using Confocal Laser Scanning Microscopy," *J. Biomed. Opt.*, vol. 2, no. 3, 1997, pp. 267–274.

894. M. Kempe, A. Z. Genak, W. Rudolph, and P. Dorn, "Ballistic and Diffuse Light Detection in Confocal and Heterodyne Imaging Systems," *J. Opt. Soc. Am. A*, vol. 14, no. 1, 1997, pp. 216–223.

895. M. Kempe, W. Rudolph, and E. Welsch, "Comparative Study of Confocal and Heterodyne Microscopy for Imaging through Scattering Media," *J. Opt. Soc. Am. A.*, vol. 13, no. 1, 1996, pp. 46–52.

896. I. V. Meglinsky, A. N. Bashkatov, E. A. Genina, D. Yu. Churmakov, and V. V. Tuchin, "The Enhancement of Confocal Images of Tissues at Bulk Optical Immersion," *Quantum Electronics*, vol. 32, no. 10, 2002, pp. 875–882.

897. I. V. Meglinsky, A. N. Bashkatov, E. A. Genina, D. Yu. Churmakov, and V. V. Tuchin, "Study of the Possibility of Increasing the Probing Depth by

the Method of Reflection Confocal Microscopy upon Immersion Clearing of Near-Surface Human Skin Layers," *Laser Physics*, vol. 13, no. 1, 2003, pp. 65–69.

898. A. N. Yaroslavsky, J. Barbosa, V. Neel, C. DiMarzio, and R. R. Anderson, "Combining Multispectral Polarized Light Imaging and Confocal Microscopy for Localization of Nonmelanoma Skin Cancer," *J. Biomed. Opt.*, vol. 10, no. 10, 2005, pp. 014011-1–6.

899. D. Huang, E. A. Swanson, C. P. Lin, J. S. Schuman, W. G. Stinson, W. Chang, M. R. Hee, T. Flotte, K. Gregory, C. A. Puliafito, and J. G. Fujimoto, "Optical coherence tomography," *Science* vol. 254, 1991, pp. 1178–1181.

900. J. A. Izatt, M. D. Kulkarni, K. Kobayashi, et al., "Optical Coherence Tomography for Biodiagnostics," *Opt. Photon. News*, vol. 8, no. 5, 1997, pp. 41–47, 65.

901. A. F. Fercher, C. K. Hitzenberger, and W. Drexler, "Ocular Partial-Coherence Interferometry," *Proc. SPIE* 2732, 1996, pp. 210–228.

902. A. F. Fercher, W. Drexler, and C. K. Hitzenberger, "Ocular Partial-Coherence Tomography," *Proc. SPIE* 2732, 2996, pp. 229–241.

903. J. M. Schmitt, "Array Detection for Speckle Reduction in Optical Coherence Microscopy," *Phys. Med. Biol.*, vol. 42, 1997, pp. 1427–1439.

904. V. M. Gelikonov, G. V. Gelikonov, N. D. Gladkova, et al., "Coherent Optical Tomograhy of Microscopic Inhomogeneities in Biological Tissues," *JETP's Lett.*, vol. 61, 1995, pp. 149–153.

905. J. M. Schmitt and A. Knüttel, "Model of Optical Coherence Tomography of Heterogeneous Tissue," *J. Opt. Soc. Am. A*, vol. 14, 1997, pp. 1231–1242.

906. G. Häusler, J. M. Herrmann, R. Kummer, and M. W. Linder, "Observation of Light Propagation in Volume Scatterers with 10^{11}-Fold Slow Motion," *Opt. Lett.*, vol. 21, 1996, pp. 1087–1089.

907. G. J. Tearny, M. E. Brezinsky, B. E. Bouma, et al., "Optical Coherence Tomography," *Science*, vol. 276, 1997, pp. 2037–2039.

908. Z. Chen, T. Milner, S. Srinivas, et al., "Noninvasive Imaging of In-Vivo Blood Flow Velocity Using Optical Doppler Tomography," *Opt. Lett.*, vol. 22, 1997, pp. 1119–1121.

909. J. G. Fujimoto, C. Pitris, S. A. Boppart, and M. E. Brezinski, "Optical Coherence Tomography: an Emerging Technology for Biomedical Imaging and Optical Biopsy," *Neoplasia*, vol. 2, 2000, pp. 9–25.

910. J. M. Schmitt, "Restoration of Optical Coherence Images of Living Tissue Using the CLEAN Algorithm," *J. Biomed. Opt.*, vol. 3, no. 1, 1998, pp. 66–75.

911. B. W. Colston, Jr., M. J. Everett, L. B. DaSilva, L. L. Otis, P. Stroeve, and H. Nathel, "Imaging of Hard- and Soft-Tissue Structure in the Oral Cavity by Optical Coherence Tomography," *Appl. Opt.*, vol. 37, no. 16, 1998, pp. 3582–3585.

912. J. M. Schmitt, S. L. Lee, and K. M. Yung, "An Optical Coherence Microscope with Enhanced Resolving Power in Thick Tissue," *Opt. Communs*, vol. 142, 1997, pp. 203–207.

913. J. M. Schmitt and S. H. Xiang, "Cross-Polarized Backscatter in Optical Coherence Tomography of Biological Tissue," *Opt. Lett.*, vol. 23, no. 13, 1998, pp. 1060–1062.

914. J. K. Barton, T. E. Milner, T. J. Pfefer, et al., "Optical Low-Coherence Reflectometry to Enhance Monte Carlo Modeling of Skin," *J. Biomed. Opt.*, vol. 2, no. 2, 1997, pp. 226–234.

915. H. Brunner, R. Lazar, and R. Steiner, "Optical Coherence Tomography (OCT) of Human Skin with a Slow-Scan CCD-Camera," OSA TOPS 6, Optical Society of America, Washington, DC, 1996, pp. 50–55.

916. E. Lankenau, J. Welzel, R. Birngruber, and R. Engelhardt, "*In vivo* Tissue Measurements with Optical Low Coherence Tomography," *Proc. SPIE* 2981, 1995, pp. 78–84.

917. C. K. Hitzenberger, W. Drexler, A. Baumgartner, and A. F. Fercher, "Dispersion Effects in Partial Coherence Interferometry," *Proc. SPIE* 2981, 1997, pp. 29–36.

918. A. G. Podoleanu, M. Seeger, G. M. Dobre, et al., "Transversal and Longitudinal Images from the Retina of the Living Eye Using Low Coherence Reflectometry," *J. Biomed. Opt.*, vol. 3, no. 1, 1998, pp. 12–20.

919. G. Häusler and M. W. Lindner, ""Coherence Radar" and "Spectral Radar"—New Tools for Dermatological Diagnosis," *J. Biomed. Opt.*, vol. 3, no. 1, 1998, pp. 21–31.

920. A. Baumgartner, C. K. Hitzenberger, H. Sattmann, et al., "Signal and Resolution Enhancements in Dual Beam Optical Coherence Tomography of the Human Eye," *J. Biomed. Opt.*, vol. 3, no. 1, 1998, pp. 45–54.

921. W. Drexler, O. Findl, R. Menapace, et al., "Dual Beam Optical Coherence Tomography: Signal Identification for Ophthalmologic Diagnosis," *J. Biomed. Opt.*, vol. 3, no. 1, 1998, pp. 55–65.

922. B. Bouma, L. E. Nelson, G. J. Tearney, et al., "Optical Coherence Tomographic Imaging of Human Tissue at 1.55 μm and 1.81 μm Using Er- and Tm-Dopted Fiber Sources," *J. Biomed. Opt.*, vol. 3, no. 1, 1998, pp. 76–79.

923. R. Walti, M. Bohnke, R. Gianotti, et al., "Rapid and Precise *In Vivo* Measurement of Human Corneal Thickness with Optical Low-Coherence Reflectometry in Normal Human Eyes," *J. Biomed. Opt.*, vol. 3, no. 3, 1998, pp. 253–258.

924. F. I. Feldchtein, G. V. Gelikonov, V. M. Gelikonov, et al., "*In vivo* OCT Imaging of Hard and Soft Tissue of the Oral Cavity," *Optics Express*, vol. 3, no. 6, 1998, pp. 239–250; www.osa.org.

925. J. M. Schmitt, "OCT Elastography: Imaging Microscopic Deformation and Strain in Tissue," *Optics Express*, vol. 3, 1998, pp. 199–211; www.osa.org.

926. X. Wang, T. Milner, Z. Chen, and J. S. Nelson, "Measurement of Fluid-Flow-Velocity Profile in Turbid Media by the Use of Optical Doppler Tomography," *Appl. Opt.*, vol. 36, no. 1, 1997, pp. 144–149.

927. Z. Chen, T. Milner, X. Wang, et al., "Optical Doppler Tomography: Imaging *In Vivo* Blood Flow Dynamics Following Pharmacological Intervention and Photodynamic Therapy," *Photochem. Photobiol.* vol. 67, no. 1, 1998, pp. 56–60.

928. J. A. Izatt, M. D. Kulkarni, and S. Yazdanfar, "*In Vivo* Bidirectional Color Doppler Flow Imaging of Picoliter Blood Volumes Using Optical Coherence Tomography," *Opt. Lett.*, vol. 22, no. 18, 1997, pp. 1439–1441.

929. D. A. Boas, K. K. Bizheva, and A. M. Siegel, "Using Dynamic Low-Coherence Interferometry to Image Brownian Motion within Highly Scattering Media," *Opt. Lett.*, vol. 23, no. 5, 1998, pp. 319–321.

930. V. G. Kolinko, F. F. M. de Mul, J. Greve, and A. V. Priezzhev, "Feasibility of Picosecond Laser-Doppler Flowmetry Provides Basis for Time-Resolved Doppler Tomography of Biological Tissue," *J. Biomed. Opt.*, vol. 3, no. 2, 1998, pp. 187–190.

931. B. Masters, "Early Development of Optical Low-Coherence Reflectometry and Some Recent Biomedical Applications,"*J. Biomed. Opt.*, vol. 4, no. 2, 1999, pp. 236–247.

932. R. K. Wang and V. V. Tuchin, "Optical Coherence Tomography: Light Scattering and Imaging Enhancement," Chapter 13 in *Coherent-Domain Optical Methods: Biomedical Diagnostics, Environmental and Material Science*, V. V. Tuchin (ed.), Kluwer Academic Publishers, Boston, vol. 2, 2004, pp. 3–60.

933. P. E. Andersen, L. Thrane, H. T. Yura, A. Tycho, and T. M. Jørgensen, "Optical Coherence Tomography: Advanced Modeling," Chapter 14 in *Coherent-Domain Optical Methods: Biomedical Diagnostics, Environmental and Material Science*, V. V. Tuchin (ed.), Kluwer Academic Publishers, Boston, vol. 2, 2004, pp. 61–118.

934. C. K. Hitzenberger, "Absorption and Dispersion in OCT," Chapter 15 in *Coherent-Domain Optical Methods: Biomedical Diagnostics, Environmental and Material Science*, V. V. Tuchin (ed.), Kluwer Academic Publishers, Boston, vol. 2, 2004, pp. 119–161.

935. A. Podoleanu, "En-Face OCT Imaging," Chapter 16 in *Coherent-Domain Optical Methods: Biomedical Diagnostics, Environmental and Material Science*, V. V. Tuchin (ed.), Kluwer Academic Publishers, Boston, vol. 2, 2004, pp. 163–209.

936. J. F. de Boer, "Polarization Sensitive Optical Coherence Tomography: Phase Sensitive Interferometry for Multi-Functional Imaging," Chapter 18 in *Coherent-Domain Optical Methods: Biomedical Diagnostics, Environmental and Material Science*, V. V. Tuchin (ed.), Kluwer Academic Publishers, Boston, vol. 2, 2004, pp. 271–314.

937. Z. Chen, "Optical Doppler Tomography," Chapter 19 in *Coherent-Domain Optical Methods: Biomedical Diagnostics, Environmental and Material Science*, V. V. Tuchin (ed.), Kluwer Academic Publishers, Boston, vol. 2, 2004, pp. 315–342.

938. S. Neerken, G. W. Lucassen, T. (A. M.) Nuijs, E. Lenderink, and R. F. M. Hendriks, "Comparison of Confocal Laser Scanning Microscopy and Optical Coherence Tomography," Chapter 19 in *Coherent-Domain Optical Methods: Biomedical Diagnostics, Environmental and Material Science*, V. V. Tuchin (ed.), Kluwer Academic Publishers, Boston, vol. 2, 2004, pp. 417–439.

939. F. Reil and J. E. Thomas, "Heterodyne Techniques for Characterizing Light Fields," Chapter 8 in *Coherent-Domain Optical Methods: Biomedical Diagnostics, Environmental and Material Science*, V. V. Tuchin (ed.), Kluwer Academic Publishers, Boston, vol. 1, 2004, pp. 319–351.

940. S. Roth and I. Freund, "Second Harmonic Generation in Collagen," *J. Chem. Phys.*, vol. 70, 1979, pp. 1637–1643.

941. I. Freund, M. Deutsch, and A. Sprecher, "Connective Tissue Polarity," *Biophys. J.*, vol. 50, 1986, pp. 693–712.

942. P. J. Campagnola, H. A. Clark, W. A. Mohler, A. Lewis, and L. M. Loew, "Second-Harmonic Imaging Microscopy of Living Cells," *J. Biomed. Opt.*, vol. 6, no. 3, 2001, pp. 277–286.

943. P. J. Campagnola, A. C. Millard, M. Terasaki, P. E. Hoppe, S. J. Malone, and W. A. Mohler, "Three-Dimensional High-Resolution Second-Harmonic Generation Imaging of Endogenous Structural Proteins in Biological Tissues," *Biophys. J.*, vol. 82, no. 2, 2002, pp. 493–508.

944. P. Stoller, B.-M. Kim, and A. M. Rubenchik, "Polarization-Dependent Optical Second-Harmonic Imaging of a Rat-Tail Tendon," *J. Biomed. Opt.*, vol. 7, no. 2, 2002, pp. 205–214.

945. P. Stoller, K. M. Reiser, P. M. Celliers, and A. M. Rubenchik, "Polarization-Modulated Second-Harmonic Generation in Collagen," *Biophys. J.*, vol. 82, no. 2, 2002, pp. 3330–3342.

946. A. T. Yeh, B. Choi, J. S. Nelson, and B. J. Tromberg, "Reversible Dissosiation of Collagen in Tissues," *J. Invest. Dermatol.*, vol. 121, 2003, pp. 1332–1335.

947. T. Yasui, Y. Tohno, and T. Araki, "Characterization of Collagen Orientation in Human Dermis by Two-Dimensional Second-Harmonic-Generation Polarimetry," *J. Biomed. Opt.*, vol. 9, no. 2, 2004, pp. 259–264.

948. M. Han, L. Zickler, G. Giese, M. Walter, F. H. Loesel, and J. F. Bille, "Second-Harmonic Imaging of Cornea after Intrastromal Femtosecond Laser Ablation," *J. Biomed. Opt.*, vol. 9, no. 4, 2004, pp. 760–766.

949. V. V. Tuchin, *Optical Clearing of Tissues and Blood*, vol. PM 154, SPIE Press, 2005.

950. A. P. Ivanov, S. A. Makarevich, and A. Ya. Khairulina, "Propagation of Radiation in Tissues and Liquids with Densely Packed Scatterers," *J. Appl. Spectrosc. (USSR)*, vol. 47, no. 4, 1988, pp. 662–668.

951. G. A. Askar'yan, "The Increasing of Laser and Other Radiation Transport through Soft Turbid Physical and Biological Media," *Sov. J. Quant. Electr.*, vol. 9, no. 7, 1982, pp. 1379–1383.

952. F. Veretout, M. Delaye, and A. Tardieu, "Molecular Basis of Lens Transparency. Osmotic Pressure and X-ray Analysis of α-Crystallin Solutions," *J. Mol. Biol.*, vol. 205, 1989, pp. 713–728.

953. R. Barer and S. Joseph, "Refractometry of Living Cells," *Q. J. Microsc. Sci.*, vol. 95, 1954, pp. 399–406.

954. B. A. Fikhman, *Microbiological Refractometry*, Medicine, Moscow, 1967.

955. E. Eppich, J. Beuthan, C. Dressler, and G. Müller, "Optical Phase Measurements on Biological Cells," *Laser Physics*, vol. 10, 2000, pp. 467–477.

956. B. Chance, H. Liu, T. Kitai, and Y. Zhang, "Effects of Solutes on Optical Properties of Biological Materials: Models, Cells, and Tissues," *Anal. Biochem.*, vol. 227, 1995, pp. 351–362.

957. H. Liu, B. Beauvoit, M. Kimura, and B. Chance, "Dependence of Tissue Optical Properties on Solute-Induced Changes in Refractive Index and Osmolarity," *J. Biomed. Opt.*, vol. 1, 1996, pp. 200–211.

958. V. V. Tuchin, "Optical Immersion as a New Tool to Control Optical Properties of Tissues and Blood," *Laser Phys.*, vol. 15, no. 8, 2005, pp. 1109–1136.

959. V. V. Tuchin, "Optical Clearing of Tissue and Blood Using Immersion Method," *J. Phys. D: Appl. Phys.*, vol. 38, 2005, pp. 2497–2518.

960. A. N. Bashkatov, V. V. Tuchin, E. A. Genina, Yu. P. Sinichkin, N. A. Lakodina, and V. I. Kochubey, "The Human Sclera Dynamic Spectra: *In Vitro* and *In Vivo* Measurements," *Proc. SPIE* 3591, 1999, pp. 311–319.

961. V. V. Tuchin, J. Culver, C. Cheung, S. A. Tatarkova, M. A. DellaVecchia, D. Zimnyakov, A. Chaussky, A. G. Yodh, and B. Chance, "Refractive Index Matching of Tissue Components as a New Technology for Correlation and Diffusing-Photon Spectroscopy and Imaging," *Proc. SPIE* 3598, 1999, pp. 111–120.

962. V. V. Tuchin (ed.), "Controlling of Tissue Optical Properties: Applications in Clinical Study," *Proc. SPIE* 4162, 2000.

963. V. V. Tuchin, "Controlling of Tissue Optical Properties," *Proc. SPIE* 4001, 2000, pp. 30–53.

964. V. V. Tuchin, "Advances in Immersion Control of Optical Properties of Tissues and Blood," *Proc. SPIE* 5254, 2003, pp. 1–13.

965. V. V. Tuchin, A. N. Bashkatov, E. A. Genina, Yu. P. Sinichkin, and N. A. Lakodina, "*In Vivo* Investigation of the Immersion-Liquid-Induced Human Skin Clearing Dynamics," *Technical Physics Lett.*, vol. 27, no. 6, 2001, pp. 489–490.

966. R. K. Wang, X. Xu, V. V. Tuchin, and J. B. Elder, "Concurrent Enhancement of Imaging Depth and Contrast for Optical Coherence Tomography by Hyperosmotic Agents," *J. Opt. Soc. Am. B*, vol. 18, 2001, pp. 948–953.

967. R. K. Wang and V. V. Tuchin, "Enhance Light Penetration in Tissue for High Resolution Optical Imaging Techniques by the Use of Biocompatible Chemical Agents," *J. X-Ray Science and Technology*, vol. 10, 2002, pp. 167–176.

968. R. K. Wang and J. B. Elder, "Propylene Glycol as a Contrast Agent for Optical Coherence Tomography to Image Gastrointestinal Tissue," *Lasers Surg. Med.*, vol. 30, 2002, pp. 201–208.

969. A. N. Bashkatov, E. A. Genina, and V. V. Tuchin, "Optical Immersion as a Tool for Tissue Scattering Properties Control" in *Perspectives in Engineering Optics*, K. Singh and V. K. Rastogi (eds.), Anita Publications, New Delhi, 2003, pp. 313–334.

970. V. V. Tuchin, "Optical Spectroscopy of Tissue," in *Encyclopedia of Optical Engineering*, R. G. Driggers (ed.), Marcel-Dekker, New York, 2003, pp. 1814–1829.

971. R. K. Wang, X. Xu, Y. He, and J. B. Elder, "Investigation of Optical Clearing of Gastric Tissue Immersed with Hyperosmotic Agents," *IEEE J. Select. Tops. Quant. Electr.*, vol. 9, 2003, pp. 234–242.

972. X. Xu and R. K. Wang, "Synergetic Effect of Hyperosmotic Agents of Dimethyl Sulfoxide and Glycerol on Optical Clearing of Gastric Tissue Studied with Near Infrared Spectroscopy," *Phys. Med. Biol.*, vol. 49, 2004, pp. 457–468.

973. M. H. Khan, B. Choi, S. Chess, K. M. Kelly, J. McCullough, and J. S. Nelson, "Optical Clearing of *In Vivo* Human Skin: Implications for Light-Based Diagnostic Imaging and Therapeutics," *Lasers Surg. Med.*, vol. 34, 2004, pp. 83–85.

974. F. Zhou and R. K. Wang, "Theoretical Model on Optical Clearing of Biological Tissue with Semipermeable Chemical Agents," *Proc. SPIE* 5330, 2004, pp. 215–221.

975. V. V. Tuchin and A. B. Pravdin, "Dynamics of Skin Diffuse Reflectance and Autofluorescence at Tissue Optical Immersion," in *Materials on European Workshop "BioPhotonics 2002,"* October 18–20, 2002, Heraklion, Crete, Foundation for Research and Technology–Hellas, Heraklion, CD-edition.

976. D. Y. Churmakov, I. V. Meglinski, and D. A. Greenhalgh, "Amending of Fluorescence Sensor Signal Localization in Human Skin by Matching of the Refractive Index," *J. Biomed. Opt.*, vol. 9, 2004, pp. 339–346.

977. Y. He, R. K. Wang, and D. Xing, "Enhanced Sensitivity and Spatial Resolution for *In Vivo* Imaging with Low-Level Light-Emitting Probes by Use of Biocompatible Chemical Agents," *Opt. Lett.*, vol. 28, no. 21, 2003, pp. 2076–2078.

978. E. I. Galanzha, V. V. Tuchin, Q. Luo, H. Cheng, and A. V. Solov'eva, "The action of Osmotically Active Drugs on Optical Properties of Skin and State of Microcirculation in Experiments," *Asian J. Physics*, vol. 10, no. 4, 2001, pp. 503–511.

979. G. Vargas, A. Readinger, S. S. Dosier, and A. J. Welch, "Morphological Changes in Blood Vessels Produced by Hyperosmotic Agents and Measured by Optical Coherence Tomography," *Photochem. Photobiol.*, vol. 77, no. 5, 2003, pp. 541–549.

980. E. I. Galanzha, V. V. Tuchin, A. V. Solovieva, T. V. Stepanova, Q. Luo, and H. Cheng, "Skin Backreflectance and Microvascular System Functioning at the Action of Osmotic Agents," *J. Phys. D: Appl. Phys.*, vol. 36, 2003, pp. 1739–1746.

981. M. Brezinski, K. Saunders, C. Jesser, X. Li, and J. Fujimoto, "Index Matching to Improve OCT Imaging through Blood," *Circulation*, vol. 103, 2001, pp. 1999–2003.

982. X. Xu, R. K. Wang, J. B. Elder, and V. V. Tuchin, "Effect of Dextran-Induced Changes in Refractive Index and Aggregation on Optical Properties of Whole Blood," *Phys. Med. Biol.*, vol. 48, 2003, pp. 1205–1221.

983. B. Grzegorzewski and E. Kowaliáska, "Optical Properties of Human Blood Sediment," *Acta Physica Polonica A*, vol. 101, no. 1, 2002, pp. 201–209.

984. B. Grzegorzewski, E. Kowaliáska, A. Gãrnicki, and A. Gutsze, "Diffraction Measurement of Erythrocyte Sedimentation Rate," *Optica Applicata*, vol. 32, no. 1, 2002, pp. 15–21.

985. A. K. Amerov, J. Chen, G. W. Small, and M. A. Arnold, "The Influence of Glucose upon the Transport of Light through Whole Blood," *Proc. SPIE* 5330, 2004, pp. 101–111.

986. A. N. Bashkatov, E. A. Genina, Yu. P. Sinichkin, V. I. Kochubey, N. A. Lakodina, and V. V. Tuchin, "Estimation of the Glucose Diffusion Coefficient in Human Eye Sclera," *Biophysics*, vol. 48, no. 2, 2003, pp. 292–296.

987. A. N. Bashkatov, E. A. Genina, Yu. P. Sinichkin, V. I. Kochubey, N. A. Lakodina, and V. V. Tuchin, "Glucose and Mannitol Diffusion in Human *Dura Mater*," *Biophysical J.*, vol. 85, no. 5, 2003, pp. 3310–3318.

988. E. A. Genina, A. N. Bashkatov, Yu. P. Sinichkin, V. I. Kochubey, N. A. Lakodina, G. B. Altshuler, and V. V. Tuchin, "*In Vitro* and *In Vivo* Study of Dye Diffusion into the Human Skin and Hair Follicles," *J. Biomed. Opt.* vol. 7, 2002, pp. 471–477.

989. V. V. Tuchin, E. A. Genina, A. N. Bashkatov, G. V. Simonenko, O. D. Odoevskaya, and G. B. Altshuler, "A Pilot Study of ICG Laser Therapy of *Acne Vulgaris*: Photodynamic and Photothermolysis Treatment," *Lasers Surg. Med.*, vol. 33, no. 5, 2003, pp. 296–310.

990. E. A. Genina, A. N. Bashkatov, G. V. Simonenko, O. D. Odoevskaya, V. V. Tuchin, and G. B. Altshuler, "Low-Intensity ICG-Laser Phototherapy of *Acne Vulgaris*: A Pilot Study," *J. Biomed. Opt.*, vol. 9, no. 4, 2004, pp. 828–834.

991. Yu. P. Sinichkin, S. R. Utz, and H. A. Pilipenko, "*In Vivo* Human Skin Spectroscopy: I Remittance Spectra," *Opt. Spectrosc.*, vol. 80, no. 2, 1996, pp. 228–234.

992. Yu. P. Sinichkin, S. R. Utz, L. E. Dolotov, H. A. Pilipenko, and V. V. Tuchin, "Technique and Device for Evaluation of the Human Skin Erythema and Pigmentation," *Radioengineering*, no. 4, 1997, pp. 77–81.

993. L. E. Dolotov, Yu. P. Sinichkin, V. V. Tuchin, S. R. Utz, G. B. Altshuler, and I. V. Yaroslavsky, "Design and Evaluation of a Novel Portable Erythema-Melanin-Meter," *Lasers Surg. Med.*, vol. 34, no. 2, 2004, pp. 127–135.

994. B. C. Wilson, M. S. Patterson, and L. Lilge, "Implicit and Explicit Dosimetry in Photodynamic Therapy: a New Paradigm," *Lasers Med. Sci.* vol. 12, 1997, pp. 182–199.

995. Y. Ito, R. P. Kennan, E. Watanabe, and H. Koizumi, "Assessment of Heating Effects in Skin During Contineous Wave Near Infrared Spectroscopy," *J. Biomed. Opt.*, vol. 5, 2000, pp. 383–390.

996. S.-j. Yeh, O. Khalil, Ch. F. Hanna, and S. Kantor, "Near-Infrared Thermo-Optical Response of the Localized Reflectance of Intact Diabetic and Nondiabetic Human Skin," *J. Biomed. Opt.*, vol. 8, 2003, pp. 534–544.

997. A. T. Yeh, B. Kao, W. G. Jung, Z. Chen, J. S. Nelson, and B. J. Tromberg, "Imaging Wound Healing Using Optical Coherence Tomography and Multiphoton Microscopy in an *In Vitro* Skin-Equivalent Tissue Model," *J. Biomed. Opt.*, vol. 9, no. 2, 2004, pp. 248–253.

998. T. Sakuma, T. Hasegawa, F. Tsutsui, and S. Kurihara, "Quantitative Analysis of the Whiteness of Atypical Cervical Transformation Zone," *J. Reprod. Med.*, vol. 30, 1985, pp. 773–776.

999. Ya. Holoubek, "Note on Light Attenuation by Scattering: Comparison of Coherent and Incoherent (Diffusion) Approximations," *Opt. Eng.*, 37, 1998, pp. 705–709.

1000. J. S. Balas, G. C. Themelis, E. P. Prokopakis, I. Orfanudaki, E. Koumantakis, and E. S. Helidonis, "*In Vivo* Detection and Staging of Epithelial Dysplasias and Malignancies Based on the Quantitative Assessment of Acetic Acid-Tissue," *J. Photochem. Photobiol.*, vol. 53, 1999, pp. 153–157.

1001. A. Agrawal, U. Utzinger, C. Brookner, C. Pitris, M. F. Mitchell, and R. Richards-Kortum, "Fluorescence Spectroscopy of the Cervix: Influence of Acetic Acid, Cervical Mucus, and Vaginal Medications," *Lasers Surg. Med.*, vol. 25, 1999, pp. 237–249.

1002. R. A. Drezek, T. Collier, C. K. Brookner, A. Malpica, R. Lotan, R. Richards-Kortum, and M. Follen, "Laser Scanning Confocal Microscopy of Servical Tissue before and after Application of Acetic Acid," *Am. J. Obstet. Gynecol.*, vol. 182, 2000, pp. 1135–1139.

1003. B. W. Pogue, H. B. Kaufman, A. Zelenchuk, W. Harper, G. C. Burke, E. E. Burke, and D. M. Harper, "Analysis of Acetic Acid-Induced Whitening of High-Grade Squamous Intraepithelial Lesions," *J. Biomed. Opt.*, vol. 6, 2001, pp. 397–403.

1004. B. Choi, T. E. Milner, J. Kim, J. N. Goodman, G. Vargas, G. Aguilar, and J. S. Nelson, "Use of Optical Coherence Tomography to Monitor Biological Tissue Freezing During Cryosurgery," *J. Biomed. Opt.*, vol. 9, 2004, pp. 282–286.

1005. G. N. Stamatas and N. Kollias, "Blood Stasis Contributions to the Perception of Skin Pigmentation," *J. Biomed. Opt.*, vol. 9, 2004, pp. 315–322.

1006. M. Rajadhyaksha, S. Gonzalez, and J. M. Zavislan, "Detectability of Contrast Agents for Confocal Reflectance Imaging of Skin and Microcirculation," *J. Biomed. Opt.*, vol. 9, 2004, pp. 323–331.

1007. J. K. Barton, N. J. Halas, J. L. West, and R. A. Drezek, "Nanoshells as an Optical Coherence Tomography Contrast Agent," *Proc. SPIE* 5316, 2004, pp. 99–106

1008. R. K. Wang and V. V. Tuchin, "Enhance Light Penetration in Tissue for High Resolution Optical Techniques by the Use of Biocompatible Chemical Agents," *Proc. SPIE* 4956, 2003, pp. 314–319.

1009. R. K. Wang, Y. He, and V. V. Tuchin, "Effect of Dehydration on Optical Clearing and OCT Imaging Contrast after Impregnation of Biological Tissue with Biochemical Agents," *Proc. SPIE* 5316, 2004, pp. 119–127.

1010. M. Lazebnik, D. L. Marks, K. Potgieter, R. Gillette, and S. A. Boppart, "Functional Optical Coherence Tomography of Stimulated and Spontaneous Scattering Changes in Neural Tissue," *Proc. SPIE* 5316, 2004, pp. 107–112.

1011. X. Xu, R. Wang, and J. B. Elder, "Optical Clearing Effect on Gastric Tissues Immersed with Biocompatible Chemical Agents Investigated by Near Infrared Reflectance Spectroscopy," *J. Phys. D: Appl. Phys.*, vol. 36, 2003, pp. 1707–1713.

1012. Y. He and R. K. Wang, "Dynamic Optical Clearing Effect of Tissue Impregnated with Hyperosmotic Agents and Studied with Optical Coherence Tomography," *J. Biomed. Opt.*, vol. 9, 2004, pp. 200–206.

1013. N. Guzelsu, J. F. Federici, H. C. Lim, H. R. Chauhdry, A. B. Ritter, and T. Findley, "Measurement of Skin Strech via Light Reflection," *J. Biomed. Opt.*, vol. 8, 2003, pp. 80–86.

1014. A. F. Zuluaga, R. Drezek, T. Collier, R. Lotan, M. Follen, and R. Richards-Kortum, "Contrast Agents for Confocal Microscopy: How Simple Chemicals Affect Confocal Images of Normal and Cancer Cells in Suspension," *J. Biomed. Opt.*, vol. 7, 2002, pp. 398–403.

1015. H. Schneckenburger, A. Hendinger, R. Sailer, W. S. L. Strauss, and M. Schmitt, "Laser-Assisted Optoporation of Single Cells," *J. Biomed. Opt.*, vol. 7, 2002, pp. 410–416.

1016. B. Grzegorzewski and S. Yermolenko, "Speckle in Far–Field Produced by Fluctuations Associated with Phase Separation," *Proc. SPIE* 2647, 1995, pp. 343–349.

1017. C.-L. Tsai and J. M. Fouke, "Noninvasive Detection of Water and Blood Content in Soft Tissue from the Optical Reflectance Spectrum," *Proc. SPIE* 1888, 1993, pp. 479–486.

1018. L. D. Shvartsman and I. Fine, "Optical Transmission of Blood: Effect of Erythrocyte Aggregation," *IEEE Trans. Biomed. Eng.*, vol. 50, 2003, pp. 1026–1033.

1019. O. Cohen, I. Fine, E. Monashkin, and A. Karasik, "Glucose Correlation with Light Scattering Patterns—a Novel Method for Non-Invasive Glucose Measurements," *Diabetes Technol. Ther.*, vol. 5, 2003, pp. 11–17.

1020. A. N. Yaroslavskaya, I. V. Yaroslavsky, C. Otto, G. J. Puppels, H. Duindam, G. F. J. M. Vrensen, J. Greve, and V. V. Tuchin, "Water Exchange in Human Eye Lens Monitored by Confocal Raman Microspectroscopy," *Biophysics*, vol. 43, no. 1, 1998 pp. 109–114.

1021. V. Tuchin, I. Maksimova, D. Zimnyakov, I. Kon, A. Mavlutov, and A. Mishin, "Light Propagation in Tissues with Controlled Optical Properties," *Proc. SPIE* 2925, 1996, pp. 118–14.

1022. A. N. Bashkatov, E. A. Genina, Yu. P. Sinichkin, N. A. Lakodina, V. I. Kochubey, and V. V. Tuchin "Estimation of Glucose Diffusion Coefficient in Scleral Tissue," *Proc. SPIE* 4001, 2000, pp. 345–355.

1023. E. A. Genina, A. N. Bashkatov, N. A. Lakodina, S. A. Murikhina, Yu. P. Sinichkin, and V. V. Tuchin "Diffusion of Glucose Solution through Fibrous Tissues: *In Vitro* Optical and Weight Measurements," *Proc. SPIE* 4001, 2000, pp. 255–261.

1024. B. O. Hedbys and S. Mishima, "The Thickness-Hydration Relationdhip of the Cornea," *Exp. Eye Res.*, vol. 5, 1966, pp. 221–228.

1025. A. I. Kholodnykh, K. Hosseini, I. Y. Petrova, R. O. Esenaliev, and M. Motamedi, *In vivo* OCT Assessment of Rabbit Corneal Hydration and Dehydration, *Proc. SPIE* 4956, 2003, pp. 295–298.

1026. X. Xu and R. K. Wang, "The Role of Water Desorption on Optical Clearing of Biotissue: Studied with Near Infrared Reflectance Spectroscopy," *Med. Phys.*, vol. 30, 2003, pp. 1246–1253.

1027. J. Jiang and R. K. Wang, "Comparing the Synergetic Effects of Oleic Acid and Dimethyl Sulfoxide as Vehicles for Optical Clearing of Skin Tissue *In Vitro*," *Phys. Med. Biol.*, vol. 49, 2004, pp. 5283–5294.

1028. V. V. Tuchin, T. G. Anishchenko, A. A. Mishin, and O. V. Soboleva, "Control of Bovine Sclera Optical Characteristics with Various Osmolytes," *Proc. SPIE* 2982, 1997, pp. 284–290.

1029. I. L. Kon, V. V. Bakutkin, N. V. Bogomolova, S. V. Tuchin, D. A. Zimnyakov, and V. V. Tuchin, "Trazograph Influence on Osmotic Pressure and Tissue Structures of Human Sclera," *Proc. SPIE* 2971, 1997, pp. 198–206.

1030. D. A. Zimnyakov, V. V. Tuchin, and K. V. Larin, "Speckle Patterns Polarization Analysis as an Approach to Turbid Tissue Structure Monitoring," *Proc. SPIE* 2981, 1997, pp. 172–180.

1031. A. N. Bashkatov, E. A. Genina, Yu. P. Sinichkin, and V. V. Tuchin, "The Influence of Glycerol on the Transport of Light in the Skin," *Proc. SPIE* 4623, 2002, pp. 144–152.

1032. V. V. Tuchin, I. L. Maksimova, A. N. Bashkatov, Yu. P. Sinichkin, G. V. Simonenko, E. A. Genina, and N. A. Lakodina, "Eye Tissues Study— Scattering and Polarization Effects," OSA TOPS, Optical Society of America, Washington, DC, 1999, pp. 255–258.

1033. V. V. Tuchin, A. N. Bashkatov, E. A. Genina, and Yu. P. Sinichkin, "Scleral Tissue Clearing Effects," *Proc. SPIE* 4611, 2002, pp. 54–58.

1034. A. V. Papaev, G. V. Simonenko, and V. V. Tuchin, "A Simple Model for Calculation of Polarized Light Transmission Spectrum of Biological Tissue Sample," *J. Opt. Technol.*, vol. 71, no. 5, 2004, pp. 3–6.

1035. S. Yu. Shchyogolev, "Inverse Problems of Spectroturbidimetry of Biological Disperse Systems: an Overview," *J. Biomed. Opt.*, vol. 4, 1999, pp. 490–503.

1036. V. V. Tuchin, X. Xu, and R. K. Wang, "Sedimentation of Immersed Blood Studied by OCT," *Proc. SPIE* 4241, 2001, pp. 357–369.

1037. V. V. Tuchin, X. Xu, R. K. Wang, and J. B. Elder, "Whole Blood and RBC Sedimentation and Aggregation Study using OCT," *Proc. SPIE* 4263, 2001, pp. 143–149.

1038. A. N. Bashkatov, E. A. Genina, I. V. Korovina, Yu. P. Sinichkin, O. V. Novikova, and V. V. Tuchin, "*In Vivo* and *In Vitro* Study of Control of Rat Skin Optical Properties by Action of 40%-Glucose Solution," *Proc. SPIE* 4241, 2001, pp. 223–230.

1039. A. N. Bashkatov, E. A. Genina, I. V. Korovina, V. I. Kochubey, Yu. P. Sinichkin, and V. V. Tuchin, "*In Vivo* and *In Vitro* Study of Control of Rat Skin Optical Properties by Acting of Osmotical Liquid," *Proc. SPIE* 4224, 2000, pp. 300–311.

1040. A. N. Bashkatov, A. N. Korolevich, V. V. Tuchin, Yu. P. Sinichkin, E. A. Genina, M. M. Stolnitz, N. S. Dubina, S. I. Vecherinski, and M. S. Belsley, "*In Vivo* Investigation of Human Skin Optical Clearing and Blood Microcirculation under the Action of Glucose Solution," *Asian J. of Physics*, vol. 15, no. 1, 2006, pp. 1–14.

1041. E. V. Cruz, K. Kota, J. Huque, M. Iwaku, and E. Hoshino, "Penetration of Propylene Glycol into Dentine," *Int. Endodontic J.*, vol. 35, 2002, pp. 330–336.

1042. A. N. Bashkatov, D. M. Zhestkov, E. A. Genina, and V. V. Tuchin, "Immersion Optical Clearing of Human Blood in the Visible and Near Infrared Spectral Range," *Opt. Spectrosc.*, vol. 98, no. 4, 2005, pp. 638–646.

1043. D. M. Zhestkov, A. N. Bashkatov, E. A. Genina, and V. V. Tuchin, "Influence of Clearing Solutions Osmolarity on the Optical Properties of RBC," *Proc. SPIE* 5474, 2004, pp. 321–330.

1044. S. P. Chernova, N. V. Kuznetsova, A. B. Pravdin, and V. V. Tuchin, "Dynamics of Optical Clearing of Human Skin *In Vivo*," *Proc. SPIE* 4162, 2000, pp. 227–235.

1045. P. L. Walling and J. M. Dabney, "Moisture in Skin by Near-Infrared Reflectance Spectroscopy," *J. Soc. Cosmet. Chem.*, vol. 40, 1989, pp. 151–171.

1046. K. A. Martin, "Direct Measurement of Moisture in Skin by NIR Spectroscopy," *J. Soc. Cosmet. Chem.*, vol. 44, 1993, pp. 249–261.

1047. K. Wichrowski, G. Sore, and A. Khaiat, "Use of Infrared Spectroscopy for *In Vivo* Measurement of the Stratum Corneum Moisturization after Application of Cosmetic Preparations," *Int. J. Cosmet. Sci.*, vol. 17, 1995, pp. 1–11.

1048. J. M. Schmitt, J. Hua, and J. Qu, "Imaging Water Absorption with OCT," *Proc. SPIE* 3598, 1999, pp. 36–46.

1049. K. F. Kolmel, B. Sennhenn, and K. Giese, "Investigation of Skin by Ultraviolet Remittance Spectroscopy," *British J. Dermatol.*, vol. 122, no. 2, 1990, pp. 209–216.

1050. H.-J. Schwarzmaier, M. P. Heintzen, W. Müller, et al., "Optical Density of Vascular Tissue before and after 308-nm Excimer Laser Irradiation," *Opt. Eng.*, vol. 31, 1992, pp. 1436–1440.

1051. R. Splinter, R. H. Svenson, L. Littman, et al., "Computer Simulated Light Distributions in Myocardial Tissues at the Nd-YAG Wavelength of 1064 nm," *Lasers Med. Sci.*, vol. 8, 1993, pp. 15–21.

1052. V. V. Tuchin, "Control of Tissue and Blood Optical Properties," in *Biophotonics—Principles and Applications*, NATO Advanced Study Institute, September 29–October 9, 2004, Ottawa, Canada, *Advances in Biophotonics*, B. W. Wilson, V. V. Tuchin, and S. Tanev (eds.), IOS Press, Amsterdam, 2005, pp. 79–122.

1053. S. Tanev, V. V. Tuchin and P. Paddon, "Light Scattering Effects of Gold Nanoparticles in Cells: FDTD Modeling," *Laser Physics Letters*, vol. 3, no. 12, 2006, pp. 594–598.

1054. S. Tanev, W. Sun, N. Loeb, P. Paddon, and V. Tuchin, "The Finite-Difference Time-Domain Method in the Biosciences: Modelling of Light Scattering by Biological Cells in Absorptive and Controlled Extra-cellular Media," in *Biophotonics—Principles and Applications*, NATO Advanced Study Institute, September 29–October 9, 2004, Ottawa, Canada, *Advances in Biophotonics*, B. W. Wilson, V. V. Tuchin, and S. Tanev (eds.), IOS Press, Amsterdam, 2005, pp. 45–78.

1055. B. Choi, L. Tsu, E. Chen, T. S. Ishak, S. M. Iskandar, S. Chess, and J. S. Nelson, "Determination of Chemical Agent Optical Clearing Potential Using *In Vitro* Human Skin," *Lasers Surg. Med.*, vol. 36, 2005, pp. 72–75.

1056. V. V. Tuchin, G. B. Altshuler, A. A. Gavrilova, A. B. Pravdin, D. Tabatadze, J. Childs, and I. V. Yaroslavsky, "Optical Clearing of Skin Using Flashlamp-Induced Enhancement of Epidermal Permeability," *Lasers Surg. Med.*, vol. 38, 2006, pp. 824–836.

1057. A. A. Gavrilova, D. Tabatadze, J. Childs, I. Yaroslavsky, G. Altshuler, A. B. Pravdin, and V. V. Tuchin, "*In vitro* optical clearing of rat skin using lattice of photoinduced islands for enhancement of transdermal permeability," *Proc. SPIE* 5771, 2005, pp. 344–348.

1058. O. F. Stumpp and A. J. Welch, "Injection of Glycerol into Porcine Skin for Optical Skin Clearing with Needle-Free Injection Gun and Determination of Agent Distribution Using OCT and Fluorescence Microscopy," *Proc. SPIE* 4949, 2003, pp. 44–50.

1059. O. F. Stumpp, A. J. Welch, T. E. Milner, and J. Neev, "Enhancement of Transdermal Skin Clearing Agent Delivery Using a 980 nm Diode Laser," *Lasers Surg. Med.*, vol. 37, 2005, pp. 278–285.

1060. M. H. Khan, S. Chess, B. Choi, K. M. Kelly, and J. S. Nelson, "Can Topically Applied Optical Clearing Agents Increase the Epidermal Damage Threshold and Enhance Therapeutic Efficacy," *Lasers Surg. Med.*, vol. 35, 2004, pp. 93–95.

1061. M. H. Khan, C. Przeklasa, B. Choi, K. M. Kelly, and J. S. Nelson, "Laser Assisted Tattoo Removal in Combination with Topically Applied Optical Clearing Agents," *25th Annual Meeting of the American Society for Laser Medicine and Surgery*, March 30–April 3, 2005, Lake Buena Vista, Florida, Abstracts, *Lasers Surg. Med.*, Suppl. 17, March 2005, p. 85.

1062. R. J. McNichols, M. A. Fox, A. Gowda, S. Tuya, B. Bell, and M. Motamedi, "Temporary Dermal Scatter Reduction: Quantitative Assessment and Imlications for Improved Laser Tattoo Removal," *Lasers Surg. Med.*, vol. 36, 2005, pp. 289–296.

1063. R. Cicchi, F. S. Pavone, D. Massi, and D. D. Sampson, "Contrast and Depth Enhacement in Two-Photon Microscopy of Human Skin *Ex Vivo* by Use of Optical Clearing Agents," *Opt. Express*, vol. 13, 2005, pp. 2337–2344.

1064. D. A. Zimnyakov, V. V. Tuchin, A. A. Mishin, et al., "*In Vitro* Human Sclera Structure Analysis Using Tissue Optical Immersion Effect," *Proc. SPIE* 2673, 1996, pp. 233–243.

1065. V. V. Tuchin, D. A. Zimnykov, I. L. Maksimova, G. G. Akchurin, A. A. Mishin, S. R. Utz, and I. S. Peretochkin, "The Coherent, Low-Śoherent and Polarized Light Interaction with Tissues undergo the Refractive Indices Matching Control," *Proc. SPIE* 3251, 1998, pp. 12–21.

1066. A. Kotyk and K. Janaček, *Membrane Transport: an Interdisciplinary Approach*, Plenum Press, New York, 1977.

1067. A. Pirie and van R. Heyningen, *Biochemistry of the Eye*, Blackwell Scientific Publications, Oxford, 1966.

1068. I. S. Grigor'eva and E. Z. Meilikhova (eds.), *Handbook of Physical Constants*, Energoatomizdat, Moscow, 1991.

1069. I. K. Kikoin (ed.), *Handbook of Physical Constants*, Atomizdat, Moscow, 1976.

1070. H. Schaefer and T. E. Redelmeier, *Skin Barier: Principles of Percutaneous Absorption*, Karger, Basel, 1996.

1071. F. Pirot, Y. N. Kalia, A. L. Stinchcomb, G. Keating, A. Bunge, and R. H. Guy, "Characterization of the Permeable Barrier of Human Skin *In Vivo*," *Proc. Natl. Acad. Sci. USA*, vol. 94, 1997, pp. 1562–1567.

1072. I. H. Blank, J. Moloney, A. G. Emslie, et al., "The Diffusion of Water Across the Stratum Corneum as a Function of its Water Content," *J. Invest. Dermatol.*, vol. 82, 1984, pp. 188–194.

1073. T. von Zglinicki, M. Lindberg, G. H. Roomans, and B. Forslind, "Water and Ion Distribution Profiles in Human Skin," *Acta Derm. Venerol. (Stockh)*, vol. 73, 1993, pp. 340–343.

1074. G. Altshuler, M. Smirnov, and I. Yaroslavsky, "Lattice of Optical Islets: a Novel Treatment Modality in Photomedicine," *J. Phys. D: Appl. Phys.*, vol. 38, 2005, pp. 2732–2747.

1075. P. Michailova, *Medical Cosmetics*, Moscow, Medicine, 1984.

1076. M. Kirjavainen, A. Urtti, I. Jaaskelainen, et al., "Interaction of Liposomes with Human Skin *In Vitro*—the Influence of Lipid Composition and Structure," *Biochem. Biophys. Acta*, vol. 1304, 1996, pp. 179–189.

1077. K. D. Peck, A.-H. Ghanem, and W. I. Higuchi, "Hindered Diffusion of Polar Molecules Through and Effective Pore Radii Estimates of Intact and Ethanol Treated Human Epidermal Membrane," *Pharmaceutical Res.*, vol. 11, 1994, pp. 1306–1314.

1078. T. Inamori, A.-H. Ghanem, W. I. Higuchi, and V. Srinivasan, "Macromolecule Transport in and Effective Pore Size of Ethanol Pretreated Human Epidermal Membrane," *Intern. J. Pharmaceutics*, vol. 105, 1994, pp. 113–123.

1079. M. Sznitowska, "The influence of Ethanol on Permeation Behavior of the Porous Pathway in the Stratum Corneum," *Int. J. Pharmacol.*, vol. 137, 1996, pp. 137–140.

1080. A. K. Levang, K. Zhao, and J. Singh, "Effect of Ethanol/Propylene Glycol on the *In Vitro* Percutaneous Absorption of Aspirin, Biophysical Changes and Macroscopic Barrier Properties of the Skin," *Int. J. Pharm.*, vol. 181, 1999, pp. 255–263.

1081. D. Bommannan, R. O. Potts, and R. H. Guy, "Examination of the Effect of Ethanol on Human Stratum Corneum *In Vivo* Using Infrared Spectroscopy," *J. Control Release*, vol. 16, 1991, pp. 299–304.

1082. C. A. Squier, M. J. Kremer, and P. W. Wertz, "Effect of Ethanol on Lipid Metabolism and Epidermal Permeability Barrier of Skin and Oral Mucosa in The Rat," *J. Oral Pathol. Med.*, vol. 32, 2003, pp. 595–599.

1083. U. Jacobi, J. Bartoll, W. Sterry, and J. Lademann, "Orally Administered Ethanol: Transepidermal Pathways and Effects on the Human Skin Barrier," *Arch. Dermatol. Res.*, vol. 296, 2005, pp. 332–338.

1084. http://www.dmso.org.

1085. J. Lademann, N. Otberg, H. Richter, H.-J. Weigmann, U. Lindemann, H. Schaefer, and W. Sterry, "Investigation of Follicular Penetration of Topically Applied Substances," *Skin Pharmacol. Appl. Skin Physiol.*, vol. 14, 2001, pp. 17–22.

1086. J. Lademann, U. Jacobi, H. Richter, N. Otberg, H.-J. Weigmann, H. Meffert, H. Schaefer, U. Blume-Peytavi, and W. Sterry, "*In Vivo* Determination of UV-Photons Entering into Human Skin," *Laser Phys.*, vol. 14, 2004, pp. 234 - 237.

1087. J. Lademann, A. Rudolph, U. Jacobi, H.-J. Weigmann, H. Schaefer, W. Sterry, and M. Meinke "Influence of Nonhomogeneous Distribution of Topically Applied UV Filters on Sun Protection Factors," *J. Biomed. Opt.*, vol. 9, 2004, pp. 1358–1362.

1088. S. Lee, D. J. McAuliffe, N. Kollias, T. J. Flotte, and A. G. Doukas "Photomechanical Delivery of 100-nm Microspheres Through the Stratum Corneum: Implications for Transdermal Drug Delivery," *Laser Surg. Med.*, vol. 31, 2002, pp. 207–210.

1089. S. Lee, T. Anderson, H. Zhang, T. J. Flotte, and A. G. Doukas, "Alteration of Cell Membrane by Stress Waves *In Vitro*," *Ultrasound Med. & Biol.*, vol. 22, 1996, pp. 1285–1293.

1090. S. Lee, D. J. McAuliffe, H. Zhang, Z. Xu, J. Taitelbaum, T. J. Flotte, and A. G. Doukas, "Stress-Waves-Induced Membrane Permiation of Red Blood Cells is Faciliated by Aquaporins," *Ultrasound Med. & Biol.*, vol. 23, 1997, pp. 1089–1094.

1091. D. J. McAuliffe, S. Lee, H. Zhang, T. J. Flotte, and A. G. Doukas, "Stress-Waves-Assisted Transport through the Plasma Membrane *In Vitro*," *Laser Surg. Med.*, vol. 20, 1997, pp. 216–222.

1092. C. L. Gay, R. H. Guy, G. M. Golden, V. H. W. Mak, and M. L. Francoeur, "Characterization of Low-Temperature (i.e., <65°C) Lipid Transitions in Human Stratum Corneum," *J. Invest. Dermatol.*, vol. 103, 1994, pp. 233–239.

1093. J. S. Nelson, J. L. McCullough, T. C. Glenn, W. H. Wright, L.-H. L. Liaw, and S. L. Jacques, "Mid-Infrared Laser Ablation of Stratum Corneum Enhances *In Vitro* Percutaneous Transport of Drugs," *J. Invest. Dermatol.*, vol. 97, 1991, pp. 874–879.

1094. J.-Y. Fang, W.-R. Lee, S.-C. Shen, Y.-P. Fang, and C.-H. Hu "Enhancement of Topical 5-Aminolaevulinic Acid Delivery by Erbium:YAG Laser and Microdermabrasion: A Comparison with Iontophoresis and Electroporation," *British J. Dermatol.*, vol. 151, 2004, pp. 132–140.

1095. U. Jacobi, E. Waibler, W. Sterry, and J. Lademann, "*In Vivo* Determination of the Long-Term Reservoir of the Horny Layer Using Laser Scanning Microscopy," *Laser Phys.*, 15, no. 4, 2005, pp. 565 – 569.

1096. E. Waibler, "Investigation of the long-term reservoir of the stratum corneum—Quantification, localization and residence time," Summary of the doctoral thesis, January 22, 2005, Charité-Universitätsmedizin, Berlin.

1097. S. Mordon, C. Sumian, and J. M. Devoisselle, "Site-Specific Methylene Blue Delivery to Pilosebaceous Structures Using Highly Porous Nylon Microspheres: An Experimental Evaluation," *Lasers Surg. Med.*, vol. 33, 2003, pp. 119–125.

1098. D. A. Zimnyakov and Yu. P. Sinichkin, "A study of Polarization Decay as Applied to Improved Imaging in Scattering Media," *J. Opt. A: Pure Appl. Opt.*, vol. 2, 2000, pp. 200–208.

1099. M. Gu, X. Gan, A. Kisteman, and M. G. Xu, "Comparison of Penetration Depth Between Two-Photon Excitation and Single-Photon Excitation in Imaging Through Turbid Tissue Media," *Appl. Phys. Lett.*, vol. 77, 2000, pp. 1551–1553.

1100. E. Beaurepaire, M. Oheim, and J. Mertz, "Ultra-Deep Two-Photon Fluorescence Excitation in Turbid Media," *Opt. Commun.*, vol. 188, 2001, pp. 25–29.

1101. B. R. Masters and P. T. C. So, "Confocal Microscopy and Multi-Photon Excitation Microscopy of Human Skin *In Vivo*," *Opt. Express*, vol. 8, 2001, pp. 2–10.

1102. K. König and I. Riemann, "High-Resolution Multiphoton Tomography of Human Skin with Subcellular Spatial Resolution and Picosecond Time Resolution," *J. Biomed. Opt.*, vol. 8, 2003, pp. 432–439.

1103. A. V. Priezzhev, O. M. Ryaboshapka, N. N. Firsov, and I. V. Sirko, "Aggregation and Disaggregation of Erythrocytes in Whole Blood: Study by Backscattering Technique," *J. Biomed. Opt.*, vol. 4, no. 1, 1999, pp. 76–84.

1104. A. H. Gandjbakhche, P. Mills, and P. Snabre, "Light-Scattering Technique for the Study of Orientation and Deformation of Red Blood Cells in a Concentrated Suspension," *Appl. Opt.*, vol. 33, 1994, pp. 1070–1078.

1105. A. V. Priezzhev, N. N. Firsov, and J. Lademann, "Light Backscattering Diagnostics of Red Blood Cells Aggregation in Whole Blood Samples," Chapter 11 in *Handbook of Optical Biomedical Diagnostics*, vol. PM107, V. V. Tuchin (ed.), SPIE Press, Bellingham, WA, 2002, pp. 651–674.

1106. S. M. Bertoluzzo, A. Bollini, M. Rsia, and A. Raynal, "Kinetic Model for Erythrocyte Aggregation," *Blood Cells, Molecules, and Diseases*, vol. 25, no. 22, 1999, pp. 339–349.

1107. S. Chien, "Physiological and Pathophysiological Significance of Hemorheology," in *Clinical Hemorheology*, S. Chien, J. Dormandy, E. Ernst, and A. Matarai (eds.), Martinus Nijhoff, Dordrecht, 1987, pp. 125–134.

1108. D. H. Tycko, M. H. Metz, E. A. Epstein, and A. Grinbaum, "Flow-Cytometric Light Scattering Measurement of Red Blood Cell Volume and Hemoglobin Concentration," *Appl. Opt.*, vol. 24, 1985, pp. 1355–1365.

1109. M. Yu. Kirillin and A. V. Priezzhev, "Monte Carlo Simulation of Laser Beam Propagation in a Plane Layer of the Erythrocytes Suspension: Comparison of Contributions from Different Scattering Orders to the Angular Distribution of Light Intensity," *Quant. Electr.*, vol. 32, no. 10, 2002, pp. 883–887.

1110. N. G. Khlebtsov and S. Yu. Shchyogolev, "Account for Particle Nonsphericity at Determination of Parameters of Disperse Systems by a Turbidity Spectrum Method. 1. Characteristic Functions of Light Scattering by Nonspherical Particle Systems in Rayleigh-Gans Approximation," *Opt. Spectrosc.*, vol. 42, 1977, pp. 956–962.

1111. A. Ya. Khairullina and S. F. Shumilina, "Determination of Size Distribution Function of the Erythrocytes According to Size by the Spectral Transparency Method," *J. Appl. Spectrosc.*, vol. 19, 1973, pp. 1078–1083.

1112. J. Beuthan, O. Minet, M. Herring, G. Mueller, and C. Dressler, "Biological Cells as Optical Phase-Filters—A Contribution to Medical Functional Imaging," *Minimal Invasive Medizin*, vol. 5, no. 2, 1994, pp. 75–78.

1113. K. V. Larin, T. Akkin, M. Motamedi, R. O. Esenaliev, and T. E. Millner, "Phase-Sensitive Optical Low-Coherence Reflectometry for Detection of Analyte Concentration," *Appl. Opt.*, vol. 43, 2004, pp. 3408–3414.

1114. L. Heinemann, U. Kramer, H. M. Klotzer, M. Hein, D. Volz, M. Hermann, T. Heise, and K. Rave, "Non-Invasive Task Force: Noninvasive Glucose Measurement by Monitoring of Scattering Coefficient During Oral Glucose Tolerance Tests," *Diabetes Technol. Ther.*, vol. 2, 2000, pp. 211–220.

1115. M. Essenpreis, A. Knüttel, D. Boecker, inventors; Boehringer Mannheim GmbH, assignee: "Method and Apparatus for Determining Glucose Concentration in a Biological Sample," US patent 5,710,630, January 20, 1998.

1116. F. O. Nuttall, M. C. Gannon, W. R. Swaim, and M. J. Adams, "Stability Over Time of Glycohemoglobin, Glucose, and Red Blood Cells Survival in Hemologically Stable People without Diabetes," *Metabolism*, vol. 53, no. 11, 2004, pp. 1399–1404.

1117. K. V. Larin, I. V. Larina, M. Motamedi, V. Gelikonov, R. Kuranov, and R. O. Esenaliev, "Potential Application of Optical Coherent Tomography for Non-invasive Monitoring of Glucose Concentration," *Proc. SPIE* 4263, 2001, pp. 83–90.

1118. D. W. Schmidtke, A. C. Freeland, A. Heller, and R. T. Bonnecaze, "Measurements and Modeling of the Transient Difference Between Blood and Subcutaneous Glucose Concentrations in the Rat After Injection of Insulin," *Proc. Natl. Acad. Sci. USA*, vol. 95, 1998, pp. 294–299.

1119. M. Han, L. Zickler, G. Giese, M. Walter, F. H. Loesel, and J. F. Bille, "Second-Harmonic Imaging of Cornea after Intrastromal Femtosecond Laser Ablation," *J. Biomed. Opt.*, vol. 9, no. 4, 2004, pp. 760–766.

1120. B. Chance, Q. Luo, S. Nioka, D. C. Alsop, and J. A. Detre, "Optical Investigations of Physiology: a Study of Intrinsic and Extrinsic Biomedical Contrast," *Phil. Trans. R. Soc. Lond. B*, vol. 352, 1997, pp. 707–716.

1121. Q. Luo, S. Nioka, and B. Chance, "Functional Near-Infrared Imager," *Proc. SPIE* 2979, 1997, pp. 84–93.

1122. S. J. Matcher and C. E. Cooper, "Absolute Quantification of Deoxyhaemoglobin Concentration in Tissue Near Infrared Spectroscopy," *Phys. Med. Biol.*, vol. 39, 1994, pp. 1295–1312.

1123. H. R. Heekeren, R. Wenzel, H. Obrig, et al., "Functional Human Brain Mapping During Visual stimulation using near-infrared light," *Proc. SPIE* 2979, 1997, pp. 847–858.

1124. M. B. Lilledahl, O. A. Haugen, M. Barkost, and L. O. Svaasand, "Reflection Spectroscopy of Atherosclerotic Plaque," *J. Biomed. Opt*, vol. 11, no. 2, 2006, pp. 021005-1–7.

1125. V. Ntziachristos, X. H. Ma, M. Schnall, A. Yodh, and B. Chance, "Concurrent Multi-Channel Time-Resolved NIR with MR Mammography: Instrumentation and Initial Clinical Results," in *Advances in Optical Imaging and Photon Migration*, J. G. Fujimoto and M. S. Patterson (eds.), OSA TOPS 21, Optical Society of America, Washington, DC, 1998, pp. 284–288.

1126. F. E. W. Schmidt, M. E. Fry, J. C. Hebden, and D. T. Delpy, "The Development of a 32-Channel Time-Resolved Optical Tomography System," in *Advances in Optical Imaging and Photon Migration*, J. G. Fujimoto and M. S. Patterson (eds.), OSA TOPS 21, Optical Society of America, Washington, DC, 1998, pp. 120–122.

1127. S. Ijichi, T. Kusaka, K. Isobe, F. Islam, K. Okubo, H. Okada, M. Namba, K. Kawada, T. Imai, and S. Itoh, "Quantification of Cerebral Hemoglobin as a Function of Oxygenation Using Near-infrared Time-resolved Spectroscopy in a Piglet Model of Hypoxia," *J. Biomed. Opt.*, vol. 10, no. 2, 2005, pp. 024026-1–9.

1128. NIM Inc., 3508 Market St., Philadelphia, PA 19104.

1129. G. G. Akchurin, D. A. Zimnyakov, and V. V. Tuchin, "Optoelectronic Module for Laser Microwave Modulation Spectroscopy and Tomography of Biological Tissues," *Critical Rev. in Biomed. Eng.*, vol. 1, no. 1, 2000, pp. 46–53.

1130. H. Y. Ma, C. W. Du, and B. Chance, "A Homodyne Frequency-Domain Instrument—I&Q Phase Detection System," *Proc. SPIE* 2979, 1997, pp. 826–837.

1131. B. Chance, E. Anday, S. Nioka, et al., "A Novel Method for Fast Imaging of Brain Function, Non-invasively, with Light," *Optics Express*, vol. 2, 1998, pp. 411–423.

1132. S. Nioka, S. Zhou, E. Anday, W. Thayer, D. Kurth, M. Papadopoulos, Y. Chen, and B. Chance, "Phazed Array Functional Imaging of Neonate's Meurological Disorders," in *Advances in Optical Imaging and Photon Migration*, J. G. Fujimoto and M. S. Patterson (eds.), OSA TOPS 21, Optical Society of America, Washington, DC, 1998, pp. 262–265.

1133. B. Chance, E. Anday, E. Conant, S. Nioka, S. Zhou, and H. Long, "Rapid and Sensitive Optical Imaging of Tissue Functional Activity, and Breast," in *Advances in Optical Imaging and Photon Migration*, J. G. Fujimoto and M. S. Patterson (eds.), OSA TOPS 21, Optical Society of America, Washington, DC, 1998, pp. 218–225.

1134. R. M. Daneu, Y. Wang, X. D. Li, et al.,"Regional Imager for Low-resolution Functional Imaging of the Brain with Diffusing Near-infrared Light," *Photochem. Photobiol.*, vol. 67, 1998, pp. 33–40.

1135. J. H. Choi, M. Wolf, V. Toronov, U. Wolf, C. Polzonetti, D. Hueber, L. P. Safonova, R. Gupta, A. Michalos, W. Mantulin, and E. Gratton, "Noninvasive Determination of the Optical Properties of Adult brain: Near-infrared Spectroscopy Approach," *J. Biomed. Opt.*, vol. 9 , no. 1, 2004, pp. 221–229.

1136. J. Zhao, H. S. Ding, X. L. Hou, C. L. Zhou, and B. Chance, "In Vivo Determination of the Optical Properties of Infant Brain using Frequency-domain Near-infrared Spectroscopy," *J. Biomed. Opt.*, vol. 10, no. 2, 2005, pp. 024028-1–7.

1137. H. Fang, M. Ollero, E. Vitkin, L. M. Kimerer, P. B. Cipolloni, M. M. Zaman, S. D. Freedman, I. J. Bigio, I. Itzkan, E. B. Hanlon, and L. T. Perelman, "Noninvasive Sizing of Subcellular Organelles with Light Scattering Spectroscopy," *IEEE J. Sel. Top. Quant. Elect.*, vol. 9, no. 2, 2003, pp. 267–276.

1138. H.-J. Schnorrenberg, R. Haßner, M. Hengstebeck, K. Schlinkmeier, and W. Zinth, "Polarization Modulation Can Improve Resolutionin Diaphanography," *Proc. SPIE* 2326, 1995, pp. 459–464.

1139. D. A. Zimnyakov and Yu. P. Sinichkin, "A Study of Polarization Decay as Applied to Improved Imaging in Scattering Media," *J. Opt. A: Pure Appl. Opt.* vol. 2, 2000, pp. 200–208.

1140. S. G. Demos, W. B. Wang, and R. R. Alfano, "Imaging Objects Hidden in Scattering Media with Fluorescence Polarization Preservation of Contrast Agents," *Appl. Opt.*, vol. 37, 1998, pp. 792–797.

1141. S. G. Demos, W. B. Wang, J. Ali, and R. R. Alfano, "New Optical Difference Approaches for Subsurface Imaging of Tissues," in *Advances in Optical Imaging and Photon Migration*, J. G. Fujimoto and M. S. Patterson (eds.), OSA TOPS 21, Optical Society of America, Washington, DC, 1998, pp. 405–410.

1142. A. Muccini, N. Kollias, S. B. Phillips, R. R. Anderson, A. J. Sober, M. J. Stiller, and L. A. Drake, "Polarized Light Photography in the Evaluation of Photoaging," *J. Am. Acad. Dermatol.*, vol. 33, 1995, pp. 765–769.

1143. Yu. P. Sinichkin, D. A. Zimnyakov, D. N. Agafonov, and L. V. Kuznetsova, "Visualization of Scattering Media upon Backscattering of a Linearly Polarized Nonmonochromatic Light," *Opt. Spectrosc.*, vol. 93, 2002, pp. 110–116.

1144. S. L. Jacques, K. Lee, and J. Roman, "Scattering of Polarized Light by Biological Tissues," *Proc. SPIE* 4001, 2000, pp. 14–28.

1145. J. S. Tyo, "Enhancement of the Point-spread Function for Imaging in Scattering Media by Use of Polarization-Difference Imaging," *J. Opt. Soc. Amer. A*, vol. 17, 2000, pp. 1–10.

1146. D. A. Zimnyakov, Yu. P. Sinichkin, and V. V. Tuchin, "Polarization Reflectance Spectroscopy of Biological Tissues: Diagnostical Applications," *Izv. VUZ Radiphysics*, vol. 47, 2005, pp. 957–975.

1147. A. P. Sviridov, V. Chernomordik, M. Hassan, A. C. Boccara, A. Russo, P. Smith, and A. Gandjbakhche, "Enhancement of Hidden Structures of Early Skin Fibrosis Using Polarization Degree Pattern and Pearson Correlation Analysis," *J. Biomed. Opt.*, vol. 10, no. 5, 2005, pp. 051706-1–6.

1148. A. Myakov, L. Nieman, L. Wicky, U. Utzinger, R. Richards-Kortum, and K. Sokolov, "Fiber Optic Probe for Polarized Reflectance Spectroscopy In Vivo: Design and Performance," *J. Biomed. Opt.*, vol. 7, no. 3, 2002, pp. 388–397.

1149. R. H. Newton, J. Y. Brown, and K. M. Meek, "Polarised Light Microscopy Technique for Quantitative Mapping Collagen Fibril Orientation in Cornea," *Proc. SPIE* 2926, 1996, pp. 278–284.

1150. S. Inoué, "Video Imaging Processing Greatly Enhance Contrast, Quality, and Speed in Polarization-based Microscopy," *J. Cell Biol.*, vol. 89, 1981, pp. 346–356.

1151. R. Oldenbourg and G. Mei, "New Polarized Light Microscope with Precision Universal Compensator," *J. Microscopy*, vol. 180, no. 2, 1995, pp. 140–147.

1152. T. T. Tower and R. T. Tranquillo, "Alignment Maps of Tissues: I. Microscopic Elliptical Polarimetry," *Biophys. J.*, vol. 81, 2001, pp. 2954–2963.

1153. T. T. Tower and R. T. Tranquillo, "Alignment Maps of Tissues: II. Fast Harmonic Analysis for Imaging," *Biophys. J.*, vol. 81, 2001, pp. 2964–2971.

1154. D. A. Yakovlev, S. P. Kurchatkin, A. B. Pravdin, E. V. Gurianov, M. Yu. Kasatkin, and D. A. Zimnyakov, "Polarization Monitoring of Structure and Optical Properties of the Heterogenous Birefringent Media: Application in the Study of Liquid Crystals and Biological Tissues," *Proc. SPIE* 5067, 2003, pp. 64–72.

1155. X. Gan, S. P. Schilders, and M. Gu, "Image Enhancement through Turbid Media under a Microscope by Use of Polarization Gating Methods," *J. Opt. Soc. Am. A*, vol. 16, 1999, pp. 2177–2184.

1156. X. Gan and M. Gu, "Image Reconstruction through Turbid Media under a Transmission-mode Microscope," *J. Biomed. Opt.*, vol. 7, no. 3, 2002, pp. 372–377.

1157. N. Huse, A. Schönle, and S. W. Hell, "Z-polarized Confocal Microscopy," *J. Biomed. Opt.*, vol. 6, no. 3, 2001, pp. 273–276.

1158. A. Asundi and A. Kishen, "Digital Photoelastic Investigations on the Tooth-bone Interface," *J. Biomed. Opt.*, vol. 6, 2001, pp. 224–230.

1159. A. Kishen and A. Asundi, "Photomechanical Investigations on Post Endodontically Rehabilitated Teeth," *J. Biomed. Opt.*, vol. 7, no. 2, 2002, 262–270.

1160. D. B. Tata, M. Foresti, J. Cordero, P. Tomashefsky, M. A. Alfano, and R. R. Alfano, "Fluorescence Polarization Spectroscopy and Time-resolved Fluorescence Kinetics of Native Cancerous and Normal Rat Kidney Tissues," *Biophys. J.*, vol. 50, 1986, pp. 463–469.

1161. A. Pradhan, S. S. Jena, B. V. Laxmi, and A. Agarwal, "Fluorescence Depolarization of Normal and Diseased Skin Tissues," *Proc. SPIE* 3250, 1998, pp. 78–82.

1162. S. K. Mohanty, N. Ghosh, S. K. Majumder, and P. K. Gupta, "Depolarization of Autofluorescence from Malignant and Normal Human Breast Tissues," *Appl. Opt.*, vol. 40, no. 7, 2001, pp. 1147–1154.

1163. N. Ghosh, S. K. Majumder, and P. K. Gupta, "Polarized Fluorescence Spectroscopy of Human Tissue," *Opt. Lett.*, vol. 27, 2002, pp. 2007–2009.

1164. Y. Aizu and T. Asakura, "Bio-speckle Phenomena and their Application to the Evaluation of Blood Flow," *Opt. Laser Technol.*, vol. 23, 1991, pp. 205–219.

1165. B. Ruth, "Measuring the Steady-state Value and the Dynamics of the Skin Blood Flow Using the Non-contact Laser Method," *Med. Eng. Phys.*, vol. 16, 1994, pp. 105–111.

1166. E. N. D. Stenov and P. Å. Öberg, "Design and Evaluation of a Fibre-optic Sensor for Limb Blood Flow Measurements," *Physiol. Meas.*, vol. 15, no. 3, 1994, pp. 261–270.

1167. S. C. Tjin, S. L. Ng, and K. T. Soo, "New Side-projected Fiber Optic Probe for *In Vivo* Flow Measurements," *Opt. Eng.*, vol. 35, no. 11, 1996, pp. 3123–3129.

1168. S. C. Tjin, D. Kilpatrick, and P. R. Johnston, "Evaluation of the Two-fiber Laser Doppler Anemometer for *In Vivo* Blood Flow Measurements: Experimental and Flow Simulation Results," *Opt. Eng.*, vol. 34, no. 2, 1995, pp. 460–469.

1169. R. R. Ansari, "Ocular Static and Dynamic Light Scattering: A Non-invasive Diagnostic Tool for Eye Research and Clinical Practice," *J. Biomed. Opt.*, vol. 9, no. 1, 2004, pp. 22–37.

1170. B. Chu, *Laser Light Scattering: Basic Principles and Practice*, Academic Press, New York, 1991.

1171. T. Tanaka and G. B. Benedek, "Observation of Protein Diffusivity in Intact Human and Bovine Lenses with Application to Cataract," *Invest. Ophthal. Vis. Sci.*, vol. 14, no. 6, 1975, pp. 449–456.

1172. S. E. Bursell, P. C. Magnante, and L. T. Chylack, "*In Vivo* Uses of Quasi-elastic Light Scattering Spectroscopy as a Molecular Probe in the Anterior Segment of the Eye," *Noninvasive Diagnostic Techniques in Ophthalmology*, B. R. Masters (ed.), Springer-Verlag, New York, 1990, pp. 342–365.

1173. L. Rovati, F. Fankhauser II, and J. Rick, "Design and Performance of a New Ophthalmic Instrument for Dynamic Light Scattering in the Human eye," *Rev. Sci. Instrum.*, vol. 67, no. 7, 1996, p. 2620.

1174. D. A. Boas and A. G. Yodh, "Spatially Varying Dynamical Properties of Turbid Media Probed with Diffusing Temporal Light Correlation," *J. Opt. Soc. Am. A.*, vol. 14, no. 1, 1997, pp. 192–215.

1175. D. A. Boas, I. V. Meglinsky, L. Zemany, et al., "Diffusion of Temporal Field Correlation with Selected Applications," SPIE CIS Selected Paper 2732, 1996, pp. 34–46.

1176. A. G. Yodh and N. Georgiades, "Diffusing-wave Interferometry," *Opt. Communs*, vol. 83, 1991, pp. 56–59.

1177. I. V. Meglinsky, D. A. Boas, A. G. Yodh, B. Chance, and V. V. Tuchin, "The Development of Intensity Fluctuations Correlation Method for Noninvasive Monitoring and Quantifying of Blood Flow Parameters," *Izvestija VUZ Applied Nonlinear Dynamics*, vol. 4, no. 6, 1996, pp. 72–81.

1178. A. Ya. Khairulina, "The Informativity of the Autocorrelation Function of the Time Domain Fluctuations of the Backscattered by Erythrocytes Suspension Radiation," *Opt. Spectrosc.*, vol. 80, no. 2, 1996, pp. 268–273.

1179. G. Yu, G. Lech, C. Zhou, B. Chance, E. R. Mohler III, and A. G. Yodh, "Time-Dependent Blood Flow and Oxygenation in Human Sceletal Muscles Measured with Noninvasive Near-Infrared Diffuse Optical Spectroscopies," *J. Biomed. Opt.*, vol. 10, no. 2, 2005, pp. 024027-1–7.

1180. K. U. Frerichs and G. Z. Feuerstein, "Laser Doppler Flowmetry: A Review of its Application for Measuring Cerebral and Spinal Cord Blood Flow," *Mol. Chem. Neuropathol.*, vol. 12, 1990, pp. 55–61.

1181. G. V. Belcaro, U. Hoffman, A. Bollinger, and A. N. Nicolaides, *Laser Doppler*, Med-Orion Publishing Company, London, 1994.

1182. E. Berardesca, P. Elsner, and H. I. Maibach (eds.), *Bioengineering of the Skin: Cutaneous Blood Flow and Erythema*, CRC Press, Roca Raton, 1995.

1183. C. E. Riva, B. L. Petring, R. D. Shonat, and C. J. Pournaras, "Scattering Process in LDV from Retinal Vessels," *Appl. Opt.*, vol. 28, 1989, pp. 1078–1083.

1184. E. R. Ingofsson, L. Tronstad, E. V. Hersh, and C. E. Riva, "Efficacy of Laser Doppler Flowmetry in Determining Pulp Vitality of Human Teeth," *Endod. Dent. Traumatol.*, vol. 10, 1994, pp. 83–87.

1185. A. V. Priezzhev, B. A. Levenko, and N. B. Savchenko, "Investigation of Blood Flow Dynamics in the Embryogenesis of *Macropodus Opercularis*," *Biophysics*, vol. 40, no. 6, 1995, pp. 1373–1378.

1186. J. K. Barton and S. Stromski, "Flow Measurements without Phase Information in Optical Coherence Tomography Images," *Optics Express*, vol. 13, no. 14, 2005, pp. 5234–5239.

1187. F. F. M. de Mul, M. H. Koelink, M. L. Kok, et al., "Laser Doppler Velocimetry and Monte Carlo Simulations on Models for Blood Perfusion in Tissue," *Appl. Opt.*, vol. 34, no. 28, 1995, pp. 6595–6611.

1188. D. Y. Zang, P. Wilder-Smith, J. E. Millerd, and A. M. A. Arrastia, "Novel Approach to Laser Doppler Measurement of Pulpal Blood Flow," *J. Biomed. Opt.*, vol. 2, 1997, pp. 304–309.

1189. E. Logean, L. F. Schmetterer, and C. E. Riva, "Optical Doppler Velocimetry at Various Retinal Vessel Depths by Variation of the Source Coherence Length," *Appl. Opt.*, vol. 39, no. 16, 2000, pp. 2858–2862.

1190. D. V. Kudinov and A. V. Priezzhev, "Numerical Simiulation of Light Scattering in a Turbid Medium with Moving Particles as Applied to Medical Optical Tomography," *Moscow Univ. Phys. Bull.*, vol. 53, no. 3, 1998, pp. 39–45.

1191. A. Serov, B. Steinacher, and T. Lasser, "Full-field Laser Doppler Perfusion Imaging and Monitoring with an Intelligent CMOS camera," *Optics Express*, vol. 13, 2005, pp. 3681–3689.

1192. A. Serov and T. Lasser, "High-speed Laser Doppler Perfusion Imaging Using an Integrating CMOS Image Sensor," *Optics Express*, vol. 13, no. 17, 2005, pp. 6416–6428.

1193. A. Serov and T. Lasser, "High-speed Laser Doppler Imaging of Blood Flow in Biological Tissue," *Proc. SPIE* 6163, 2006, pp. 00–00.

1194. Y. Aizu, K. Ogino, T. Sugita, et al., "Evaluation of Blood Flow at Ocular Fundus by Using Laser Speckle," *Appl. Opt.*, vol. 31, 1992, pp. 3020–3029.

1195. B. Zang, C. M. Pleass, and C. S. Ih, "Feature Information Extraction from Dynamic Biospeckle," *Appl. Opt.*, vol. 33, 1994, pp. 231–237.

1196. P. Zakharov, S. Bhat, P. Schurtenberger, and F. Scheffold, "Multiple Scattering Suppression in Dynamic Light Scattering Based on a Digital Camera Detection Scheme," *Appl. Opt.* vol. 45, 2006, pp. 1756–1764.

1197. A. K. Dunn, H. Bolay, M. A. Moskowitz, and D. A. Boas, "Dynamic Imaging of Cerebral Blood Flow using Laser Speckle," *J. Cereb. Blood Flow Metab.*, vol. 21, 2001, pp. 195–201.

1198. S. Yuan, A. Devor, D. A. Boas, and A. K. Dunn, "Determination of Optimal Exposure Time for Imaging of Blood Flow Changes with Laser Speckle Contrast Imaging," *Appl. Opt.*, vol. 44, 2005, pp. 1823–1830.

1199. K. R. Forrester, C. Stewart, J. Tulip, C. Leonard, and R. C. Bray, "Comparison of Laser Speckle and Laser Doppler Perfusion Imaging: Measurement in Himan Skin and Rabbit Articulat Tissue," *Med. Biol. Eng. Comput.*, vol. 40, 2002, pp. 687–697.

1200. K. R. Forrester, J. Tulip, C. Leonard, C. Stewart, and R. C. Bray, "A Laser Speckle Imaging Technique for Measuring Tissue Perfusion," *IEEE Trans. Biomed. Eng.*, vol. 51, 2004, pp. 2074–2084.

1201. Q. Liu, Z. Wang, and Q. Luo, "Temporal Clustering Analysis of Cerebral Blood Flow Activation Maps Measured by Laser Speckle Contrast Imaging," *J. Biomed. Opt.*, vol. 10, no. 2, 2005, pp. 024019-1-7.

1202. A. C. Völker, P. Zakharov, B. Weber, F. Buck, and F. Scheffold, "Laser Speckle Imaging with an Active Noise Reduction Scheme," *Optics Express*, vol. 13, no. 24, 2005, pp. 9782–9787.

1203. B. Weber, C. Burger, M. T. Wyss, G. K. von Schulthess, F. Scheffold, and A. Buck, "Optical Imaging of the Spatiotemporal Dynamics of Cerebral Blood Flow and Oxidative Metabolism in the Rat Barrel Cortex," *Europ. J. Neurosci.*, vol. 20, 2004, pp. 2664–2671.

1204. A. Serov, W. Steenbergen, and F. de Mul, "Prediction of the Photodetector Signal Generated by Doppler-induced Speckle Fluctuations: Theory and Some Validations," *J. Opt. Soc. Am. A*, vol. 18, 2001, pp. 622–639.

1205. R. Bonner and R. Nossal, "Model for Laser Doppler Measurements of Blood Flow in Tissue," *Appl. Opt.*, vol. 20, 1981, pp. 2097–2107.

1206. H. Bolay, U. Reuter, A. K. Dunn, Z. Huang, D. A. Boas, and A. M. Moskowitz, "Intrinsic Brain Activity Triggers Trigeminal Meningeal Afferernts in a Migraine Model," *Nat. Med.*, vol. 8, 2002, pp. 136–142.

1207. Z. Wang, Q. M. Luo, H. Y. Cheng, W. H. Luo, H. Gong, and Q. Lu, "Blood flow activation in rat somatosensory cortex under sciatic nerve stimulation revealed by laser speckle imaging," *Prog. Nat. Sci.*, vol. 13, 2003, pp. 522–527.

1208. A. C. Ngai, J. R. Meno, and H. R. Winn, "Simultaneous Measurements of Pial Arteriolar Diameter and Laser-Doppler Flow during Somatosensory Stimulation," *J. Cereb. Blood Flow Metab.*, vol. 15, 1995, pp. 124–127.

1209. T. Matsuura and I. Kanno, "Quantitative and Temporal Relationship between Local Cerebral Blood Flow and Neuronal Activation induced by Somatosensory Stimulation in Rats," *Neurosci. Res.*, vol. 40, 2001, pp. 281–290.

1210. H. Y. Cheng, Q. M. Luo, S. Q. Zeng, S. B. Chen, J. Cen, and H. Gong, "Modified Laser Speckle Imaging Method with Improved Spatial Resolution," *J. Biomed. Opt.*, vol. 8, no. 3, 2003, pp. 559–564.

1211. M. Henning, D. Gerdt, and T. Spraggins, "Using a Fiber-Optic Pulse Sensor in Magnetic Resonance Imaging," *Proc. SPIE* 1420, 1991, pp. 34–40.

1212. S. M. Khanna, R. Danliker, J.-F. Willemin, et al., "Cellular Vibration and Motility in the Organ of Corti," *Acta Oto-laryngologica*, Suppl. 467, 1989.

1213. S. M. Khanna, C. J. Koester, J. F. Willemin, et al., "A Noninvasive Optical System for the Study of the Function of Inner Ear in Living Animals," *Proc. SPIE* 2732, 1996, pp. 64–81.

1214. N. Stasche, H.-J. Foth, K. Hoermann et al, "Middle Ear Transmission Disorders—Tympanic Membrane Vibration Analysis by Laser-Doppler-Vibrometry," *Acta Oto-laryngologica*, vol. 114, 1994, pp. 59–63.

1215. M. Maeta, S. Kawakami, T. Ogawara, and Y. Masuda, "Vibration Analysis of the Tympanic Membrane with a Ventilation Tube and a Perforation by Holography," *Proc. SPIE* 1429, 1991, pp. 152–161.

1216. H. D. Hong and M. Fox, "Noninvasive Detection of Cardiovascular Pulsations by Optical Doppler Techniques," *J. Biomed. Opt.*, vol. 2, no. 4. 1997, pp. 382–390.

1217. J. Hast, R. Myllylä, H. Sorvoja, and J. Miettinen, "Arterial Pulse Shape Measurement Using Self-Mixing Effect in a Diode Laser," Quantum Electron., vol. 32, no. 11, 2002, pp. 975–980.

1218. J. Hast, Self-Mixing Interferometry and its Applications in Noninvasive Pulse Detection, PhD Dissertation, Oulu University Press, Oulu, Finland, 2003.

1219. V. Tuchin, A. Ampilogov, A. Bogoroditsky, et al., "Laser Speckle and Optical Fiber Sensors for Micromovements Monitoring in Biotissues," *Proc. SPIE* 1420, 1991, pp. 81–92.

1220. S. Yu. Kuzmin, S. S. Ul'yanov, V. V. Tuchin, and V. P. Ryabukho, "Speckle and Speckle-Interferometric Methods in Cardiodiagnostics," *Proc. SPIE* 2732, 1996, pp. 82–99.

1221. M. Conerty, J. Castracane, E. Saravia, et al., "Development of Otolaryngological Interferometric Fiber Optic Diagnostic Probe," *Proc. SPIE* 1649, 1992, pp. 98–105.

1222. S. S. Ul'yanov and V. V. Tuchin, "The Analysis of Space-Time Projection of Differential and Michelson-type Output Signal for Measurement," *Proc. SPIE* 1981, 1992, pp. 165–174.

1223. R. Berkovits and S. Feng, "Theory of Speckle-Pattern Tomography in Multiple-Scattering Media," *Phys. Rev. Lett.*, vol. 65, 1990, pp. 3120–3123.

1224. D. A. Zimnyakov and V. V. Tuchin, "Fractality of Speckle Intensity Fluctuations," *Appl. Opt.*, vol. 35, 1996, pp. 4325–4333.

1225. D. A. Zimnyakov, V. V. Tuchin, and A. A. Mishin, "Visualization of Biotissue Fractal Structures Using Spatial Speckle-Correlometry Method," *Izvestiya VUZ, Appl. Nonlinear Dynamics*, vol. 4, 1996, pp. 49–58.

1226. D. A. Zimnyakov, V. V. Tuchin, S. R. Utz, and A. A. Mishin, "Speckle Imaging Methods Using Focused Laser Beams in Applications to Tissue Mapping," *Proc. SPIE* 2433, 1995, pp. 411–420.

1227. D. A. Zimnyakov and V. V. Tuchin, "About "Two-Modality" of Speckle Intensity Distributions for Large-Scale Phase Scatterers," *Lett. J. Technical Phys.*, vol. 21, 1995, pp. 10–14.

1228. D. A. Zimnyakov and V. V. Tuchin, ""Lens-like" Local Scatterers Approach to the Biotissue Structure Analysis," *Proc. SPIE* 2647, 1995, pp. 334–342.

1229. D. A. Zimnyakov, V. P. Ryabukho, and K. V. Larin, ""Micro-Lens" Effect Manifestation in Focused Beam Diffraction on Large Scale Phase Screens," *Lett. J. Technical Phys.*, vol. 20, 1994, pp. 14–19.

1230. D. A. Zimnyakov, "Scale Effects in Partially Developed Speckle Structure. The Case of Gaussian Phase Screens," *Opt. Spectrosc.*, vol. 79, 1995, pp. 155–162.

1231. D. A. Zimnyakov, V. V. Tuchin, and S. R. Utz, "Human Skin Epidermis Structure Investigation Using Coherent Light Scattering," *Proc. SPIE* 2100, 1994, pp. 218–224.

1232. D. A. Zimnyakov, V. V. Tuchin, S. R. Utz, and A. A. Mishin, "Human Skin Image Analysis Using Coherent Focused Beam Scattering," *Proc. SPIE* 2329, 1995, pp. 115–125.

1233. D. A. Zimnyakov, V. V. Tuchin, A. A. Mishin, and K. V. Larin, "Correlation Dimension of Speckle Fields for Scattering Structures with Fractal Properties," *Izvestija VUZ. Applied Nonlinear Dynamics*, vol. 3, no. 6, 1995, pp. 126–134; English translation: *Proc. SPIE* 3177, 1997, pp. 158–164.

1234. S. J. Jacques and S. Kirkpatrick, "Acoustically Modulated Speckle Imaging of Biological Tissues," *Opt. Let.*, vol. 23, no. 11, 1998, pp. 879–881.

1235. S. Kirkpatrick and M. J. Cipolla, "High Resolution Imaged Laser Speckle Strain Gauge for Vascular Applications," *J. Biomed. Opt.*, vol. 5, no. 1, 2000, pp. 62–71.

1236. D. D. Duncan and S. Kirkpatrick, Processing Algorithms for tracking Speckle Shifts in Optical Elastography of Biological Tissues," *J. Biomed. Opt.*, vol. 6, no. 4, 2001, pp. 418–426.

1237. A. Kishen, V. M. Murukeshan, V. Krishnakumar, and A. Asundi, "Analysis on the Nature of Thermally Induced deformation in Human dentine by Electronic Speckle Pattern Interferometry (ESPI)," *J. of Dentistry*, vol. 29, 2001, pp. 531–537.

1238. J. Lademann, H.-J. Weigmann, W. Sterry, V. Tuchin, D. Zimnyakov, G. Müller, and H. Schaefer, "Analysis of the penetration Process of Drigs and Cosmetic Products into the Skin by Tape Strippings in Combination with Spectroscopic Measurements," *Proc. SPIE* 3915, 2000, pp. 194–201.

1239. P. Zaslansky, J. D. Currey, A. A. Friesem, and S. Weiner, "Phase Shifting Speckle Interferometry for Determination of Strain and Young's Modulus of Mineralized Biological Materials: A Study of Tooth Dentin Compression in Water," *J. Biomed. Opt.*, vol. 10, no. 2, 2005, pp. 024020-1–13.

1240. V. P. Tychinsky, "Coherent Phase Microscopy of Intracellular Processes," *Physics—Uspekhi*, vol. 44, 2001, pp. 617–629.

1241. V. P. Tychinsky, "Microscopy of Subwave Structures," *Physics—Uspekhi*, vol. 166, 1996, pp. 1219–1229.

1242. C. Dressler, E. V. Perevedentseva, J. Beuthan, O. Minet, E. Balanos, G. Graschew, and G. Mueller, "Research on Human Carcinoma Cells in Different Physiological States Using the Laser Phase Microscopy," *Proc. SPIE* 3726, 1999, pp. 397–402.

1243. P. Corcuff, C. Bertrand, and J. L. Leveque, "Morphometry of Human Epidermis *In Vivo* by Real-Time Confocal Microscopy," *Arch. Dermatol. Res.*, vol. 285, 1993, pp. 475–481.

1244. V. B. Karpov, "Study of Biological Samples with a Laser Fourier Holographic Microscopy," *Laser Physics*, vol. 4, 1994, pp. 618–623.

1245. S. C. W. Hyde, N. P. Barry, R. Jones, J. C. Dainty, and P. M. W. French, "Sub-100 pm Depth-Resolution Holographic Imaging through Scattering Media in the Near-Infrared," *Opt. Lett.*, vol. 20, 1996, pp. 2320–2322.

1246. C. Dunsby and P. French, "Techniques for Depth-Resolved Imaging through Turbid Media Including Coherence-Gated Imaging," *J. Phys. D: Appl. Phys.*, vol. 36, no. 14, 2003, pp. R207–R227.

1247. P. French, "Low-Coherence Holography," in *Coherent-Domain Optical Methods: Biomedical Diagnostics, Environmental and Material Science*, V. V. Tuchin (ed.), Kluwer Academic Publishers, Boston, vol. 1, 2004, pp. 199–234.

1248. V. V. Tuchin, V. V. Ryabukho, D. A. Zimnyakov, et al., "Tissue Structure and Blood Microcirculation Monitoring by Speckle Intereferometry and Full-Field Correlometry," *Proc. SPIE* 4251, 2001, pp. 148–155.

1249. V. V. Tuchin, L. I. Malinova, V. P. Ryabukho, et al., "Optical Coherence Techniques for Study of Blood Sedimentation and Aggregation," *Proc. SPIE* 4619, 2002, pp. 149–156.

1250. S. N. Roper, M. D. Moores, G. V. Gelikonov, F. I. Feldchtein, N. M. Beach, M. A. King, V. M. Gelikonov, A. M. Sergeev, and D. H. Reitze, "*In vivo* Detection of Experimentally Induced Cortical Dysgenesis in the Adult Rat Neocortex Using Optical Coherence Tomography," *J. Neurosci. Meth.*, vol. 80, 1998, pp. 91- 98.

1251. W. Drexler, U. Morgner, F. X. Kartner, C. Pitris, S. A. Boppart, X. D. Li, E. P. Ippen, and J. G. Fujimoto, "*In Vivo* Ultrahigh Resolution Optical Coherence Tomography," *Opt. Lett.*, vol. 24, 1999, pp. 1221–1223.

1252. M. Wojtkowski, R. Leitgeb, A. Kowalczyk, T. Bajraszewski, and A. F. Fercher, "*In Vivo* Human Retinal Imaging by Fourier Domain Optical Coherence Tomography," *J. Biomed. Opt.*, vol. 7, 2002, pp. 457–463.

1253. R. Leitgeb, C. K. Hitzenberger, and A. F. Fercher, "Performance of Fourier Domain vs. Time Domain Optical Coherence Tomography," *Opt. Express*, vol. 8, 2003, pp. 889–894.

1254. M. A. Coma, M. V. Sarunic, C. Yang, and J. A. Izatt, "Sensitivity Advantage of Swept Source and Fourier Domain Optical Coherence Tomography," *Optics Express*, vol. 11, no. 18, 2003, pp. 2183–2189.

1255. A. Maheshwari, M. A. Choma, and J. A. Izatt, "Heterodyne Swept-Source Optical Coherence Tomography for Complete Complex Conjugate Ambiguity Removal," *Proc. SPIE* 5690, 2005, pp. 91–95.

1256. R. V. Kuranov, V. V. Sapozhnikova, I. V. Turchin, E. V. Zagainova, V. M. Gelikonov, V. A. Kamensky, L. B. Snopova, and N. N. Prodanetz, "Complementary Use of Cross-Polarization and Standard OCT for Differential Diagnosis of Pathological Tissues," *Opt. Express*, vol. 10, 2002, pp. 707–713.

1257. S. J. Matcher,C. P. Winlove, and S. V. Gangnus, "Collagen Structure of Bovine Intervertebral Disc Studied Using PolarizationSensitive Optical Coherence Tomography," *Phys. Med. Biol.*, vol. 49, 2004, pp. 1295–1306.

1258. N. Ugrumova, D. P. Attenburrow, C. P. Winlove, and S. J. Matcher, "The collagen Structure of Equine Articular Cartilage, Characterized Using Polarization-Sensitive Optical Coherence Tomography," *J. Phys. D: Appl. Phys.*, vol. 38, 2005, pp. 2612–2619.

1259. L. Vabre, A. Dubois, and A. C. Boccara, "Thermal-Lifgt-Full-Field Optical Coherence Tomography," *Opt. Lett.*, vol. 27, 2002, pp. 530–533.

1260. A. Dubois, K. Grieve, G. Moneron, R. Lecaque, L. Vabre, and A. C. Boccara, "Ultra-High Resolution Full-FieldOptical Coherence Tomography," *Appl. Opt.*, vol. 43, 2004, pp. 2874–2883.

1261. H.-W. Wang, A. M. Rollins, and J. A. Izatt, "High Speed, Full Field Optical Coherence Tomography," *Proc. SPIE* 3598, 1999, pp. 204–212.

1262. A. M. Sergeev, V. M. Gelikonov, G. V. Gelikonov, F. I. Feldchtein, R. V. Kuranov, N. D. Gladkova, N. M. Shakhova, L. B. Snopova, A. V. Shakhov, I. A. Kuznetzova, A. N. Denisenko, V. V. Pochinko, Yu. P. Chumakov, and O. S. Streltzova, "*In Vivo* Endoscopic OCT Imaging of Precancer and Cancer States of Human Mucosa," *Opt. Express*, vol. 1, 1997, pp. 432–440.

1263. A. Eigensee, G. Häusler, J. M. Herrmann, and M. W. Lindner, "A New Method of Short-Coherence Interferometry in Human Skin (*In Vivo*) and in Solid Volume Scatterers," *Proc. SPIE* 2925, 1996, pp. 169–178.

1264. L. Poupinet and G. Jarry, "Heterodyne Detection for Measuring Extinction Coefficient in Mammalian Tissue," *J. Optics (Paris)*, vol. 24, 1993, pp. 279–285.

1265. G. Jarry, L. Poupinet, J. Watson, and T. Lepine, "Extinction Measurements in Diffusing Mammalian Tissue with Heterodyne Detection and a Titanium: Sapphire Laser," *Appl. Opt.*, vol. 34, 1995, pp. 2045–2050.

1266. H. Inaba, "Photonic Sensing Technology is Opening New Fronties in Biophotonics," *Opt. Rev.*, vol. 4, 1997, pp. 1–10.

1267. B. Devaraj, M. Usa, K. P. Chan, T. Akatsuka, and H. Inaba, "Recent Advances in Coherent Detection Imaging (CDI) in Biomedicine. Laser Tomography of Human Tissue *In Vivo* and *In Vitro*," *IEEE J. Selected Topics Quant. Electron.*, vol. 2, 1996, pp. 1008–1016.

1268. B. Devaraj, M. Takeda, M. Kobayashi, et al., "*In Vivo* Laser Computed Tomographic Imaging of Human Fingers by Coherent Detection Imaging Method Using Different Wavelengths in Nar Ifrared Rgion," *Appl. Phys. Lett.*, vol. 69, 1996, pp. 3671–3673.

1269. V. Prapavat, J. Mans, R. Schutz, et al., "*In Vivo*-Investigations on the Detection of Chronical Polyarthritis Using a CW-Transillumination Method in Interphalangeal Joints," *Proc. SPIE* 2626, 1995, pp. 360–366.

1270. A. Wax and J. E. Thomas, "Optical Heterodyne Imaging and Wigner Phase Space Distributions," *Opt. Lett.*, vol. 21, 1996, pp. 1427–1429.

1271. M. Friebel, A. Roggan, G. Müller, and M. Meinke, "Determination of Optical Properties of Human Blood in the Spectral Range 250 to 1100 nm Using Monte Carlo Simulation with Hematocrit-Dependent Effective Scattering Phase Functions," *J. Biomed. Opt.*, vol. 11, no. 3, 2006, pp. 034021-1–10.

1272. S. C. Gebhart, W. C. Lin, and A. Mahadevan-Jansen, "*In Vitro* Determination of Normal and Neoplastic Human Brain Tissue Optical Properties Using Inverse Adding-Doubling," *Phys. Med. Biol.*, vol. 51, 2006, pp. 2011–2027.

1273. E. Salomatina, B. Jiang, J. Novak, and A. N. Yaroslavsky, "Optical Properties of Normal and Cancerous Human Skin in the Visible and Near Infrared Spectral Range," *J. Biomed. Opt.* (accepted for publication).

1274. A. L. Clark, A. Gillenwater, R. Alizadeh-Naderi, A. K. El-Naggar, and R. Richards-Kortum, "Detection and Diagnosis of Oral Neoplasia with an Optical Coherence Microscope," *J. Biomed. Opt.*, vol. 9, no. 6, 2004, pp. 1271–1280.

1275. A. M. Zysk, E. J. Chaney, and S. A. Boppart, "Refractive Index of Carcinogen-Induced Rat Mammary Tumours," *Phys. Med. Biol.*, vol. 51, 2006, pp. 2165–2177.

1276. L. Oliveira, A. Lage, M. Pais Clemente, and V. V. Tuchin, "Concentration Dependence of the Optical Clearing Effect Created in Muscle Immersed in Glycerol and Ethylene Glycol," *Proc. SPIE* 6535, 2007.

1277. X. Ma, J. Q. Lu, H. Ding, and X. H. Hu, "Bulk Optical Parameters of Porcine Skin Dermis Tissues at Eight Wavelengths from 325 to 1557 nm," *Opt. Lett.*, vol. 30, 2005, pp. 412–414.

1278. H. Ding, J. Q. Lu, K. M. Jacobs, and X. H. Hu, "Determination of Refractive Indices of Porcine Skin Tissues and Intralipid at Eight Wavelengths between 325 and 1557 nm," *J. Opt. Soc. Am. A*, vol. 22, 2005, pp. 1151–1157.

1279. H. Ding, J. Q. Lu, W. A. Wooden, P. J. Kragel, and X. H. Hu, "Refractive Indices of Human Skin Tissues at Eight Wavelengths and Estimated Dispersion Relations between 300 and 1600 nm," *Phys. Med. Biol.*, vol. 51, 2006, pp. 1479–1489.

1280. D. Chan, B. Sculz, M. Rübhausen, S. Wessel, and R. Wepf, "Structural Investigations of Human Hairs by Spectrally Resolved Ellipsometry," *J. Biomed. Opt.*, vol. 11, no. 1, 2006, pp. 014029-1–6.

1281. M. Friebel and M. Meinke, "Determination of the Complex Refractive Index of Highly Concentrated Hemoglobin Solutions Using Transmittance and Reflectance Measurements," *Biomed. Opt.*, vol. 10, 2005, pp. 064019.

1282. M. Friebel and M. Meinke, "Model Function to Calculate the Refractive Index of Native Hemoglobin in the Wavelength Range of 250–1100 nm Dependent on Concentration," *Appl. Opt.*, vol. 45, no. 12, 2006, pp. 2838–2842.

1283. J. J. J. Dirckx, L. C. Kuypers, and W. F. Decraemer, "Refractive Index of Tissue Measured with Confocal Microscopy," *J. Biomed. Opt.*, vol. 10, no. 4, 2005, pp. 044014-1–8.

Index

Valery V. Tuchin was born February 4, 1944. He received his degrees MS in Radiophysics and Electronics (1966), Candidate of Sciences in Optics (PhD, 1973), and Doctor of Science in Quantum Radiophysics (1982) from Saratov State University, Saratov, Russia. He is a Professor and holds the Optics and Biomedical Physics Chair, and he is a director of Research-Educational Institute of Optics and Biophotonics at Saratov State University. Prof. Tuchin also heads the Laboratory of Laser Diagnostics of Technical and Living Systems of Precision Mechanics and Control Institute, Russian Academy of Sciences. He was dean of the Faculty of Physics of Saratov University from 1982 to 1989.

His research interests include biomedical optics, biophotonics and laser medicine, nonlinear dynamics of laser and biophysical systems, physics of optical and laser measurements. He has authored more than 300 peer-reviewed papers and books, including his latest, *Handbook of Optical Biomedical Diagnostics* (SPIE Press, Vol. PM107, 2002; Translation to Russian, Fizmatlit, Moscow, 2007), *Coherent-Domain Optical Methods for Biomedical Diagnostics, Environmental and Material Science* (Kluwer Academic Publishers, Boston, USA, vol. 1 & 2, 2004), *Optical Clearing of Tissues and Blood* (SPIE Press, Vol. PM 154, 2005), and *Optical Polarization in Biomedical Applications*, Springer, 2006 (Lihong Wang, and Dmitry Zimnyakov – co-authors). He is a holder of more than 25 patents.

Prof. Tuchin currently teaches courses on optics, tissue optics, laser and fiber optics in biomedicine, optical measurements in biomedicine, laser dynamics, biophysics and medical physics for undergraduate and postgraduate students. Since 1992, Prof. Tuchin has been the instructor of SPIE and OSA short courses on biomedical optics for an international audience of engineers, students and medical doctors; and he is an editorial board member of *J. of Biomedical Optics* (SPIE), *Lasers in the Life Sciences*, *J. of X-Ray Science and Technology*, *J. of Biophotonics*, *J. on Biomedical Photonics* (China), and the Russian journals *Izvestiya VUZ, Applied Nonlinear Dynamics, Quantum Electronics* and *Laser Medicine*. He is a member of the Russian Academy of Natural Sciences and the International Academy of Informatization, a member of the board of SPIE/RUS and a fellow of SPIE. He has been awarded the title and scholarship "Soros Professor" (1997–1999), and the Russian Federation Scholarship for the outstanding scientists (1994–2003); and Honored Science Worker of the Russian Federation (since 2000). Since 2005 Prof. Tuchin is a vice-president of Russian Photobiology Society. In 2007 he has been awarded by SPIE Educator Award.